COST STUDIES OF BUILDINGS

By the same author:
Contractual Procedures in the Construction Industry
Civil Engineering Contractual Procedures
Added Value in Design and Construction (with K. Hogg)

COST STUDIES OF BUILDINGS

FIFTH EDITION

ALLAN ASHWORTH

Allan Ashworth DUniv (Hon) MSc MRICS HMI
is visiting professor at the University of Salford where
he is involved with the Centre for Education in the
Built Environment (CEBE). He is also a visiting
professor in quantity surveying at Unitec in
New Zealand and UTAR in Malaysia.

Prentice Hall
is an imprint of

Harlow, England • London • New York • Boston • San Francisco • Toronto • Sydney • Singapore • Hong Kong
Tokyo • Seoul • Taipei • New Delhi • Cape Town • Madrid • Mexico City • Amsterdam • Munich • Paris • Milan

Pearson Education Limited
Edinburgh Gate
Harlow
Essex CM20 2JE
England

and Associated Companies throughout the world

Visit us on the World Wide Web at:
www.pearsoned.co.uk

First published under the Addison Wesley imprint 1988
Second Edition 1994
Third Edition 1999
Fourth Edition under the Prentice Hall imprint 2004
Fifth Edition 2010

ISBN 978-0-273-72895-5

British Library Cataloguing-in-Publication Data

A catalogue record for this book is
available from the British Library.

Library of Congress Cataloging-in-Publication Data

Ashworth, A. (Allan)
    Cost studies of buildings / Allan Ashworth. -- 5th ed.
       p. cm.
    Includes bibliographical references and index.
    ISBN 978-0-273-72895-5 (pbk.)
    1. Building--Cost effectiveness.   2. Construction industry--Finance.   I. Title.
    TH438.15.A83   2010
    690.068′1–dc22

                                                                    2009046199

10 9 8 7 6 5 4 3 2 1
14 13 12 11 10

Set by 35 in 10/12pt Ehrhardt
Printed and bound in Great Britain by
Ashford Colour Press Ltd, Gosport, Hampshire

CONTENTS

Preface to the fifth edition   xiii
Preface to the first edition   xv
Acknowledgements   xvi
List of abbreviations used   xviii

SECTION 1   COST CONTROL

1   INTRODUCTION   3
        Learning outcomes   3
    1.1   What this book is about   3
    1.2   The purpose of cost control   4
    1.3   The importance of cost control   5
    1.4   Cost, price and value   6
    1.5   Value for money   7
    1.6   Cost advice   10
    1.7   Contractual arrangements and their effects on costs   11
    1.8   Construction economics   13
    1.9   Economic analysis objectives   16
    1.10 Design method   17
    1.11 Building economic theories   21
    1.12 Environmental impact of the construction project   21
    1.13 Knowledge management   23
        Self assessment questions   24
        Bibliography   24

2   HISTORY OF BUILDING ECONOMICS   25
        Learning outcomes   25
    2.1   Introduction   25
    2.2   Building economics   26
    2.3   Standard texts   26
    2.4   Trends in building economics   28
    2.5   Reasons for change   30
    2.6   Historical context   30

	Self assessment questions	35
	Bibliography	36
3	THE CONSTRUCTION INDUSTRY	37
	Learning outcomes	37
3.1	Introduction	37
3.2	The importance of the construction industry	38
3.3	The construction industry as an economic regulator	43
3.4	The development process	43
3.5	Construction firms	44
3.6	The professions	45
3.7	Research and innovation	47
	Self assessment questions	47
	Bibliography	48

SECTION 2   COST INFORMATION

4	COST DATA	51
	Learning outcomes	51
4.1	Introduction	51
4.2	Characteristics of cost data	53
4.3	The hierarchical structure of cost data	53
4.4	Accuracy and consistency	56
4.5	Cost feedback	57
4.6	Price books	60
4.7	Public sector price guides	65
4.8	Monthly cost information	66
4.9	Priced bills of quantities	66
4.10	Price analysis	68
4.11	Cost analysis	68
4.12	The Building Cost Information Service	80
4.13	Market intelligence	84
4.14	International comparison of construction costs	86
	Self assessment questions	86
	Bibliography	87
5	DESIGN ECONOMICS	88
	Learning outcomes	88
5.1	Introduction	88
5.2	Client's requirements	89
5.3	Value for money	90
5.4	Factors to consider	91
5.5	Whole-life costing	105
5.6	Sustainable economic design	106
5.7	Element evolution	107
	Conclusions	109
	Self assessment questions	110
	Bibliography	110

6 THE ECONOMICS OF QUALITY   112
       Learning outcomes   112
   6.1 Introduction   112
   6.2 What is quality?   113
   6.3 Quality systems   114
   6.4 Quality control   116
   6.5 The 'M' factors affecting quality   117
   6.6 Quality standards   117
   6.7 Categories of quality costs   119
   6.8 Economic assessment   120
   6.9 Quality considerations   124
        Conclusions   124
        Self assessment questions   125
        Bibliography   126

7 INDICES AND TRENDS   127
       Learning outcomes   127
   7.1 Introduction   127
   7.2 Constructing an index   128
   7.3 Uses of index numbers   128
   7.4 Laspeyre's and Paasche's indices   130
   7.5 Difficulties in the use of indices   130
   7.6 Cost and price indices   132
   7.7 Other published indices   135
   7.8 Regional variation   137
   7.9 Measuring trends   138
   7.10 Key performance indicators and benchmarking   143
        Self assessment questions   146
        Bibliography   146

8 NEW RULES OF MEASUREMENT (NRM)   147
       Learning outcomes   147
   8.1 Introduction   147
   8.2 Standard methods of measurement   148
   8.3 NRM (New Rules of Measurement)   149
   8.4 Part 1: General   151
   8.5 Part 2: Measurement rules for order of cost estimating   153
   8.6 Part 3: Measurement rules for elemental cost planning   160
   8.7 Part 4: Tabulated rules of measurement for elemental
        cost planning   161
        Self assessment questions   162
        Bibliography   162

9 COST ANALYSIS   163
       Learning outcomes   163
   9.1 Introduction   163
   9.2 Definitions   165

9.3   Preparation of a cost analysis   166
9.4   Principles of analysis   167
9.5   Elemental cost studies   168
9.6   Preliminaries   169
9.7   Substructure   171
9.8   Superstructure   176
9.9   Internal finishes   199
9.10   Fittings and furnishings   203
9.11   Engineering services   203
9.12   External works and drainage   210
9.13   Whole-building cost analyses   211
       Conclusions   214
       Self assessment questions   214
       Bibliography   215

10   PROCUREMENT OF CONSTRUCTION PROJECTS   216
       Learning outcomes   216
10.1   Historical perspectives   216
10.2   Employer's essential requirements   218
10.3   Procurement systems   219
10.4   Partnering   220
10.5   Public private partnerships   221
10.6   Best value   224
10.7   Audit Commission   225
10.8   Constructing Excellence   225
10.9   Performance indicators   226
10.10 Other industry comparisons   227
10.11 Procurement and the cost studies of buildings   228
       Self assessment questions   228
       Bibliography   229

11   TAXATION, GRANTS AND INVESTMENT   230
       Learning outcomes   230
11.1   Taxation   230
11.2   Value-added tax   232
11.3   Corporation tax   234
11.4   Depreciation of assets   238
11.5   Tax-efficient design   241
11.6   Financial assistance for development   243
11.7   The effects of taxation and grants on cash flow   248
11.8   Investment   249
11.9   Property investment and equities   251
11.10 Property investment   252
11.11 Carbon tax   255
       Self assessment questions   256
       Bibliography   256

12   COST RESEARCH AND INNOVATION   257
        Learning outcomes   257
12.1   Introduction   257
12.2   Cost research philosophy   258
12.3   Collaboration with practice   259
12.4   Developments in quantity surveying   262
12.5   The research process   263
12.6   Research methods   266
12.7   Research organisations in the construction industry   269
12.8   Research dissemination   275
12.9   Construction Research and Innovation Strategy Panel
        (CRISP)   276
12.10   Linking construction industry needs and construction
        research outputs   277
12.11   The future   279
        Conclusions   282
        Self assessment questions   282
        Bibliography   283

SECTION 3   COST PRACTICE

13   DEVELOPMENT APPRAISAL   287
        Learning outcomes   287
13.1   Introduction   287
13.2   Development value   288
13.3   General determinants of value   288
13.4   Investment appraisal   289
13.5   Methods of valuation   290
13.6   Valuation tables   293
13.7   Developers' budgets   300
13.8   Conventional methods of investment appraisal   308
13.9   Discounting methods   311
13.10   Optimal investment criterion   312
13.11   Sensitivity analysis   312
13.12   Cost–benefit analysis   313
        Self assessment questions   317
        Bibliography   317

14   PRE-TENDER PRICE ESTIMATING   318
        Learning outcomes   318
14.1   Introduction   318
14.2   Estimate classification   319
14.3   Methods   319
14.4   General considerations   329
14.5   Other factors   331
14.6   Preparing the estimate   339

Conclusions 343
Self assessment questions 343
Bibliography 343

15 COST PLANNING 345
Learning outcomes 345
15.1 Introduction 345
15.2 History and development 347
15.3 The cost planning process 348
15.4 Cost limits 354
15.5 Elemental cost synthesis 357
15.6 Example of a cost plan 363
15.7 Construction budget 379
Self assessment questions 380
Bibliography 380

16 COST MODELLING 381
Learning outcomes 381
16.1 Introduction 381
16.2 Types of cost models 384
16.3 Purpose of cost models 385
16.4 Classification of cost models 386
16.5 Trends in cost modelling 387
16.6 Empirical methods 387
16.7 Regression analysis 388
16.8 Simulation 391
16.9 Heuristics 397
16.10 Expert systems 399
16.11 Value-for-money considerations 401
16.12 Development of models 402
Conclusions 402
Self assessment questions 404
Bibliography 404

17 WHOLE-LIFE COSTING 1: FACTORS TO CONSIDER 406
Learning outcomes 406
17.1 Introduction 406
17.2 The importance of long-term forecasting 408
17.3 A building's life 409
17.4 Deterioration and obsolescence 410
17.5 Physical deterioration 412
17.6 Obsolescence 414
17.7 Component life 418
17.8 Problems with component life data 421
17.9 Targeting the major elements of costs-in-use 423
17.10 Inflation 425
17.11 Discount rate 427

17.12   Rate of interest   429
17.13   Taxation   429
17.14   Accounting for uncertainty   430
17.15   Whole-life value   433
17.16   Whole-life Cost Forum (WLCF)   434
17.17   Whole-life cost management   434
        Self assessment questions   436
        Bibliography   436

18   WHOLE-LIFE COSTING 2: PRACTICE   439
        Learning outcomes   439
18.1   Whole-life costing applications   439
18.2   Whole-life cost plan   442
18.3   Costs-in-use   443
18.4   Calculations   447
18.5   Forecasting change   466
18.6   Historical perspectives   468
        Conclusions   469
        Self assessment questions   470
        Bibliography   470

19   VALUE MANAGEMENT   472
        Learning outcomes   472
19.1   Introduction   472
19.2   Origins of value engineering   473
19.3   Terminology   473
19.4   Methodology   475
19.5   Value management workshops   476
19.6   Cost–value reductions   477
19.7   The 'what?' scenarios   477
19.8   Unnecessary costs   479
19.9   Creativity   480
19.10   Other issues to consider   482
19.11   Securing value   483
        Self assessment questions   484
        Bibliography   485

20   RISK ANALYSIS AND MANAGEMENT   486
        Learning outcomes   486
20.1   Introduction   486
20.2   Risk and uncertainty   486
20.3   The risk management process   487
20.4   Types of risk   489
20.5   Dealing with risk   490
20.6   Some techniques for dealing with risk and uncertainty   491
20.7   Risk analysis using probabilities   494
        Conclusions   496

	Self assessment questions	496
	Bibliography	496
21	**POST-CONTRACT COST CONTROL**	498
	Learning outcomes	498
21.1	Introduction	498
21.2	Measurement contracts	499
21.3	Cost-reimbursable contracts	501
21.4	Budgetary control	502
21.5	Client's financial reports	503
21.6	Client's cash flow	506
21.7	Contractor's cost control	507
21.8	Contractor's cash flow	509
21.9	Cost comparison	511
21.10	Project cash flow	512
21.11	Cost management	513
	Self assessment questions	515
	Bibliography	515
22	**FACILITIES MANAGEMENT**	517
	Learning outcomes	517
22.1	Introduction	517
22.2	Overview of facilities management	518
22.3	Relevance of building costs	522
22.4	Operations management	523
22.5	Space management example	531
	Conclusions	535
	Self assessment questions	535
	Bibliography	536
23	**THE FUTURE DIRECTION OF COST STUDIES OF BUILDINGS**	537
	Learning outcomes	537
23.1	Introduction	537
23.2	A paradigm shift in practices	539
23.3	Added value	540
23.4	Adding value in construction	541
23.5	Lean construction	544
23.6	Benchmarking	546
23.7	Rethinking the construction industry	548
23.8	Sustainable construction	549
23.9	Procurement	552
23.10	Modern methods of construction	552
23.11	Research	553
	Conclusions	554
	Self assessment questions	555
	Bibliography	555
	Index	556

PREFACE TO THE FIFTH EDITION

"The cynic knows the price of everything and the value of nothing."
(Oscar Wilde)
"The price of inaction is far greater than the cost of making a mistake."
(Meister Eckhart)

The construction industry is one of the most important activities in any economy in the world. A large part of the natural resources are usually used in the construction and maintenance of buildings, and these play an important part both in production and in providing services to the community. Its activities are usually represented by anything up to 8 per cent of the gross domestic product and it employs about 10 per cent of the working population both directly in the activities that it carries out and in providing employment for many others. The products of the construction industry, i.e. the built environment, provide a sense of well-being amongst society and when the industry is doing well then so, most probably, is the rest of society. We all spend a lot of time in buildings, making use of the infrastructure or looking at buildings. A high-quality built environment helps to encourage optimism and this yields confidence for investment by governments, private firms and individuals. This message is so important in times like the present.

Clearly it is important that good value should be obtained from the resources that are provided. It is not surprising therefore that the built environment is subject to frequent criticism from different directions. Sometimes it also receives praise! Unfortunately most of this criticism is unfounded and ill-informed and promulgated by the media to gain attention. Projects that over-run their budgets especially appear in the headlines of local and national newspapers and the trade and professional journals. Perhaps the real reason they are newsworthy is that they are the exception rather than the rule? The large numbers of projects that are constructed annually, which run into the thousands, are completed satisfactorily on time, within budget and at the right quality. Unfortunately, it is some of the prestige projects that hit the headlines. Some of these are of an innovative nature, without prototypes, and hence things can and do go wrong. The buildings for the Olympic Games in the UK in 2012 are a good example of this. There are reasons why the budget seems to have got out of control, but this is mainly because the client was unsure of what was required and, when this was known, the financial position began to look grim. Perhaps all too frequently the complex factors relating to erecting buildings are under-estimated and as a result many of the recommendations are of little value or even damaging. Some methods of procurement introduced over the past twenty-five years have suffered the same fate.

We need to apply the best knowledge and techniques that are available to help us use the limited resources more effectively and more efficiently. It is no use speaking of a socially acceptable price in order to get a scheme approved by a client, if in the end the final result bears little resemblance to what was required or agreed. This is a recipe for disaster. We need to be clear at the outset of the need to provide transparent and trustworthy outcomes. The industry also urgently needs to focus its attention on sustainability and to ensure that resources which cannot be replenished are used as sparingly as possible.

I remain grateful for the positive comments that I receive about this book from colleagues and students in many universities and colleges from all over the world. I have been fortunate to meet some of these on their home ground and to see for myself that this book is on their reading lists *and* is being used by their students.

I am, of course, especially grateful to my wife Margaret, who has had to contend with me during these bouts of writing for most of our married life together. These are frequently intensive in their nature.

Professor Allan Ashworth
DUniv (Hon) MSc MRICS HMI
York, 2010

PREFACE TO THE FIRST EDITION

The costs of buildings and other structures are an important factor to be considered by anyone associated with a construction project. It is one of the trio of fundamental needs of the industry's clients. The Royal Institution of Chartered Surveyors has in many of its recent publications identified the need for the greater understanding of construction costs. Both in the changing scene and the evolving role, construction costs and their forecasting, analysing, planning, controlling and accounting are seen as a priority and a distinctive role of the quantity surveyor. The importance and growth of this knowledge in recent years have been considerable, and essential for the proper financial management of the construction project. Other surveyors, notably building surveyors and other professionals employed in the industry such as architects, builders and engineers, must also take account of the costs of construction and this book should also be of interest and relevance to them. It is primarily intended for students on university undergraduate courses and for those taking final level professional examinations. It will also be of use to students taking appropriate units at higher technician level.

The book has been structured so that each chapter is largely free-standing, and this will allow the student to make easy reference to the material. In an attempt to assist the more inquisitive reader, each chapter contains a select bibliography.

The order of the chapters begins by considering construction economics and its relationship to the study of building costs. It offers a background to enable building costs to be studied in context. This is followed by the broad view of development, the importance of finance and the sources of cost information. Chapter 5 concludes this general study by examining the various methods which can be used for pre-tender estimating.

Chapters 6–10 consider building costs in more detail by looking at the effects of design and examining the individual elements of a building. To these are added the importance of quality standards and the need to reduce cost information to a common timescale through the use of indices. This section includes both the theory and a worked example of a cost plan.

The third section describes the concept of total cost control through attempting to take into account costs-in-use. The importance of taxation and other government measures on building costs, together with the modern developments of cost modelling and cost research, complete the chapters.

Allan Ashworth
University of Salford 1988

ACKNOWLEDGEMENTS

We are grateful to the following for permission to reproduce copyright material:

Figures

Figure 3.1 adapted from *Construction Statistics Annual*, No. 9, Palgrave Macmillan (Office for National Statistics 2008), Crown Copyright material is reproduced with permission under the terms of the Click-Use Licence; Figure 4.1 from 'Principles of Estimating', *Estates Gazette* (Wood, R.D. 1982), Reprinted by permission; Figure 4.3 from *Cost Planning and Computers*, HMSO (Bennett, J. *et al.* 1981), Crown Copyright material is reproduced with permission under the terms of the Click-Use Licence; Figure 10.1, adapted from PFI: Strengthening long-term partnerships, *PFI/PPP Factsheet* (HM Treasury 2006), Crown Copyright material is reproduced with permission under the terms of the Click-Use Licence; Figure 10.2 from 'PFI: Strengthening long-term partnerships', *PFI/PPP Factsheet* (HM Treasury 2006), Crown Copyright material is reproduced with permission under the terms of the Click-Use Licence; Figure 13.2 from *The Decision to Build*, HMSO (1974), Crown Copyright material is reproduced with permission under the terms of the Click-Use Licence; Figure 17.7 from An integrated framework for life cycle costing in buildings, *Proceedings of RICS COBRA* (Kishk, M. and Al-Hajj, A. 1999); Figure 21.1 from Seaden Partnership, W. Roodhouse (FRICS), Reprinted by permission of W. Roodhouse (FRICS); Figure 23.1 from *Contractual Procedures in the Construction Industry*, 5th edition (Ashworth, A. 2006) p. 176; Fig. 13.1 from Barclays Bank (1974).

Tables

Table 3.1 from *Construction Statistics Annual*, No. 9, Palgrave Macmillan (Office for National Statistics 2008), Crown Copyright material is reproduced with permission under the terms of the Click-Use Licence; Table 3.2 from Number of fatal injuries as notified to HSE and local authorities (Health and Safety Executive), http://www.hse.gov.uk/statistics/fatalquarterly.htm, Crown Copyright material is reproduced with permission under the terms of the Click-Use Licence;

Table 3.4 from *Construction Statistics Annual*, No. 9, Palgrave Macmillan (Office of National Statistics 2008), Crown Copyright material is reproduced with permission under the terms of the Click-Use Licence; Table 7.3 from *Construction Statistics Annual*, The Stationery Office (Department for Business, Enterprise and Regulatory Reform 2007), Crown Copyright material is reproduced with permission under the terms of the Click-Use Licence; Table 7.10 adapted from *Construction Statistics Annual*, 9, Palgrave MacMillan (Office for National Statistics 2008), Crown Copyright material is reproduced with permission under the terms of the Click-Use License; Table 8.1 adapted from Royal Institute of British Architects (2008), The RIBA Outline Plan of Work 2007 (amended November 2008) reproduced with permission of the Royal Institute of British Architects.; Table 9.3 from Construction Markets for Corus, Corus (formerly British Steel), Reprinted by permission of Construction Markets for Corus (formerly British Steel); Tables 9.20, 9.21, 9.22, 9.23 and 9.24 from 'The finishing touch', *Chartered Quantity Surveyor*, October (1992), Dearle & Henderson; Table 9.30 from *Building*, October (Davis Langdon 2008), United Business Media; Table 10.3 from *Rethinking Construction* (Egan, J. 1998), Crown Copyright material is reproduced with permission under the terms of the Click-Use Licence; Tables 11.4, 11.5 from *UK Pension Fund Service Annual Review*, The World Markets Company PLC (The World Markets Company 2008), Figures extracted from the WM UK Pension Funds Annual Review. Copyright 2009 The World Markets company plc ("WM") a State Street Business. All statistics quoted are sourced by the State Street Investment Analytics Division unless otherwise stated. No representation is given in respect of the information shown above, it is for general purposes only, and it should not be used or reproduced without the permission of WM; Table 17.3 adapted from Preliminary Results From A Survey Of Building Surveyor's Views, *Life Expectancies of Building Components*, Research Paper No.11 (Royal Institution of Chartered Surveyors and the Building Research Establishment 1992), Reprinted by permission of Royal Institute of Chartered Surveyors.

In some instances we have been unable to trace the owners of copyright material, and we would appreciate any information that would enable us to do so.

LIST OF ABBREVIATIONS USED

ACCA	Association of Chartered Certified Accountants
ARCOM	Association of Researchers in Construction Management
AUBEA	Australian Universities Building Educators Association
BCIS	Building Cost Information Service
BCSA	British Constructional Steelwork Association
BERR	Department of Business, Enterprise and Regulatory Reform
BMI	Building Maintenance Information
BRE	Building Research Establishment
BVR	Best value review
CABE	Commission for Architecture and the Built Environment
CCPI	Co-ordinating Committee for Project Information
CEBE	Centre for Education in the Built Environment
CESSM	Civil engineering standard method of measurement
CIB	International Council for Building Research Studies
CIOB	Chartered Institute of Building
CIRIA	Construction Industry Research and Information Association
CNBR	Co-operative Network for Building Researchers
CPI	Co-ordinated project information
EPSRC	Engineering and Physical Sciences Research Council
EUQ	Element unit quantity
EUR	Element unit rate
FAST	Functional analysis system techniques
GIFA	Gross internal floor area
KPI	Key performance indicator
MMC	Modern methods of construction
MTC	Measured-term contract
NIA	Net internal area
NPV	Net present value
OGC	Office of Government Commerce
POA	Post-occupancy analysis
PSA	Property Services Agency
RICS	Royal Institution of Chartered Surveyors
SFCA	Standard form of cost analysis
SMM	Standard method of measurement
TQM	Total quality management

COST CONTROL

CHAPTER 1

INTRODUCTION

LEARNING OUTCOMES

After reading this chapter, you should have an understanding of what this book is about. You should be able to:

■ Understand the purpose and importance of cost control
■ Identify the nature of cost advice
■ Appreciate the nature of construction economics and economic analysis objectives
■ Appreciate the main components of design method
■ Appreciate the environmental impact of construction projects

1.1 WHAT THIS BOOK IS ABOUT

Cost Studies of Buildings is about the understanding and application of costs to building and other structures. One of its aims is to ensure that scarce and limited resources are used to best advantage. It is about ensuring that clients receive the best value for money for the projects that they construct. As buildings have become more complex and clients have become better informed the techniques and tools available have become more extensive. The use of information technology has also provided a new array of possibilities particularly in the ease of modelling different design and construct solutions.

The book has been divided into three sections. The first of these provides a context for the material that follows later. It includes a simplified analysis of the construction industry since building costs cannot be studied in a vacuum but need to be considered within the industry to which they are applied. A more detailed study of the industry can be found in *The Construction Industry of Great Britain* by Roger Harvey and Allan Ashworth (Butterworth-Heinemann 1997). It has also been thought appropriate to include a brief history so far of the subject of building economics. The subject material in general is provided under a number of different titles and descriptors. It has variously been described as building economics, cost planning (although this definition now means something completely different) and cost control.

The second section is about the different sorts of cost information that are required to undertake an effective study of building costs. These include the traditional sources of cost information such as material prices and measured rates for different kinds of construction work as well as the more applicable cost analyses. Cost information can also come in many guises such as indices of cost, taxation and sources of funding and design data. While the latter is not strictly cost information, the design of a project has a particular influence on costs and the economics of design are influenced by a large number of factors. The importance of research ends this section. Without this, the subject remains sterile and innovation in building costs does not take place. Cost innovation comes from many directions including designers, constructors and manufacturers.

The third section is concerned with the practice of cost studies. These use a range of techniques that can be applied to each individual project in turn. A selection of these techniques will be adopted for all projects depending upon the aims and objectives sought by the client. Even in the simplest case some form of early price estimate will be required but this on its own is insufficient for modern-day clients. The practices being used are constantly being extended and improved as Chapter 2 will identify. These include a range of cost and value techniques from the inception through to the in-use phase. The whole aspect of the study of building costs has shifted the emphasis towards value for money. This shift has included the following:

- Development appraisal
- Elemental analysis
- Application of cost planning
- Introduction of cost limits and allowances
- Educational research and practice
- Alternative procurement systems
- Cost–value reductions
- Whole-life costing
- Value engineering
- Facilities management
- Risk analysis
- The future directions of cost studies of buildings

Some will attempt to argue that a few of these techniques merely limit expenditure and apply a range of cost control practices, i.e. they are restricted to cost reduction mechanisms. In practice they do much more through refocusing the design and construction teams by adding value to the project.

1.2 THE PURPOSE OF COST CONTROL

The purpose of cost control can be generally identified as follows:

- To limit the client's expenditure to within the amount agreed. In simple terms this means that the tender sum and final account should approximately equate with the budget estimate.

- To achieve a balanced design expenditure between the various elements of the buildings.
- To provide the client with a value-for-money project. This will probably necessitate the consideration of a total-cost approach.

The client may stipulate the maximum initial cost expenditure, or provide a detailed brief to the design team who will then determine the cost. Most schemes are a combination of these two extremes.

1.3 THE IMPORTANCE OF COST CONTROL

There has in recent years been a great need for an understanding of construction economics and cost control, particularly during the design stage of projects. The importance of this is due largely to the following:

- The increased pace in society in general has resulted in clients being less likely to tolerate delays caused by redesigning buildings when tenders are too high.
- The client's requirements today are more complex than those of their Victorian counterparts. A more effective system of control is therefore desirable from inception up to the completion of the final account, and thereafter during costs-in-use.
- The clients of the industry often represent large organisations and financial institutions. This is a result of takeovers, mergers and some public ownership. Denationalisation has often meant that these large organisations remain intact as a single entity. There has thus been an increased emphasis on accountability in both the public and the private sectors of industry. The efficiency of these organisations at construction work is only as good as their advisers.
- There has been a trend towards modern designs and new techniques, materials and methods of construction. The designer is able to choose from a far wider range of products and this has produced variety in construction. The traditional methods of estimating are unable to cope in these circumstances to achieve value for money and more balanced designs.
- Several major schemes in the UK and abroad in construction and other industries have received adverse publicity on estimated costs. Even after allowing for inflationary factors, the existing estimating procedures have been very inadequate (see Chapter 14). It is not a valid diversion to suggest that projects in other industries such as the Nimrod Early Warning System, Concorde or space exploration have produced considerably more inaccurate estimates than those in the construction industry.
- Contractors' profit margins have in real terms been reduced considerably during the past decade. This has resulted in their greater cost-consciousness in an attempt to redress possible losses.
- There has, in general, been a move towards the elimination of waste, and a greater emphasis on the use of the world's scarce resources. This has necessitated a desire for improved methods of forecasting and control of costs.

- There is a general trend towards greater cost-effectiveness, and thus a need to examine construction costs not solely in the context of initial costs but in terms of whole-life costs, or total-cost appraisal.
- World recession has generally produced a shortage of funds for capital purposes and construction in general. This has been coupled with high inflation and interest charges, resulting in the costs of construction soaring to high levels. Although the relative costs compared with other commodities may be similar, the apparent high costs have resulted in greater caution, particularly on the part of clients.

1.4 COST, PRICE AND VALUE

The terms cost, price and value will represent different interpretations to different people. Their particular meaning generally lies in the context in which they are being used. It must also be remembered that much of the terminology used in the construction industry has a special interpretation appropriate only to this industry. Cost, to the building contractor, represents all those items included under the heading of his expenditure. His price is the amount charged for the work he carries out, and when this is received it becomes his income. The difference between the two is his profit. Cost is therefore reasonably clearly defined within this context. It relates largely to manufacture, whereas price relates to selling. The term 'cost price' really means selling at cost. The price, however, that the building contractor charges the building owner for doing the work is to the latter his building costs. The Building Cost Information Service (BCIS) was designed and developed on the basis of the building owner's costs. These are in reality the tender price from building contractors. A tender price index therefore attempts to measure the building contractor's prices (the building owner's costs) whereas a building cost index measures the building contractor's costs. Although there is some relationship between the two, they are not identically correlated.

It is not surprising, therefore, to realise how easy it can be to confuse these two terms if used incorrectly. To adapt the famous quotation, 'one person's (builder's) price increase is another person's (building owner's) cost increase'.

Value is a much more subjective term than either price or cost. In the economic theory of value, an object must be scarce relative to demand to have a value. Where there is an abundance of a particular object and only a limited demand for it, then, using the economic criteria, it has little or no value attributed to it. Value constitutes a measure, therefore, of the relationship between supply and demand. An increase in the value of an object can therefore be obtained through either an increase in demand or a decrease in supply.

Aristotle identified seven classes of value that are still relevant to our modern society. These classifications of value can be summarised as: economic, moral, aesthetic, social, political, religious and judicial. These bear some resemblance to the way in which we identify building life as shown in Table 17.2. Economic value may be seen as the more objective consideration, since it is measurable in terms of money. The remainder are seen as being more subjective. Maximum value is assumed to be found when a required service or function is attained and when the cost of providing that service or function is at a minimum. Value in this context can be measured objectively, but any solution found through such a procedure risks sub-optimisation. Any increase above the required level of either service or function, for a small extra cost, would often be perceived by clients as better value. A more meaningful approach when applied to the built environment considers the following four components that when aggregated combine to provide a clearer picture of value:

- **Use value.** This is the benefit attached to the function for which the item is designed
- **Esteem value.** This attribute measures the attractiveness or aesthetics of the item
- **Cost value.** This represents the costs to produce or manufacture the item and to maintain it over its period of possession or life. This relates very much to the issues surrounding whole-life costing (see Chapters 17 and 18)
- **Exchange value.** This is the worth of an item as perceived by others who are primarily interested in its acquisition.

1.5 VALUE FOR MONEY

Value is a comparative term expressing the worth of an item or commodity, usually in the context of other similar or comparable items. Cheapness in itself is of no virtue. It is well worthwhile to pay a little more if the result in gain in value exceeds the extra costs involved. For example, it has been shown that sometimes to spend an extra sum initially on construction costs can have the effect of reducing recurring or future costs and hence the overall sum (or whole-life cost) spent on an investment. This may, of course, not always be the case since high initial costs may require high recurring costs for the item's upkeep or maintenance.

Value for money is an easy concept to understand but a difficult one to explain. It is in part subjective in its assessment in that different individuals assess different things in different ways. The appearance of buildings or engineering structures will always largely be subjective, even though a framework of rules may be devised for its evaluation. However, the opinions or judgements of others cannot be entirely disregarded. Designers have developed some rules or guidelines for the assessment or judgement of aesthetics based upon shape, form, colour, proportion, etc. The assessment of aesthetic design is difficult, since personal choice and taste are factors that need to be considered.

The engineering aspects of design, such as function or performance, are in part able to be judged against more objective criteria. On the face of it, these appear to be easier to assess and to make comparisons in terms of value for money. Criteria are frequently provided to designers in a client's brief and can be realistically compared with other similar structures. For example, the judgement of the spatial layout represents the adequacy of the internal space arrangement and can be related to the extent to which it facilitates the desired functions to be performed in the building. The structural components and the environmental comforts that are created can be judged in a similar manner, often using simple numerical data and analysis.

The third factor to consider is that of cost and value. The obvious approach is to put all the measurable components on one side of the equation in the form of cost, and to set these against the subjective and objective value judgements in an attempt to determine value for money. The determination of the best solution will never be an exact science, since it will always rely upon judgements set against a client's own value judgements that may be expressed as aims and objectives or outcomes. A part of value for money is quantitative in its analysis; other aspects will always remain qualitative by definition. Value for money is the start of the process of added value. It is the principle of doing more with less, a feature that has become common in all walks of life.

A major theme at the start of the 21st century remains value for money, now more appropriately described and defined as added value. It was John Ruskin (1819–1900) who said, 'It is not the cheaper things in life that we want to possess, but the expensive things that cost less.' There are now many recent reports on the construction industry that have indicated that clients in the future will insist upon increased value for the money that is expended on their capital projects. The principle involves reducing the relative costs of construction by designing, procuring and constructing the work in a different way than at the present time. The construction industry was, of course, responding to this challenge throughout the latter part of the twentieth century, and with some success. It involves doing more for less by removing unnecessary costs. It aims to meet the perceived needs associated with efficiency, effectiveness and economy.

1.5.1 Efficiency, effectiveness and economy

The successful accomplishment of a task reflects effectiveness, while performing tasks to produce the best outcome at the lowest cost from the same resources used is efficiency. Effectiveness is doing the right things, efficiency is doing these things better. The best performance maximises both effectiveness and efficiency.

A building decision is effective from the point of view of the client where this achieves positive outcomes. These may be outcomes in respect of financial measures that a developer may expect, or they may be outcomes that are socially effective and may be measured using techniques such as cost–benefit analysis. In a broad sense they represent good decisions.

Efficiency results in maximising the effectiveness of a project. For example, if a building design enhances productivity whilst costing no more in resources than competing designs, it is described as an efficient user of these resources. Where a building project offers the same level of performance as its alternatives and costs less then this too is an efficient solution.

In order to be effective the project must meet the objectives set by the client. Where these objectives can be achieved for less or achieve more than what was expected for the same budget then an efficient solution is achieved. Sometimes the terms 'economically efficient' and 'cost-effectiveness' are used loosely to describe the same thing.

Economic optimisation is the process that is used to determine the most economic in terms of both efficiency and effectiveness. In the real world it is a goal that is rarely achievable, since possible solutions that might achieve better goals against this objective may be unknown and therefore undetected. Optimum solutions are therefore restricted to current knowledge scenarios. Optimum solutions also militate against the theory of continuous improvement that is a required part of added value expectations and its general philosophy.

1.5.2 Best value

The discernible shift that has occurred in the focus of activities from cost to value is very much embedded in the Best Value movement. Best value is a concept that has come out of the Local Government Act (1999), which sets out the requirements that are expected. The key phrase in the Act is: 'A Best Value authority must make arrangements to secure continuous improvement in the way in which its functions are exercised, having regard to a combination of economy, efficiency and effectiveness.' The concept of best value applies equally as well in the private as in the public sector. None of us wants to possess the cheaper things in life, but to possess 'the expensive things that cost less'. Best value aims to achieve a cost-effective service, ensuring competitiveness and keeping up with the best that others have to offer. It embraces a cyclical review process with regular monitoring as an essential part of its ethos.

Best value extends the concepts of value money that have been identified for a long time within both construction and property. Rethinking Construction, for example, defines value in terms of zero defects, delivery on time, to budget and with a maximum elimination of waste. In order to show that best value and added value are being achieved, it becomes essential to benchmark performance including costs. It is also necessary to benchmark the overall cost of the scheme so that improved performance in the design can be assessed against its cost. The sharing of information underpins the whole best practice process. Even the leaders in an industry need to benchmark against their competitors in order to maintain that leading edge. Whilst the aspiration of best value is both admirable and essential, its demonstration in practice presents the challenge.

The best value concept for local authorities is being managed for the Department Communities and Local Government by the Audit Commission. It is a challenging performance framework placed on local authorities by central government and, since April 2000, all local authorities in England and Wales have had a duty to plan to provide their services under the principles of best value. Each service review must show that the local authorities have applied the four 'Cs' of best value to the service, and show that it is:

- **Challenging:** Why and how the service is provided
- **Comparing** the performance with others, including non-local government providers
- **Competing:** The authority must show that it has embraced the principles of fair competition in deciding who should deliver the service
- **Consulting** local service users and residents on their expectations about the service

Local authorities are required to show that they are continuously improving the way in which their services are delivered.

The Audit Commission has set up a new inspection service to guide the work of best value. In common with other forms of inspection services and benchmarking, best value seeks out best practices and uses this to help all local authorities to improve their general levels of performance. Best practices today are unlikely to be best practices tomorrow, since the achievement of improvements in quality is always a journey and never a destination. The enhancement of quality remains the long-term goal. Best value inspectors use a simple framework of six questions to make sure that they collect the right information and evidence to support their judgements:

- Are the authority's aims clear and challenging?
- Does the service meet its aims?
- How does its performance compare?
- Does the best value review (BVR) drive improvement?
- How good is the improvement plan?
- Will the authority deliver the improvements?

Best value is a concept that is important not only to the public sector but also within private sector organisations, however it is achieved.

1.6 COST ADVICE

Throughout the development cycle the quantity surveyor will be called on to advise the client on matters of cost. This cost advice will be necessary regardless of the method used for contractor selection or tendering purposes. The advice is particularly crucial at the early stages of project inception. It is at this time that major decisions, often affecting the size and quality of the works, are determined, if only in outline form. It is important therefore that the cost advice given be as reliable as possible, so that clients can proceed with the greatest amount of confidence. Quantity surveyors are widely recognised within the construction industry as the most appropriate cost advisers. Their skills in the measurement and valuation of construction work are without equal.

The type of cost advice required will vary depending on the individual circumstances and the nature of the design and specification information available. A designer who is either unable or unwilling to provide quantifiable and qualitative information must therefore expect that the design cost advice will, by necessity, be vague. If we are unable to tell contractors what we require them to construct, then we should not expect them to price such work. Surveyors must also realise the importance of providing realistic cost advice, which will contribute to the overall success of the project. In this context they must become more familiar with design method and construction organisation and management.

The types of cost advice which may be required at the different stages of the development cycle may include a combination of the following:

- Budget estimating based on a client's brief
- Cost advice on different tendering and contractual arrangements
- Pre-tender price estimating
- Comparative costs of alternative design solutions
- Elemental target costs for cost planning
- Whole-life cost planning
- Tender analysis, reconciliation and recommendation
- Interim payments and financial statements
- Final accounting
- Cost analysis of accepted tenders
- Costs-in-use
- Taxation and insurance considerations

1.7 CONTRACTUAL ARRANGEMENTS AND THEIR EFFECTS ON COSTS

The needs of the client in connection with a proposed construction project are unique. A majority of projects are of a bespoke design even where ideas may be copied from other projects or standard components are used. Even in the case of an 'off-the-shelf' project the site characteristics, and most likely the weather conditions under which the project is constructed, will be unique to the project. This means that achieving the right price for the project within the right timescale is a challenge.

Many of the clients of the construction industry are not regular purchasers of projects. A successful outcome is achievable only where the complexity of the processes used is carefully understood and dealt with appropriately. Such clients, through their networks, will paint a glowing picture of the industry where a project has gone well in their eyes. They will thus act as ambassadors for the industry in encouraging associated clients to modernise their premises or to build new premises.

The establishment of a clear procurement strategy is the key to a successful outcome. This strategy will assist those who are involved in planning and co-ordinating projects to prioritise key outcomes as well as reflecting on risk and establishing how the process will be managed and controlled. Six key steps have been identified by the Strategic Forum *Accelerating Change* report. These are as follows:

- **Statement of business needs.** Priorities, outcomes, stakeholder and constraints
- **Business case.** Encompassing all business requirements
- **Strategic brief.** Expressed in client's terminology and an absence of industry jargon
- **Selection of the team.** This may incorporate the early selection of the contractor and other specialist firms
- **Delivery of the business solution.** Clearly focused upon by the client and the team
- **Capture learning.** Confirm benefits and inform future projects

The execution of construction work on site necessitates the awarding of a contract to a constructor. The promoter or client has many different options available for this purpose. A successful contractual arrangement, however, will generally require the adoption of some recognised procedures. The construction industry is constantly examining new ways of contractor selection to combat bad press reports on construction time, cost and performance. Methods used in other countries were supposed to reduce the contract period. Further investigation and research should enable performance generally to be improved. The quantity surveyor's skills are aimed at producing more economic designs and solutions to construction work on site. For a descriptive treatment of the various methods available, students should refer to *Contractual Procedures in the Construction Industry* by Allan Ashworth (Pearson Prentice Hall 2006).

In addition to other objectives, clients prefer to pay as little as possible for the construction projects. They also wish to know in advance, wherever this is possible, the expected price they will be required to pay. Contracts can be broadly classified as either measurement or cost reimbursement. The former type provides for a reasonably accurate cost prediction, the latter does not. However, if the cost reimbursement contracts can be shown to be less expensive then clients will often be prepared to forgo the specific price prediction. Factors such as the cost–risk element, which is greater to the contractor in the measurement contract, must be balanced with the cost control capabilities which are recognised as being weak with cost reimbursement contracts. The following are generally held opinions based on rule-of-thumb guidelines alone:

- In the absence of any form of competition, tender prices are likely to be higher than where several firms may be seeking the contract.
- Negotiated tenders, under normal conditions, are typically 5% higher than a comparable selective tender.
- Open tendering should achieve the lowest possible tender sum for a project.
- Unorthodox or unusual methods of tendering and contractual arrangements generally incur higher costs.
- There is an optimum contract period in terms of cost; where this is varied tender prices will generally increase.
- Fixed price tenders do not necessarily mean lower final accounts. Where they over-anticipate inflation they will produce higher final accounts than the comparable fluctuating price tender.

The economics of the contractual arrangements need to be measured in terms of the total cost to client, inclusive of professional fees associated with cost. It must be remembered, as in other types of evaluation, that economics is only one factor to consider. Open-ended contracts which may produce the lowest final accounts can cause immense anxieties for a client, who may prefer to pay for peace of mind. It is in reality very difficult to make realistic comparisons, even where it is possible to examine two similar projects being constructed on different contractual bases.

1.7.1 Duty of care

Reasonable care must be exercised during the preparation of estimates, tenders or quotations. A building owner cannot be held responsible or liable for the mistakes in tenders prepared by a building contractor. They can, of course, to some extent protect themselves against the unfortunate effects of these errors, by assessing a tender against an approximate estimate or the tenders from other contractors. They will not, knowing that a tenderer's price is dubiously low, want to enter into a contract with that firm, for fear of the contract being uncompleted or difficult to execute.

Professional surveyors present themselves as being qualified to do the work entrusted to them. If they do not possess the level of skill or experience which is usual in the profession, or if they neglect to use the skill which they in fact possess, they will be guilty of negligence. Although surveyors may be sued for negligence under their contract of employment with the building owner or under common law in tort, the owner would first need to prove the following:

- That the surveyor owed a duty of care. If the surveyor was receiving a fee for services then the point would be established. Even in circumstances where a fee was not charged, this would not necessarily remove the duty of care.
- That the surveyor's error was carelessly made, and performance of duty was done in a reasonable manner. It would need to be shown that under similar circumstances other surveyors were able to provide more accurate forecasts of the future prices for a proposed project.
- In order to recover damages for negligence, it is necessary then to prove the amount of damages suffered.

1.8 CONSTRUCTION ECONOMICS

Construction economics consists of the application of the techniques and expertise of economics to construction projects. Economics in general is about the choice of the way in which scarce resources are and ought to be allocated between all their possible uses. Construction economics is a small part of a much larger subject of environmental economics. This is concerned with the study of man's needs in connection with shelter and the suitable and appropriate conditions in which to live. It seeks to ensure the efficient use of resources available to the industry, and to

increase the rate of growth of construction work in the most efficient manner.
It includes a study of the following.

A client's requirements   This involves a study of the client's wants and needs,
and ensuring that the design of the project is kept within the available funds to
be provided by the client. The client's fundamental needs can be summarised as
follows: satisfaction that the building meets their needs, that it is available for
occupation on the specified date for completion, that the final account closely
resembles the estimate and that the construction project can be maintained at
reasonable cost.

The possible effects on the surrounding areas if the development is carried out
This considers the wider aspects associated with planning and the general amenities
affected by proposed new construction projects.

The relationship of space and shape   This evaluates the cost implications of the
design variables, and considers those aspects of a particular design and their effects
on cost. It does not seek to limit the designer's skill or the aesthetic appearance of
the project, but merely to inform the designer and the client of the influence of their
design on the overall cost.

The assessment of the initial cost   This factor seeks to establish an initial estimate
that is sufficiently accurate for advice purposes and can be used for comparison
purposes throughout the building process.

The reasons for, and methods of, controlling costs   One of the client's main
requirements in respect of any construction project is the assessment of its expected
cost. The methods used for controlling the costs will vary, depending on the type
of project and the nature of the client. The methods adopted should be reasonably
accurate but flexible enough to suit the individual client's requirements.

The estimation of the life of buildings and materials   The emphasis on the
initial construction costs has moved to consider the whole-life costs of a project.
The spending of a little more initially may result in a considerable saving over the
life of the building. However, the estimation of building material life, interest rates
and the economic life of a project can be difficult in practice. The influence of
taxation can have a substantial effect on whole-life costs.

Consideration must also be given to the wider aspects of the subject of construction
economics in respect of the industry in general. The economic aspects of the
following are worthy of note:

- The role of the surveyors, architects, engineers and builders employed in the
  industry
- The division of the industry between the design and construction processes

- The size of the industry, its relationship to other industries and the national economy
- The types of development undertaken
- The types and sizes of construction firms, and the availability of specialist contractors
- The variations in building costs and factors that influence these variations – market conditions, regional location etc.

The construction industry has characteristics which separate it from other industries. These characteristics can be classified as follows:

- The physical nature of the product
- The structure of the industry
- The organisation of the construction process
- The method of price determination

The final product is often large and expensive and may be required over a wide geographical area. Buildings and other structures are for the most part specially made to the requirements of each individual customer, although there is scope for

Inception

Appraisal
Establishing the needs and wants of the project
Strategic briefing
Financial and economic considerations
Whole-life considerations

Design

Sketch design
Detailed design
Tender preparation

Construction

Construction methods
Project planning
Time, cost, quality

In-use

Commissioning
Repair, replacement, refurbishment
Demolition

Facilities management

The project and its use

Sustainability

Environmental considerations

Fig. 1.1   Construction economics objectives

some speculative work, particularly in housing. The nature of the product also means that each contract often represents a large proportion of the work of a single contractor in any year, causing substantial discontinuity to the production functions.

Project cost control is the application of economic principles to the construction project. It examines not only the costs appropriate to a specific project, but also the factors and influences of the determinants of this cost (Figure 1.1).

1.9 ECONOMIC ANALYSIS OBJECTIVES

The primary objective of economic analysis is to secure cost-effectiveness for the client. In order that this can be achieved, it is necessary both to identify and to evaluate the probable economic outcome of a proposed construction project. The analysis will be required from the viewpoint of the owner of the project for his competing proposals. The analysis may be evaluated in the following terms:

- To achieve maximum profitability from the project concerned
- To minimise construction costs within the criteria set for design, quality and space
- To maximise any social benefits
- To minimise risk and uncertainty
- To maximise safety, quality and public image

1.9.1 Procedure to be followed

Economic analysis comprises four processes:

1. Preparation, which includes understanding the project, defining the client's objectives and collecting the appropriate data
2. Analysis, which requires an interpretation of the available data and the formulation of alternative solutions
3. Evaluation, which is a combination of the assessment of the suggested alternatives and the identification of the optimum solution
4. Decision-making, which involves choosing to proceed with the course of action now identified

These processes are now briefly discussed.

Understanding the project   Prior to economic analysis, the aims of the promoter of the project should be clearly understood.

Defining the objectives   The failure to identify clearly the client's needs is a main source of client dissatisfaction. The objectives must be clearly stated and understood, and be compatible with each other. These may include:

- To provide an industrial building of 30,000 m^2 required for a specific manufacturing process
- To have the building ready for occupation in three years' time

■ To have an initial cost not exceeding £20m and to provide for whole-life cost savings in the immediate ten-year period

The client's objectives and criteria must be specified and defined quantitatively. It will be necessary to define more precisely the extent of the whole-life cost measures to be incorporated.

Collecting data   A review of all known information regarding the type of project to be constructed should be collected. Data regarding the nature of the construction site and the availability of facilities should be obtained.

Analysis of data   This is the process of converting the developed data into something meaningful and useful which is capable of achieving the desired objectives. Some of this information can be handled by the computer in order to generate as much information as possible.

Interpretation of results   This will occur on completion of the analysis and will determine both the feasibility and the viability of the project under examination. The results should be well organised and comprehensive in order that they may be properly utilised in the evaluation phase.

Formulation of alternative solutions   Different solutions will be available which may lead to the same objectives. These alternatives need to be fully explored in terms of the client's evolving needs.

Evaluation of alternatives   The criteria selected for comparison should enable the optimum solution to be selected. The correct balance between initial costs and the necessity to reduce future costs should consider all criteria.

Identification of correct project   If management is to be able to make the right decisions, it must do so on the basis of all the known information and a correct economic analysis as outlined above.

1.10 DESIGN METHOD

Efficient cost advice from the quantity surveyor depends on at least some understanding on his or her part of the design method used by architects or engineers. Some effort must therefore be made to appreciate what the designer is attempting to achieve, and the method used during this process.

The design of a construction project is a combination of three facets (see Figure 1.2).

1.10.1 Function

Function includes an understanding of the way in which the project will eventually be used. It may also be necessary to consider the ways in which this use may change

Fig. 1.2   Design considerations

over the lifespan of the project, and how it may be adapted to suit these changes. If the project does not serve its function properly, then it is likely that it will prove to be an irritation to its users or owners. In this case it may need to be quickly altered or perhaps even abandoned in favour of an alternative structure. In the extreme case projects which are non-functional may prove to be nothing more than a designer's folly.

1.10.2 Technology

A good designer should be fully conversant with the materials and methods of construction which may be available. Technological aspects also take into account the manufacture and assembly of the various components. The designer needs to be aware of the resources of men, machines, materials and money which will be required for the design, and that the required technology is already available. The designer must take note of the buildability aspects which will affect performance, function and cost. A truly good design, one presupposes, is one that is in the correct sense well built. Sound building is an essential ingredient of good architecture or engineering. As well as understanding building construction the designer should also be familiar with production technology.

1.10.3 Aesthetics

Aesthetics is largely a combination of the building purpose itself together with the location of the project within the built environment. The vocabulary of visual aesthetics includes unit, texture, form, colour, proportion, balance, symmetry, character etc. Although aesthetics are to some extent value judgements, and therefore highly subjective, there are some readily defined rules which can only be ignored at one's peril.

Excellence in building and construction is therefore attained only where appearance, soundness of construction and usefulness have been developed together in a fully integrated manner. The quantity surveyor will wish to add to these criteria the importance of cost and value. In today's economy, excellence in design must be achieved at a reasonable level of cost, both in respect of initial costs and also during use. The provision of cost advice on a quantity basis can fairly readily be provided using single-price methods of estimating or by a form of analysis. New technological solutions can also be evaluated by the latter method. The costing of the qualitative aspects of the design is not so easy and relies heavily on experience and opinion and value judgements. However, since these aspects are generally fully integrated it is almost impossible to consider one without the others. Furthermore, it is necessary that such cost advice be provided throughout the various stages of both the design and construction process, and also once the project is in use.

Design methods also need to work within the general constraints imposed by the technical, legal, functional and economic framework. The technical constraints impose limitations in respect of the characteristics of materials available, the skill and craftsmanship of labour, the structural form required, the necessity of integrating engineering services into the project, and the capability of the constructional processes which might be used. For example, it may be possible to design large building units which 'clip' together, and while the theory may be sound, the idea may be limited due to access and location of the site which allow only the delivery of small components. Designers may assume that errors on site can be reduced almost to nothing. The practical aspects of construction operations today indicate that this is incorrect, and where a design relies on this assumption it will fail. The technical constraints must also take into account the general requirements of buildability. This may result in modifications to the design to ensure good building method. Buildability must not be confused with convenience building, which is not a criterion of good building design.

Construction works must also be designed within the legal framework appropriate to the country concerned. While it is possible to design a building which will exclude these requirements, in practice this will not be allowed. Within the UK, and this is common in most countries, some form of legal constraint is necessary to protect both adjoining owners and third parties at large. These legal restrictions include easements, restrictive covenants, building regulations, planning laws, ownership considerations, safety and health acts, and impositions required by the form of contract. Restrictive covenants impose conditions which govern the use of land, while an easement relates to the rights of land usage which include the right to light, right of support, right of way and right to drainage. Functional constraints are imposed by the client, and may restrict designers in their work. For example, the designer must be fully aware of the intentions of the user. This is why a purpose-built project is likely to be more satisfactory than one which was constructed by a developer for speculative reasons or 'off the peg'. The function will also impose restrictions on the aesthetics which the designer may consider to be most appropriate. The designer needs to consider both anthropometrics (which relates to the measurement of the human body) and ergonomics (which introduces the idea that fatigue should be minimised). The designer will need to consider the relationship between the different room or location uses by means of a circulation diagram. This will allow the correct positions of the various rooms to be identified. The function of the building will also dictate to some extent the needs for the provision of environmental services, and the comfort criteria which may be required. The designer must take all of these into account when formulating a design solution. A final consideration of cost or economics in the solution will also constrain designers in their work. They will need to be conscious all the time that, however grand their ideas may be for the project, these must come within the client's overall budget. A designer may be able to persuade a client to adopt some particular favourite part of the design, but in general the cost constraints will be imposed in line with the approximate estimate which was delivered to the client at inception. The purpose of economic constraints is not to provide a cheap building, but to create an acceptable solution within the funding availability of the client.

The designer's first meeting with a client will generally identify the building type, its size, the funds available and the location of the site. The designer's first duty will often be to visit the site, and to attempt to visualise at this stage some idea of the project on the site. Aspect, horizontal and vertical forms, size, shape and position will need to be considered. This is a difficult task and requires a vivid imagination on the part of the designer. The designer must attempt to utilise the full benefits of the site to the best advantage. 'Best' in this context will often be based on rule-of-thumb considerations alone, with little attention paid to its verification in practice. Clients often require their projects to fulfil a function that is related to spatial factors such as the numbers in a school, since these often determine some form of building cost criteria. The priorities for space should be ranked so that they can be given their due importance in solving the design problem in accordance with the client's requirements. Priorities will also exist in the choice of materials and methods of construction. The choice is often difficult to make, but a scarcity of financial resources makes it inevitable. These decisions are at the centre of good cost control.

The design approach will also vary depending on the nature of the designer. The mention of a proposed scheme such as a church or house or office block to a designer may immediately create an architectural form in mind. This approach, although often denied by architects, is known as designing from the outside in. It is a Wild West approach, where the façade is really all that matters. After all, this is often the only part of a building that is seen by the majority of the public and possible building clients. Before attempting to be too critical of this method, let us remember that when 'house-hunting' a poor elevational appearance may deter one from further viewing. The alternative method, which is generally recognised as being more satisfactory, is to design from the inside out. Choose the correct spatial arrangements, and then decide on the elevational treatment that is the most appropriate. This need not imply that the external appearance does not matter, but seeks to select priorities in the design.

There will often be many different ways in which to organise the spatial arrangements. The available alternatives will influence the cost–shape factor. It is not until this stage has been reached in the design process that some form of realistic cost target can be applied for elemental analysis purposes. The elemental cost targets cannot be fixed until these major decisions have been finalised. Once the spatial arrangements have been accepted then some attempt can be made to satisfy the form of the structure as may originally have been envisaged. The choice of constructional methods and materials will greatly influence the final solution. This will involve selecting the correct specification to fulfil the purpose of the scheme. The role of the quantity surveyor in identifying the costs of the various available choices is invaluable to the designer.

1.11 BUILDING ECONOMIC THEORIES

The study of economics applied to construction projects has resulted in the formation of some building economic theories. However, due largely to the infancy of this academic discipline, more theories have yet to emerge. The theories provide a broad indication of the cost implications of building design. Perhaps the best known of these theories is the wall-to-floor ratio, where the implication is that the lower this ratio, the less expensive will be the cost of building. Many such theories can be expressed in a mathematical equation form. Further research should enable us to achieve a better understanding of the determinants of building cost. At the moment only a small amount of analytical work has been carried out, and therefore our advice is often based on opinion and assumption, albeit of an expert nature. Other theories which have emerged are as follows.

$$\text{Plan shape index} = \frac{g + \sqrt{(g^2 - 16r)}}{g - \sqrt{(g^2 - 16r)}}$$

where g is the sum of the perimeters of each floor divided by the number of floors and r is the gross floor area divided by the number of floors. This is a development of the length/breadth index devised by Mr D. Banks of the then Polytechnic of the South Bank. It aims to measure the plan shape efficiency of a building.

$$\text{Optimum envelope area} = n\sqrt{N} = \frac{x\sqrt{f}}{2S}$$

where N is the optimum number of storeys, x is the roof unit cost divided by the wall unit cost, f is the total floor area (m^2) and S is the storey height (m). The envelope of a building comprises the external walling area and the roof. This theory, using the above formula, aims to select the appropriate number of storeys for a building based on roof and wall costs.

In each of the above theories only a few of the major elements are taken into account in measuring building economy. Other theories have also been developed, and some of these are explained more fully in Chapter 5.

1.12 ENVIRONMENTAL IMPACT OF THE CONSTRUCTION PROJECT

The impact of the construction industry on the environment is substantial. During the extraction and manufacture of construction materials, their transportation, the process of construction and the use of buildings, large quantities of energy are used. Major contributions are made to the overall production of carbon dioxide which exacerbates the 'greenhouse' effect. The environmental impact is global but, during the construction process, communities and individuals are affected.

Society is becoming increasingly concerned with the effect of human activity on the environment. In recent years there has been greater pressure on clients to state all the likely direct and indirect effects of their projects on the life and amenities of surrounding areas. It has been the practice to supply environmental impact assessments with planning applications for major projects. The assessment requires a statement of the impact of the project on the surrounding area. It should also include details of work which will limit the impact, e.g. soundproofing in the case of a noisy transportation system. To involve the public more closely the assessment statement should be jargon-free. A further requirement is that promoters should consider the detailed impact of their project early in the planning process and undertake wide consultations involving the public and environmental groups. Despite the requirements specified by the directive, there is concern that there is a lack of definition regarding the scope of the assessments and the level of detail required. Assessments frequently tend not to look beyond the confines of the project concerned.

The concept of a green building is an elusive one. The definition is broad and being green in a professional sense may merely come down to a change in attitude. Most buildings in the UK are designed to cope with the deficiencies of a light loose structure, designed to meet the Building Regulations thermal transmittance standards and no more. Given that about 56% of energy consumed, both nationally and internationally, is used in buildings, designers have opportunities and responsibilities to reduce global energy demand. There is a need to make substantial savings in the way that energy is used in buildings, but there is also a need to pay attention to the energy used in the manufacture and fixing in place of a building's components and materials. For a new building this can be as high as five times the amount of energy that the occupants will use in the first year.

In the 1950s and 1960s building maintenance and running costs were largely ignored at the design stage of new projects. Today the capital energy costs which are expended to produce the building materials and to transport them and fix them in place are often ignored in our so-called energy-efficient designs. In any given year the energy requirements to produce one year's supply of building materials is a small but significant proportion (5%–6%) of total energy consumption, and typically about 10% of all industry energy requirements. The building materials industry is relatively energy-intensive, second only to iron and steel. It has been estimated that the energy used in the processing and manufacture of building materials accounts for about 70% of all the energy requirements for the construction of a building. Of the remaining 30%, about half is energy used on site and the other half is attributable to transportation and overheads. Research in the USA has shown that 80 separate industries contribute most of the energy requirements of construction and five key materials account for over 50% of the total embodied energy of new buildings. This is very significant since considerable savings in the energy content of new buildings can be achieved by concentrating on reducing the energy content in a small number of key materials.

1.13 KNOWLEDGE MANAGEMENT

Knowledge management describes the process which can enable an organisation to exploit the knowledge and learning of its people. This can result in increased efficiency in project implementation and reductions in waste and their associated costs, and can foster greater innovations.

Knowledge is not simply information. Information is often passive: what is important is how that information is received and what interpretations and responses the receiver can make. Knowledge involves the ability to make judgements. Knowledge is typically accumulated through experience or education, but there is also a range of management techniques that can be applied to make this more effective, especially in its recall at the appropriate moment.

Knowledge management is concerned with much more than information and communication technology. It is also important to remember that a large amount of knowledge resides in people and is thus not easy to transfer to information systems. To use this tacit knowledge requires people-orientated activities such as mentoring, knowledge-sharing meetings, technical communities and networking. This may require a culture of change within an organisation in order to encourage knowledge sharing rather than discourage it.

When the Building Cost Information Service (BCIS) was first established in the 1960s there was some resistance then to the sharing of information that would assist competing firms with their businesses. During its early years, access to that service was restricted to chartered quantity surveyors. It was eventually opened up to any organisation on a subscription basis with the avowed intent of sharing information on a reciprocal basis. In this way all of the firms involved were able to benefit and to improve their professional services to clients.

Another aspect of knowledge management is its close relationship with innovation. Knowledge creation, which is a vogue word for innovation, is important in the current era with all of its uncertainties and rapid changes in ideas and practices. Innovation includes the introduction of new services and products to clients and improvements in business processes.

Knowledge management in project-based firms, which are typical in industries like the construction industry, is a particular challenge. Difficulties can arise in project-to-project knowledge transfer as well as project-to-business transfer. These difficulties are exacerbated by the discontinuities between projects and the teams that work on them. On the completion of projects teams will always, to some extent, disperse. Companies do not routinely and formally reflect on the successes and failures of a project in order to improve performance next time.

SELF ASSESSMENT QUESTIONS

1. What is the purpose of cost control and why is this of major importance to clients in the construction industry?

2. Why is an understanding of building design method relevant to the cost study and control of building projects?

3. Suggest how the quantity and quality of advice on the costs of construction are likely to evolve in the coming decade.

BIBLIOGRAPHY

Ashworth, A., *Building Economics and Cost Control – Worked Solutions*. Butterworth 1983.

Ashworth, A., *Contractual Procedures in the Construction Industry*. Pearson Prentice Hall 2006.

Ashworth, A., *The Education and Training of Quantity Surveyors*. Construction Information File, Chartered Institute of Building 1994.

Ashworth, A., *Precontract Studies: Development economics, tendering and estimating*. Blackwell Publishing 2008.

Ashworth, A., 'Value rather than cost'. *Journal of the Malaysian Institution of Surveyors*, 1996.

Ashworth, A. and Hogg, K., *Added Value in Design and Construction*. Longman 2000.

Bon, R., *Timing of Space: Some Thoughts on Building Economics*. Habitat International 1986.

Davis, Langdon and Everest, *Buildings That Work for Your Business*. Director Publications Ltd 2001.

Egan, J., *Accelerating Change*. Rethinking Construction 2002.

Harvey, R.C. and Ashworth, A., *The Construction Industry of Great Britain*. Butterworth-Heinemann 1997.

Hillebrandt, P.M., *Economic Theory and the Construction Industry*. Macmillan 2000.

Kelly, J., Morledge, R. and Wilkinson, S., *Best Value in Construction*. Blackwell Publishing 2002.

Langsten, C., *Sustainable Practices: ESD and the Construction Industry*. Envirobook Publishing 2001.

Powell, C., *The Challenge of Change*. Royal Institution of Chartered Surveyors 1998.

Raftery, J., *Principles of Building Economics*. BSP Professional Books 1991.

Royal Institution of Chartered Surveyors, *Surveying in the Eighties*. RICS 1980.

Royal Institution of Chartered Surveyors, *The Future of the Chartered Quantity Surveyor*. RICS 1983.

Royal Institution of Chartered Surveyors, *Quantity Surveying 2000: The Future of the Chartered Quantity Surveyor*. RICS 1991.

CHAPTER 2

HISTORY OF BUILDING ECONOMICS

LEARNING OUTCOMES

After reading this chapter, you should have an understanding of the history and development of building economics within the construction industry. You should be able to:

- Realise why the subject was developed
- Identify the main sources of textual materials
- Understand the trends that are evident in the subject development
- Identify the current issues
- Understand the factors that influence the subject discipline

2.1 INTRODUCTION

Building economics is as a subject barely 50 years old. It has largely developed since the middle of the twentieth century with the building boom that followed the ending of the Second World War. While the subject matter is not now entirely in the province of quantity surveying, its origins can be traced to quantity surveying practice. It can also be reasonably argued that authors of this subject have frequently had a background in this discipline.

The impetus for the subject's development is probably twofold. First there was huge public spending on construction works in the 1950s and 1960s. Houses, schools, hospitals, roads etc. were all required to meet a rapidly expanding population. In order to plan this spending properly and achieve value for money in the various projects, something additional needed to be done other than simply to measure and value the works. Second the expansion of post-compulsory education and the introduction of undergraduate courses for quantity surveyors needed full-time lecturers, some of whom carried out research in the subject matter of building economics.

The then Ministry of Public Building and Works and the Royal Institution of Chartered Surveyors began to develop systems of financial control and evaluation for new buildings. These systems required a greater understanding of building cost relationships and patterns. They laid the foundations for a fuller exploration, understanding and application of the principles involved in both theory and practice.

2.2 BUILDING ECONOMICS

Economics in general is about the choice in the way that scarce or limited resources are and ought to be allocated between their possible uses. It employs accepted principles and procedures to determine the real cost to the community or to an organisation. It is less concerned with earnings or revenues necessary to meet various obligations than with the principles necessary to justify the selection of a particular project or activity.

Building or construction economics is a branch of general economics. It consists of the application of the principles associated with general economic theories to the particular needs and requirements of the construction industry. It is concerned with a study of the industry and its place within the economy, the construction firm, the roles of designers and constructors, the processes employed and the final product of buildings and other structures. This is the broad definition.

Building economics, in the context of this chapter, is a much narrower definition. It is really a subset of construction economics as described above. The emphasis is largely concerned with the building product, and in attempting to make this more efficient, effective and economic. It is not concerned with costing and accounting nor in examining the contributions and roles of designers and constructors other than in the deliberate sense that they affect building economy.

2.3 STANDARD TEXTS

In 1957, the then Ministry of Education published *Building Bulletin No. 4 – Cost Study*. The Bulletin introduced to the construction industry a new method of working and the principles of cost analysis and cost planning. It was a milestone in practice. The aims of this process were carefully itemised to:

- Identify building elements, i.e. parts of a building having the same function
- Reveal the distribution of costs between the elements in a building
- Relate the cost of any constituent element to its importance as a necessary part of the whole building
- Compare the costs of the same element in different buildings
- Obtain and use cost data in planning other schools
- Ensure a proper balance, within the appropriate cost limits, between the superficial area per place and the cost per square metre

A list of the other relevant texts is shown in Table 2.1. The first of these books (Nisbet 1961) was written by an association of practising quantity surveyors. This preceded, by one year, the inauguration of the Building Cost Advisory

Table 2.1 Building economics texts

Editions

First	Latest	Author	Title and publisher
1957	1972		*Building Bulletin No. 4 – Cost Study*. HMSO
1961	1961	Nisbet, J. (ed.)	*Estimating and Cost Control*. Batsford
1961	1961	Browning, C.D.	*Building Economics*. Batsford
1964	2007	Ferry, D.J., Brandon, P.S. and Ferry, J.	*Cost Planning of Buildings*. Blackwell
1966	1983	Stone, P.A.	*Building Economy*. Pergamon Press
1968	1997	Flanagan, R. and Tate, B.	*Cost Control in Building Design*. Collins
1972	1995	Seeley, I.H.	*Building Economics*. Macmillan
1973	1980	Bathurst, P.E. and Butler, D.A.	*Building Cost Control Techniques and Economics*. Heinemann
1983	1983	Ashworth, A.	*Building Economics and Cost Control*. Butterworth
1984	1984	Williams, B.	*Design Economics for Building Services in Offices*. Building Economics Bureau
1988	2009	Ashworth, A.	*Cost Studies of Buildings*. Pearson Prentice Hall
1991	1991	Raftery, J.	*Principles of Building Economics*. BSP Professional Books
1999	1999	Best, R. and De Valance, G.	*Building in Value*. Arnold

Service (1962) that was formed by the Royal Institution of Chartered Surveyors. This was later to become the Building Cost Information Service (BCIS). Nisbet's book was written by practitioners for practitioners in an era when it would have been difficult to find many full-time surveying lecturers. While much of what the book was expounding is now assumed common practice, at the time it was somewhat revolutionary, presenting a new set of ideas and procedures that were yet to be exploited and tested in practice. Browning (1961) and much later Bathurst and Butler (1973) also helped to contribute to and expand the subject knowledge and techniques. All of these were practitioners. Bathurst, like Nisbet, had been one of the parties instrumental in the development of cost planning of building projects in the public sector. Williams (1984) much later added to the knowledge base with a specialised book on building economics and engineering services.

D.J. Ferry in 1964 prepared the first edition of *Cost Planning of Buildings* which was subsequently enlarged by Brandon in 1980 and then by Kirkham in 2007. The book developed many of the ideas of Nisbet (1961) at a time when amongst the profession generally there was a huge interest in developing such techniques in practice. The two clearly identified techniques of designing to cost (elemental cost planning) and costing a design (comparative cost planning) quickly became established as a single set of procedures. The originators of these techniques, the then Ministry of Education and the Royal Institution of Chartered Surveyors, agreed to share their ideas in a common framework.

In 1968, the Directorate of Building Management, Ministry of Public Building and Works, produced one of the few programmed learning texts for the building industry, entitled *Cost Control in Building Design*. This has only recently been revised and brought up to date by Flanagan and Tate (1997). The book is in two parts. The first part covers the principles of cost control and the second part the techniques of cost control. The late Ivor Seeley, perhaps the most prolific of quantity surveying authors, prepared the first edition of *Building Economics* in 1972. This book offered a broader approach than other authors by providing more information on the subject of economics generally. *Cost Studies of Buildings* was first published in 1988 following the earlier publication of *Building Economics and Cost Control* in 1983. The latter text was limited to a question and answer approach. Cost studies is a title that probably reflects what the subject of building economics means in theory and practice.

Two books are also worthy of this list in contributing to the subject knowledge. Stone's (1983) book, first published in 1966, on *Building Economy* provides the reader with a general background knowledge that is required for an understanding of building economics. The book provides a wide-ranging analysis of construction and its place in the economy. Much later Raftery (1991) provides a more focused book covering the more essential economic principles of supply and demand as applied to the construction industry.

Each of the above has helped to explain and expand the subject of building economics. The subject has grown rapidly during the latter half of the twentieth century and represents a major aspect of many building and surveying courses. In addition there have been numerous articles and papers written on this subject. An important source of material is most notably found in the Conference Proceedings CIB W55 Committee Building Economics (International Council for Building Research Studies and Documentation).

2.4 TRENDS IN BUILDING ECONOMICS

Table 2.2 suggests some of the trends in building economics that occurred during the latter part of the twentieth century. The emphasis throughout has been on improving the quality of advice in order to allow clients to make better decisions. Coupled with these developments has been an increase in knowledge about the behaviour of costs and in the use and application of information technology. The rapid retrieval of data and the ease by which models can be updated to take into account design decisions have allowed such improved advice to be provided.

The trends have swung between a heavy reliance on the importance of experience and judgement and a rationale that construction costs can all be analysed in simple (or complex) formulae. There is now a genuine belief that costs are a combination of each of these aspects. Today the emphasis is towards providing design and construction solutions that seek to resolve the economic choice while still meeting the specific needs of clients. Value for money is seen as a process of adding value

Table 2.2   Chronology of developments in building economics

Date	Building economics	Other developments	Practice
Pre-1960s	*Building Bulletin: Cost Study* (1957) Building price books RICS Cost Research Panel	Post-war building boom	Approximate estimating Bills of quantities Final accounts
1960s	Cost studies of elements Cost limits and allowances Value for money in building Building Cost Information Service The Wilderness Group	Cost–benefit analysis	Elemental bills Operational bills Cut and shuffle Cost planning Standard phraseology
1970s	Costs-in-use Cost modelling Contractor's estimating Cost control	Measurement conventions Data coordination Building maintenance information Buildability Value-added tax/taxation Bidding strategies Computer applications Undergraduate surveying degrees	Computer bills Formula methods of price adjustment Cash flow forecasting Engineering and construction
1980s	Life-cycle costing Cost data explosion Cost engineering techniques Accuracy in forecasting Value engineering	Coordinated project information Procurement systems European comparisons Construction industry analysis Postgraduate education Single-point responsibility	Project management Post-contract cost control Contractual procedures Contractual claims Design and build
1990s	Value management Risk analysis Quality systems Expert systems	Facilities management Commercial revolution Single European market Building sustainability Information technology	Fee competition Diversification Blurring of professional boundaries Development appraisal
2000s	Benchmarking Added value in building and design Whole-life costing	IT in construction Knowledge management	Rethinking construction Lean construction Facilities management

to the project. Incorporated within this economic choice is the importance of the whole-life costs associated with the project.

2.5 REASONS FOR CHANGE

The wide interest in building economics is a result of many different factors.

2.5.1 Knowledge

The acquisition of increased knowledge has occurred in all walks of life and in all academic disciplines. This is a significant indicator. This knowledge is increasing at an exponential rate. It presents difficulties for those who design courses of study in that it has been tempting to add all of this knowledge onto the already overloaded curriculum. Courses have now become more discriminative in the way in which they select their material for students to study. The need for Masters programmes to encapsulate the material has provided one solution to this problem.

2.5.2 Information technology

Information technology is perhaps the biggest single indicator that has helped us unlock techniques and practice that would otherwise not have been available using manual practices. It allows us, for example, to search for information via the Internet. Many of the previously repetitive tasks, which were frequently time-consuming, can now be stored in the computer's memory for future use. Data used in building economics can be easily and rapidly changed or updated. This facility alone has allowed for a greater range of advice to be provided. Design decisions can more easily be incorporated.

2.5.3 Increased awareness of other industries and professions

The academic base has allowed the different professions to understand each other's work more easily. The mystique associated with other professional practices has partially been removed. The whole subject area of surveying was once defined as a number of discrete disciplines, but these have started to overlap with each other. This phenomenon has also occurred between the different professions as they have sought to expand and to claim the prime ground for themselves.

2.6 HISTORICAL CONTEXT

The ending of the Second World War (1939–45) followed the period of high unemployment of the 1930s. The British economy was in a serious condition since many of its overseas assets had been used to pay for the war and the country was heavily in debt. There was a backlog of building which had been postponed for the six years of war and there was a new government's social and economic policies.

Four important pieces of legislation were to follow:

1. The 1944 Education Act had to be implemented which, with the increase in school-leaving age, required many new schools.
2. The 1948 National Health Service Act came into force which resulted in a programme of new hospitals and health buildings.
3. The Robbins Report of 1963 on higher education created a demand for new universities with the subsequent requirement for capital spending on new buildings.
4. The Housing Acts of 1957, 1959 and 1961 dealt with the elimination of slum property.

During the immediate post-war years, several major factors helped to create conditions of an economic boom in the construction and property industries:

- The desire to renew much outdated property, much of which had fallen into decay through neglect
- The need to replace the considerable amount of property that had been damaged by the ravages of war
- The need to provide new buildings for homes, employment, education, health etc. for a rapidly growing population and the then needs of the welfare state
- The need to provide new buildings to meet the aspirations of a new generation

Coupled with these factors were changes in manufacturing industry, with the newer industries requiring investment in new premises to suit changes in technology, particularly automation and mass production. There was also the switch that was taking place in society with a greater proportion of jobs being created in offices than on the shop floor. In addition, there was the major development of the motor car and the need to build better roads and highways. These all helped to create the need for building and construction development and a relatively prosperous and expanding construction industry.

2.6.1 Early years

Building economics in the middle of the twentieth century confined itself almost to the forecasting of contractors' tenders through approximate estimating. The techniques used were often referred to as single-price methods although in many cases they used more than a single quantified description. Approximate estimates were required to provide clients with a budget. They also provided an overall cost within which architects would then complete the working design drawings for the project. They were methods that allowed for a simple and quick quantification of the building but methods that required significant expertise and judgement in their pricing. These forecasts frequently resulted in the preparation of addendum bills of quantities.

The reasons for the discrepancies between approximate estimates and tender sums were and are not difficult to find. Changes in design and specification, a failure on the part of the architect to keep to the client's brief, increased costs, lack of accurate cost data and information and poor estimating were just some of the reasons. The initial studies of building economics encouraged the development of a set of procedures that would help, at least, to minimise the variability between early price estimates and tender sums. The attempted reconciliation between estimates, tender sums and final accounts came much later.

During these early years implications for value for money were also raised. Why, for example, did projects constructed for the same purpose and occupancy rates cost different amounts to build? The answers were not difficult to detect. In most cases it was due simply to providing more space or constructing to a better quality.

The then Ministry of Education studied the costs of school buildings, realising that the limited budgets were not being spent in the best possible way. Some form of planning building expenditure was required. Cost planning was introduced with the twin aims of costing different design solutions and designing within an overall cost framework. This was the introduction to value for money in buildings and a principle that more could be done with less. This is a philosophy that has taken some time to develop in other areas of work. The principles of evaluating building designs in this way quickly spread to other sectors of the construction industry, particularly among those clients who carried out a large amount of building works.

The Wilderness Study Group, a group of quantity surveyors, began investigations into the design–cost relationships of a large number of hypothetical steel-framed buildings of equal floor area and similar specification. This was one of the earliest forms of cost study ever undertaken.

It was also recognised at this time that more extensive and reliable cost data would be required to allow clients to be properly informed of design decisions made on their behalf. In 1956 the Royal Institution of Chartered Surveyors had set up its own Cost Research Panel which was later instrumental in developing the Building Cost Advisory Service. This was later to become the Building Cost Information Service (BCIS). By the mid-1980s access to this information was provided electronically.

2.6.2 1970s and 1980s

The middle years of building economics were dominated by two themes. The first was concerned with the ability to forecast contractors' tender sums more accurately. There was much evidence to show that early price estimates were often a poor indicator of final costs. Cost modelling was introduced, initially based upon the two different techniques of regression analysis and simulation. Both of these techniques offered a radically different approach to building cost forecasting than had previously been the case. These represented a paradigm shift to what was being employed on a routine basis in practice. They attempted to capitalise on the widening availability of computers that were becoming a more common tool in the industry. While the introduction of these techniques did not achieve expectations, they did offer new perspectives on how building costs are calculated and determined and began to challenge the perceived wisdom of the traditional methods.

The second concerned itself with value for money in building, even though there was no agreed definition of what was meant by this term. Several different possibilities were explored. Several studies had indicated that clients were not always obtaining value for money in the procurement of their buildings. Different designs, meeting the same client's brief, often resulted in different costs. Different designs frequently resulted in different construction methods and techniques being adopted on site and these also generated different costs to the client. Both aspects were considered in the pursuit of value for money. The theme of value for money, which was sometimes misunderstood and difficult to explain, would be a theme that the construction industry embraced for the remaining part of the twentieth century.

Building economics also tended to polarise the researchers involved, resulting in two different perspectives. The fundamentalists wanted to find a rationale for the practices used by the profession. They believed that many of the practices being carried out had no sound basis for their justification. In some cases long-held myths were exposed. Some of the procedures being used were flawed since it could be shown that in many cases they did not work properly. Practitioners were sometimes intolerant of these views and wanted to justify their largely pragmatic processes. The modernists were at the other extreme, wanting to take practice into untried and untested waters. They sought to capitalise on the latest ideas from other research areas. As new ideas were developed the previous all-important techniques or procedures would be disowned. These two approaches had all the classic symptoms of embryonic research practices.

2.6.3 The later years

In more recent years building economics research and development has built upon past studies and has been informed by changes in other parts of society. Criticism about the performance of buildings, sometimes unfair, has required the industry as a whole to respond with new ideas and proposals. Value for money remains a constant companion. This has encouraged developments in value engineering and management techniques to examine ways of meeting clients' longer-term objectives. In this context there has also been a shift away from evaluating buildings on the basis of their initial costs and values alone to looking at the longer-term perspective. While whole-life costing cannot adequately predict what will happen in the future (nothing can!) it has encouraged designers, manufacturers and constructors to re-examine their own design and construction philosophies.

Much of the early research into building economics has been done with little support or enthusiasm from practice. Practitioners often have their own agendas and sadly these did at one time seem incompatible with academic research. The surveying profession does not even now have a strong research ethos. Many practitioners also failed to embrace and exploit the new ideas and technologies that were becoming available. There are of course exceptions and some surveying practices are able to boast about their research departments. The best research is always likely to be a combination of theory and practice. The more recent development of expert systems relies upon a strong input from those in practice. They are, after all, the experts. These systems have also allowed practitioners some ownership and direction of the developments in building economics.

The major theme at the start of the twenty-first century remains value for money, now more correctly described as added value (see *Added Value in Design and Construction* by Allan Ashworth and Keith Hogg, Longman, 2000). Recent reports have indicated that clients of the future will require increased value for the money that is expended on their capital projects. The principle involves reducing the relative costs of construction by designing, procuring and constructing the work in a different way. The construction industry has, of course, been responding to this challenge for the past 50 years and with some success. It involves doing more for less by removing unnecessary costs. It aims to meet the perceived needs of efficiency, effectiveness and economy.

The emphasis throughout has been on improving the quality of advice in order to allow clients to make better decisions. Coupled with these developments has been an increase in knowledge about the behaviour of building costs and in the use and application of information technology. The rapid retrieval of data and the ease by which economic forecasts can be updated to take into account design decisions have allowed such improved advice to be provided.

2.6.4 The future

Historically, much of the important and valuable work that has been undertaken to improve the economics of construction has focused on the building product and especially the design and details being developed during the early stages of the project design. It is easy to demonstrate the cost and value gains that have been achieved through the better understanding of building economics and a reflection and evaluation of what the designer is contemplating. Difficulties occur where the process is left until a later stage and there is then the reasonable desire not to extensively redesign what has already been developed into a workable solution.

The better understanding by the whole design team of the cost implications of construction is to be welcomed, although there is always scope for further improvements in attempting to eliminate waste in design and construction solutions. However, such gains may be considered to be small when set in the context of a changing construction industry and the drive by some to reduce initial building costs by upwards of 30% without adversely affecting the whole-life costs of the project. Whilst the emphasis so far has quite rightly been on the product, it is now time to examine the process of construction more clearly.

Matters such as the impact of procurement and industry practices, and organisations and their impact upon cost, are now being carefully examined in order to reduce costs and add value. The industry has already done a considerable amount of rethinking through appointing the contractor as the lead consultant, through partnering and through supply chain management.

Added value is a key concept in this process. Added value is a term that is used to describe the contribution that a process makes to the development of its products. The construction industry uses plant, materials, people and other resources. The completed projects are usually greater than the sum of the various parts. This represents added value. It is now generally believed that where benefits, or added value, cannot be demonstrated then whatever the practice or service provided, it will disappear. An added value analysis is a management tool that assists in the identification of strengths and weaknesses by the interpretation of the issues raised by analysis. This is considered further in Chapter 23.

The future for building economics can be summarised as follows:

- Construction costs will always need to be forecast, budgeted, controlled, accounted and evaluated. The latter aspect will become more important.
- The principle of doing more for less, which is a principle adopted throughout many facets of industry and society, will remain a topical theme. The focus of the industry is an attempt to do its work quicker, better and less expensively.
- Seeking out better or more economic (not necessarily the same thing) methods is likely to remain high on our agendas. The study and application of building economics have a huge part to play in this process.
- The use of information technology in the capture and use of data and knowledge will become more interactive.
- Best practices, achieved through benchmarking, will set the standards to be adopted and exploited.
- Building economics will become more of a whole-industry approach rather than just the province of a particular discipline. But within this there will be specialists who will spend more of their time dealing with these issues.
- The presence of advanced communication systems promises to make countries more alike and to reduce the importance of national boundaries. Building economics is an international subject discipline. However, it has been developed from different traditions reflecting the variations in theory and practice. Approaches to problems are likely to be more consistent with each other although solutions may not be uniformly applied.
- There is a tendency to overestimate what may happen in the next two years in terms of building economics, but underestimate what may happen over the next ten years.
- The future is always going to be difficult to predict. The answer to the question, 'Where are you likely to be in five or ten years' time?' is that if I knew I would probably be there already!

SELF ASSESSMENT QUESTIONS

1. Describe what is understood by the term building economics within the context of cost studies of buildings.

2. Explain why building economics has achieved greater prominence in the construction industry over the last 50 years and suggest what influences are likely to direct its future.

3. A range of new techniques and practices are now used in connection with the forecasting and control of building costs. Critically evaluate these and identify those that are likely to be the most significant in the long term.

BIBLIOGRAPHY

Ashworth, A., *Building Economics and Cost Control*. Butterworth 1983.

Ashworth, A. and Hogg, K., *Added Value in Design and Construction*. Longman 2000.

Bathurst, P.E. and Butler, D.A., *Building Cost Control Techniques and Economics*. Heinemann 1980.

Best, R. and De Valance, G., *Building in Value*. Arnold 1999.

Browning, C.D., *Building Economics*. Batsford 1961.

Ferry, D.J., Brandon, P.S. and Ferry, J., *Cost Planning of Buildings*. Blackwell 2007.

Flanagan, R. and Tate, B., *Cost Control in Building Design*. Collins 1997.

Ministry of Education, *Building Bulletin No. 4 – Cost Study*. HMSO 1972.

Nisbet, J. (ed.), *Estimating and Cost Control*. Batsford 1961.

Raftery, J., *Principles of Building Economics*. BSP Professional Books 1991.

Seeley, I.H., *Building Economics*. Macmillan 1995.

Spencer, N.C. and Winch, G.M., *How Buildings Add Value for Clients*. Thomas Telford 2002.

Stone, P.A., *Building Economy*. Pergamon Press 1983.

Williams, B., *Design Economics for Building Services in Offices*. Building Economics Bureau 1984.

CHAPTER 3

THE CONSTRUCTION INDUSTRY

LEARNING OUTCOMES

After reading this chapter, you should have an understanding of the nature of the construction industry in order to place cost studies of buildings within this context. You should be able to:

- Identify the characteristics of the construction industry
- Appreciate its size and complexity
- Understand the development process

3.1 INTRODUCTION

A sound understanding of construction economics requires a good understanding of the size and shape of the construction industry. The industry is important in all countries of the world since it helps to add to the quality of life both through its architecture and also through the facilities it provides to users, both internally and externally.

The construction industry includes building, civil engineering and process plant engineering, but the demarcation between these areas is blurred. It includes planning, regulation, design, manufacture, installation and maintenance of buildings and other structures. Construction work includes a wide variety of activities, depending on the size and type of projects undertaken and the professional and trade skills required. Projects can vary from work worth a few hundred pounds to major schemes costing several million pounds; and some in excess of a billion pounds. While the principles of execution are similar, the scale, complexity and intricacy can vary enormously. The larger the project the more specialist will be those involved at every level.

The industry is also responsible for a significant amount of work undertaken overseas on behalf of British consultants and constructors. About 10%–15% of the annual turnover of the major contractors is undertaken overseas. The construction industry has characteristics that separate it from all other industries. These are:

- The physical nature of the product
- The fact that the product is normally manufactured on the client's premises, i.e. the construction site
- The fact that many of its projects are one-off designs, with no prototype model available
- The arrangement of the industry, where design is normally separate from construction
- The organisation of the construction process
- The methods used for price determination

The final product is often large and expensive, and can represent a client's largest single capital outlay. Buildings and other structures are for the most part bespoke, designed and manufactured to suit the individual needs of each customer, although there is provision for repetitive and speculative work, particularly in the case of housing. The nature of the work also means that an individual project can often represent a large proportion of the turnover of a single contractor in any year. There are usually no prototypes as is the case in aeronautical engineering and shipbuilding, but projects learn lessons from the past and it is important that this information is collated and used by design teams in order to avoid similar mistakes being repeated.

3.2 THE IMPORTANCE OF THE CONSTRUCTION INDUSTRY

The importance of the construction industry in the economy of any country derives from the following factors.

3.2.1 Scale of the construction industry

The construction industry in Great Britain accounts for about 8% of GDP. This percentage will fluctuate with the economy, since construction is one of the first industries to enter a recession and one of the earliest to recover. Construction provides over half of the fixed capital investment in the economy (ships, vehicles, aircraft, plant, etc.). The industry experienced rapid growth in the late 1980s, but the recession in the early 1990s had severe repercussions, with output plummeting. The impact was cushioned by the volume of work already in progress, and the fact that a number of major projects were already under construction. The industry was especially buoyant at the beginning of the twenty-first century, but this would be short-lived as the world faced a recession that was unprecedented in perhaps the past 50 years. This was formed from a banking crisis and toxic debts especially in the USA, but which affected all the four corners of the earth.

The output of the construction industry in the UK increased from about £55m during the mid-1980s to in excess of £120m by 2007. This is a remarkable increase in real terms, taking into account the relatively low levels of inflation over this period and especially since the start of the new millennium. These figures do not take into account the severe recession facing the industry in 2008–9. New work

Table 3.1 Value of construction output

Sector	Value of output (£m)						
	1980	1985	1990	1995	2000	2005	2007
New housing							
Public	1 711	843	934	1 656	1 320	2 680	4 235
Private	2 585	3 797	5 746	5 470	8 665	18 384	20 088
Infrastructure	1 762	2 254	4 965	5 647	6 372	6 498	6 967
Other public	1 762	2 611	4 414	4 650	4 854	10 191	10 403
Industrial	2 806	2 159	3 394	2 996	3 717	4 291	5 030
Commercial	2 430	3 642	11 310	6 208	12 652	17 369	23 206
Repair/maintenance	8 997	14 358	25 444	25 900	32 096	47 594	52 364
Total	22 053	29 664	56 207	52 527	69 676	107 007	122 294
Percentage new	59%	52%	56%	51%	54%	56%	57%
Percentage public	40%	30%	26%	36%	36%	33%	31%

Source: Office for National Statistics (2008)

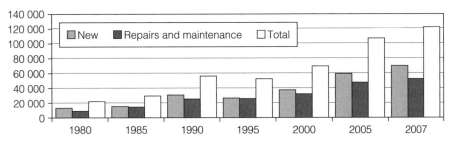

Fig. 3.1 New work, repairs and maintenance and total output compared
Source: Office for National Statistics (2008)

represents just over half of the total output, with repairs and maintenance representing the remainder (Table 3.1 and Figure 3.1). These reflect a fairly static proportion over the last twenty years. These figures exclude the considerable market for DIY in the UK, which is now described as one of the nation's favourite pastimes. They also of course exclude the unregulated economy which is reckoned by some to account for a further £10bn.

The British construction industry is the fourth largest in Europe, representing about 10% of the total output of work. It is exceeded only by Germany (32%), France (14%) and Italy (12%). Germany's share has swelled since the unification of East and West and the increase in building activity in the East. Europe is, however, dwarfed by the construction industries of the USA, China and Japan. The USA's construction industry, for example, is over five times larger than that of the UK (Davies Langdon 2007).

3.2.2 Sectors

In 2007 the house building sector accounted for 35% of the construction industry's output of new work. By far the greatest slice of this was in private sector housing which accounted for 29% or almost one-third of the industry's output for new construction works. This statistic identifies the very serious situation of the industry in 2009, since house building was at record low levels. Whilst there was a suggestion that investment in social (public sector) housing would be encouraged, at the time of writing this had yet to materialise. Public sector housing in 1980 represented 40% of the housing market (see Table 3.1) but by 2007 this had shrunk to just 17%. This reflects the rapid increase in private housing development at the start of the 21st century coupled with a reduced reliance on public sector accommodation.

Infrastructure accounted for 10% of all new construction work in 2007. This was a considerably lower figure than, for example, in 2000 when its proportionate worth was 16%. Road building is the main source of infrastructure work. Construction prices naturally fall during a recession and government should take advantage of this and at the same time help to smooth out falls in the industry workload.

Other public sector work includes schools and hospitals but also numerous other types of project such as libraries, swimming pools and public sector office buildings. In total these accounted for 15% of the new construction workload in 2007. These figures show an increase since the turn of the decade, largely accounted for by new school and college buildings and buildings for the National Health Service.

Industrial construction is the smallest of the basic sectors and in 2007 this accounted for just 7%. By comparison, commercial construction work accounted for 33% of all new construction work in 2007. This represented the largest sector.

The definition of building materials includes the raw materials, goods, components and fixtures. There has been in recent years considerable rationalisation in the sector, with the creation of several multinational conglomerates. Building materials can represent as much as 60% of the total construction output. This figure is likely to increase as more attention is given to off-site manufacture followed by assembly on site.

3.2.3 Employment

The construction industry directly employs about 1.5 million people. A large number of others are employed indirectly by materials and components manufacturers and plant and vehicle builders. A wide range of secondary employment relies on a prosperous construction industry. Therefore, the industry accounts directly for about 4.5% of the employed labour force. This figure can fluctuate widely, and may be as much as 25% higher in times of boom in the industry. Within these figures there are about 700,000 operatives, although due to the changing nature of employment this figure declined steadily throughout the 1990s. During the same period of time, the numbers of the self-employed in the construction industry more than doubled, to over 700,000. A further 300,000 are employed in administration, professional, technical and clerical occupations. The number of people employed in the repair and maintenance sector alone is considerably greater than those in agriculture, coal mining, shipbuilding or many of the other traditional industries, even when employment within these was at its peak.

3.2.4 Investment

The construction industry produces goods which are predominantly of an investment nature. Its products are not wanted for their own sake but for the goods or services they can help to create.

3.2.5 Government

Whilst the public sector has declined in size, largely due to some of its departments being privatised and the influences of the public–private partnerships, it still remains the major employer of the construction industry. Table 3.1 indicates that government output accounts for about one-third of the total output during the period indicated by these figures and especially since 1995. The output of government, although affected by economic influences, is much more stable as a sector in terms of its construction output.

3.2.6 Private Finance Initiative

The Private Finance Initiative (PFI) is currently one of the government's favourite ways of funding major new public sector construction projects, such as hospitals, schools, prisons and roads. Private consortiums, which usually involve large construction firms, are contracted to design, build and then manage the project after completion. Contracts typically last for 30 years. The completed project is not publicly owned but leased by a public authority, such as a council or health trust, from the private consortium. The private consortium raises the cash to build the project and this is then paid back, with interest, by the government through regular payments over the period of the contract. In common with other forms of hire purchase, buying in this way is more expensive than paying for it up front. The Edinburgh Royal Infirmary, for example, cost £180m to construct but will eventually cost £900m. Unions are opposed to PFI since they claim that one of the main ways that private companies profit from the PFI is by staffing the buildings as cheaply as possible. However, since the introduction of PFI schemes the construction industry has become a much more stable environment without the boom and bust conditions that have previously prevailed. Companies that engage in PFI can expect to make between three and ten times more than on traditional contracts.

The UK government defends PFI by its use of the 'public sector comparator'. This identifies whether or not a privately financed scheme offers better value for money than conventional funding. However, the main problem with this is that the government has provided an accounting device called 'risk costing' which has meant that private firms generally emerge as winners. When a consortium of private companies agrees to build a project for a public body, it acquires the risk that the project might fail. This risk is 'costed' and becomes a key component of the value-for-money calculations. According to the Association of Chartered Certified Accountants (ACCA), this risk is often exaggerated.

3.2.7 Health and safety

Whilst Britain's construction industry is the country's largest industry, it is also one of the most dangerous. In the last 25 years, over 2,800 people have died from injuries they received as a result of construction work. Many more have been injured or made ill. The construction industry has the largest number of fatal injuries of the main industry groups. In 2007–8 there were 72 fatal injuries, giving a rate of 3.4 per 100,000 workers. The rate of fatal injuries has over the past decade shown a downward trend, although the rate has shown little change in the most recent years. Despite this falling trend, the rate of major injuries in construction is the highest of any main industry group (599.2 per 100,000 employees in 2007–8). Relative to other industries, a higher proportion of reported injuries were caused by falls from height, falling objects, and contact with moving machinery (see Table 3.2). More information can be obtained from the website of the Health and Safety Executive (www.hse.gov.uk), which provides statistics on fatal injuries as notified to HSE and local authorities. These highlight the dangerous world of construction.

3.2.8 International

The construction industry is an important industry worldwide. Even in the poorer countries the net output of construction as a percentage of GNP is between 3% and 6%. The British construction industry works in the world market, being involved with work in developing countries. Overseas work typically accounts for a turnover of some £10bn. The British industry's important markets are in the Middle East, the Americas and the European Community. British professionals such as architects, engineers and surveyors are employed widely in these countries.

Table 3.2   Number of fatal injuries as notified to HSE and local authorities in the period April–December 2008

	Agriculture	Extractive/ utilities	Manufacturing	Construction	Services	Totals
Employees	7	5	26	29	30	97
Self-employed	12	0	0	17	11	40
Workers	**19**	**5**	**26**	**46**	**41**	**137**
Members of the public	4	1	2	2	71	80
Total fatalities	**23**	**6**	**28**	**48**	**112**	**217**

Source: Health and Safety Executive (www.hse.gov.uk)

3.3 THE CONSTRUCTION INDUSTRY AS AN ECONOMIC REGULATOR

The government remains a major client of the construction industry. It is tempting to argue that it uses the construction industry as an economic regulator. While the industry is damaged by the stop–go nature of its activities, there is really only scant evidence that government effectively turns the tap on or off in order to regulate economic performance. It may defer or cancel construction projects for other reasons, such as to reduce the public sector borrowing requirement, which in turn often creates a knock-on effect. Cuts in public expenditure may sometimes have a high construction consequence, but are often accompanied by other measures, and it is debatable whether these can be cited as an example of regulation. In times of economic famine, capital projects will be postponed and in times of prosperity they will be encouraged.

Government can intervene in the construction market in three ways, through finance, legislation or regulation and provision, as follows:

- Intervention in the market through finance by grants, benefits, subsidies and taxation
- Grants for the construction of industrial or commercial premises in areas of high unemployment
- Incentives for the construction of certain types of project, such as private housing
- Taxation relief against profits for the annual maintenance of building projects
- Changing regulations, such as in town and country planning, which can create opportunities for construction development, e.g. by allowing a wider range of projects in restricted areas to stimulate development
- Since government is a major client it has considerable scope to influence construction activity through the development, repair or maintenance of projects

3.4 THE DEVELOPMENT PROCESS

The development process applied to a construction project commences at inception and ends with demolition, when redevelopment of the site may occur. Table 3.3 indicates the different stages of development, although in practice these are not discrete activities. The traditional view considers the project from inception through to the stage of handing over to the client. This might more correctly be termed the capital development process. It is an outmoded view. Greater emphasis is now being placed on total- or whole-life analysis and project involvement, the designer or developer thus taking a longer-term interest, and often advising the client on maintenance planning and facilities management throughout the life of the project. This links the design to use, makes the designer more accountable and should result in a feedback loop so that problems are not repeated on future schemes.

Table 3.3   The development cycle

Stage	Phase	Typical time duration (years)
Inception	Brief	1
	Feasibility	
	Viability	
Design	Outline proposals	1
	Sketch design	
	Detail design	
	Contractual documentation	
	Procurement	
Construction	Project planning	3
	Installation	
	Commissioning	
In use	Maintenance	80
	Repair	
	Modification	
Demolition	Replacement	

3.5 CONSTRUCTION FIRMS

An estimated 210,000 construction firms were operating in Britain at the start of the 1990s. By the mid-1990s the recession in the industry had reduced this to 190,000. These ranged from sole proprietors to the huge conglomerates of businesses employing several thousands in their workforce. Construction firms start up, grow, merge with other firms, break up and sometimes die. They can be grouped and organised in many different ways according to their activities, their location, the number of their employees, the size of their annual turnover, their capital resources, or in several other ways, and in any combination or permutation that might be thought desirable (see *The Construction Industry of Great Britain* by Harvey and Ashworth). There are difficulties of classification, due in part to the individuality of the different firms and to the diverse nature of some of their activities. In 1970, there were 70,000 construction firms, of which 0.28% had a workforce of over 600 employees. Twenty-five years later, by 1995, the total number of firms had increased threefold but of these less than 0.05% had more than 600 employees. The larger firms with over 1,200 employees had also declined in number over the same period. Conversely, there had been a dramatic increase in the number of small firms. These figures help to explain the rapid and widespread increase in the subcontracting business that occurred over this period of time. In 1970 the one-man business represented 28% of the total number of construction firms. By 1995 it represented 48%. These figures had grown not just by number but also by proportion. Of the extra 135,000 firms which came into existence between 1970 and 1995, over 80,000 were mainly one-man businesses.

Table 3.4 Classification of firms by size

Size	1990	1995	2000	2005
1	91 223	99 099	87 712	73 117
2–13	68 444	89 146	69 147	97 354
14–599	11 877	5 741	6 481	12 052
600–1199	69	51	51	65
1200	47	33	35	56
Totals	171 660	194 070	163 426	182 644
General builders	88 496	79 843	59 708	45 706
Subcontractors	83 164	114 227	103 718	136 938

Source: Office for National Statistics (2008)

Table 3.4 provides an indication of the size distribution of the different construction firms in the UK. In 1970 there 73 420 firms compared with almost 200,000 by the mid-1990s. Whilst the number of the largest firms was declining during the latter half of the twentieth century, the start of the current century has seen a reversal of this trend, due, to some extent, to mergers and realignments of business operations and practices.

Of the total number of construction firms in 2005, almost 25% described themselves as general builders. Proportionately, this represents a small decline over the two decades in respect of the total number of firms in business. The figures are partly masked by size and the aspirational descriptions some firms may employ, in order to attempt to secure what they believe may be the more lucrative work.

3.6 THE PROFESSIONS

The professions have been one of the fastest growing sectors of the occupational structure in Britain. At the beginning of the twentieth century they represented about 4% of the employed population. In the early 1970s this had risen to over 11%, and the trend accelerated as the service sectors increased their importance during the 1980s and manufacturing either became more automated or generally declined. The temporary lull in the expansion of the professions, due to the recession in 2008, has caused much discussion of their benefits to society. A similar trend in comparable groups is evident in all Western capitalist societies. Several reasons are given for the rapid growth of the professions, such as an increasing complexity of commerce and industry, the need for more scientific and technical knowledge and a desire for greater accountability.

3.6.1 The built environment professions

The built environment professions in Britain are many and varied. They are a distinctive feature of the construction industry, and a matter for much debate.

There are about 350,000 members and students in the eight main chartered professional bodies that work in the construction industry. The RICS is the largest, with a membership of over 110,000. About one-third are quantity surveyors, one-third general practice surveyors and the remaining third comprise building surveyors, land surveyors and other smaller disciplines. It is sometimes argued that the difficulties which arise in the industry are due, at least in part, to the many different professional groups involved, a division that is not evident in mainland Europe. Others argue that the services the British construction industry provides have now become so specialised that one or two professional groups would be inadequate to cope with the complexities of the British construction process. There is no commonality of practices across the world and considerable differences exist even across mainland Europe. Such differences also exist in many different subject areas outside the construction industry.

There are wide cultural considerations to be taken into account in any comparison between the construction industry professions in Britain and those in other parts of the world, notably Europe, the USA and Japan. Historically, practices developed differently. In much of the rest of the world, architects and engineers dominate the construction industry, but the proportions vary in different countries. The various professional disciplines in Britain are not mirrored elsewhere, other than in Commonwealth and ex-Commonwealth countries, but these in themselves are considerable. The role of the professional bodies also varies. In Britain a professional qualification is needed in order to practise. In Europe a professional body is more of an exclusive club, to which relatively few of those engaged in practice are members. In the USA there is the emerging discipline of construction management alongside those of architect and engineer.

3.6.2 The future of the built environment professions

The built environment professional bodies grew steadily both in membership and number throughout the twentieth century. The number of professional bodies has continued to increase in spite of the amalgamation and mergers that have taken place. It can be argued that there is a proliferation of professional bodies in Britain. The future of the professions in Britain is influenced by:

- The effects of the Single European Market, since the industry structure in mainland Europe is different from that in Britain
- The diversification and blurring of professional boundaries, with the non-built environment professions such as those involved with the law and finance often included
- Their role as learned societies
- The education structure of courses in the built environment
- The pressure groups both within and outside the construction industry
- The desire in some quarters for the formation of a single construction institute, to unify all professionals in the construction industry

3.7 RESEARCH AND INNOVATION

Technical change is accelerating and progressive businesses tend to adopt new techniques and applications quickly. Research and innovation are inseparable from the well-being and prosperity of a country and of the businesses within the country. To make direct comparisons between countries' respective inventive strengths, it is usual to concentrate on the US patent statistics. Comparisons suggest that a dramatic decline has taken place in Britain's innovative performance since the Second World War compared with that of other developed countries. The relative strengths of Germany and Japan are clear.

Expenditure on research increased during the past decade in Britain, but still amounts to only 0.65% of the construction output. Construction companies contribute 10% to this sum, which is approximately one-third of that of their competitors in France, Germany and Japan. By way of comparison, the Kajima Corporation in Japan produces a wide range of high-technology products and services designed to accelerate the overall innovation of construction systems. In respect of expenditure on research, the British construction industry lags far behind both competitors overseas and other British industries. For example, all British industries spend some 2.3% of turnover on research. Although the public sector accounts for only 25% of all construction in Britain, it is often believed that public sector funds should pay for a substantial portion of construction research.

The value of research to the construction industry cannot be overestimated. Research is necessary to maintain international competitiveness and success, particularly as the craft-based traditions of construction diminish and the technological base expands. As the construction market shifts periodically across the world, new conditions and constraints relating to environment and materials must be taken into account. This background of constant change and challenges demands an effective research base to introduce change effectively and efficiently. A process of trial and error by the industry is slow and expensive. Clients would be reluctant to accept new and unproven techniques without the reassurance that they have been rigorously researched to guarantee performance. The cost–benefits of research projects are difficult to assess.

SELF ASSESSMENT QUESTIONS

1. Identify and describe those characteristics that help to make the construction industry unique amongst industries.

2. Explain why a vibrant construction industry has overall implications in a country's economy.

3. Using the data from this chapter and information from other sources, describe the nature, size and structure of the construction industry and assess how the industry might change over the next decade.

BIBLIOGRAPHY

BCIS, *Quarterly Review of Building Prices*. RICS.

Davis Langdon, *World Construction 2007–08*. Davis Langdon 2007.

Harvey, R.C. and Ashworth, A., *The Construction Industry of Great Britain*. Butterworth-Heinemann 1997.

Health and Safety Executive website: www.hse.gov.uk

Morton, R., *Construction UK: Introduction to the Industry*. Blackwell Publishing 2002.

Murray, M. and Langford, D., *Construction Reports 1944–98*. Blackwell Publishing 2002.

Office for National Statistics, *Construction Statistics Annual No. 9*. Palgrave Macmillan 2008.

'The top fifty contractors' (annual survey), *Building*. United Business Media.

'The top 500 European companies' (annual survey), *Building*. United Business Media.

COST INFORMATION

CHAPTER 4

COST DATA

LEARNING OUTCOMES

After reading this chapter, you should have an understanding of cost data and their importance in the study of building costs. You should be able to:

- Identify the characteristics of cost data
- Understand the hierarchical structure of cost data
- Understand issues of accuracy and consistency
- Identify the sources and contents of published cost information
- Develop an understanding of the vagaries of tendering and the role of feedback
- Objectively select cost data in practice

4.1 INTRODUCTION

The collection, analysis, publication and retrieval of cost and price information together comprise a very important facet of all sectors of the construction industry. Contractors and surveyors will tend, wherever possible, to use their own generated data in preference to commercially published data, since the former incorporate those factors which are inherent to themselves. Published data will therefore be used for back-up purposes. The existence of a wide variety of published data leads one to suppose, however, that they are much more greatly relied on than is sometimes admitted.

Construction costs in the context that follows is a broad term and can be interpreted to mean costs of any sort to anyone associated with construction works. Contractor's costs, however, literally mean the contractor's expenditure on labour, materials and plant and include all the items shown in Figure 4.1, with the exception of profit. The inclusion of profit, which is influenced by many considerations, not least of which are market conditions, converts contractor's costs to contractor's prices. This coincides with client's costs. Generally speaking, it should be obvious from the context in which these terms are used what their meaning is intended to be (see Chapter 1).

The use of cost data cannot be restricted to capital construction costs alone. This information is also relevant to costs-in-use, and in practice similar sources of

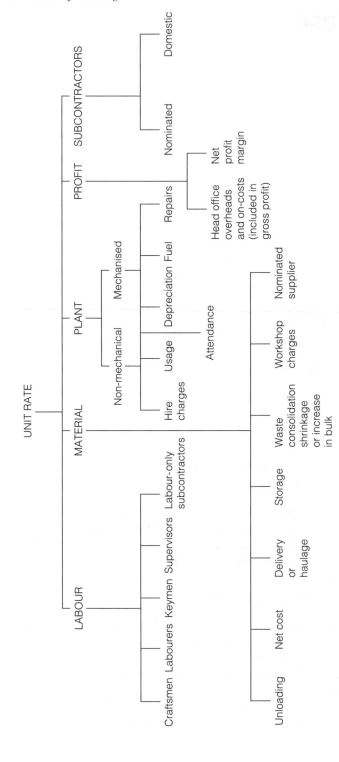

Fig. 4.1 Diagrammatic representation of a unit rate
Source: Wood (1982)

information are referred to for this purpose. The only major differences which need to be considered are the life expectancy of materials and systems, and the discounting rates to be used. These are discussed in Chapters 17 and 18 under the title of whole-life costing.

4.2 CHARACTERISTICS OF COST DATA

Cost data are required at various levels of sophistication for the theory and practice of building economics. They are required during the inception stage of the design process in order to provide clients with an indication of possible costs associated with a proposed construction project. They will also be required at the various levels of detail as the project proceeds through the design and construction stages. In all cases, however, a reliable estimate of cost will be needed.

Cost data, whether they are applied to items based on an analysis of a standard method of measurement (e.g. SMM7) or builders' quantities, are based on what are believed to be the determinants of cost. However, very little research of an analytical nature has been undertaken to establish precisely what these determinants are. We continue to rest on assumptions. We base our costs, for example, on the cost per square metre of gross internal floor area (GIFA), because we believe that this provides a broad indication of total cost. In reality this approach is far too subjective, since the variables of shape, height and quality, for example, will have a major influence on our choice of a rate.

Cost data during the early stages of the design process can be usefully related to function and design. The degree of reliability, however, is highly suspect and requires an appraisal of several variable factors. During the later stages of the design process, cost is apparently more related to quantity and specification. Both of the above are the traditional approach. An alternative view is to suggest that costs are process-determined, i.e. that the methods, equipment and plant selected by the contractor will determine costs. The generally accepted viewpoint is that building costs are quantity-related, whereas civil engineering costs are more process-determined. The latter viewpoint supports the fact that civil engineering costs have in the past been rarely published since they are of more limited value unless the process used is known.

4.3 THE HIERARCHICAL STRUCTURE OF COST DATA

The construction industry has adopted a hierarchical structure for its cost data as shown in Figure 4.2. This structure consists of eight levels of analysis using total cost as the lowest level of detail. This total cost is often used in conjunction with one of the single-price methods of estimating. It is generally realised that the variability in construction projects, even of a similar type, is such that it is not possible to consider total cost as anything more than a general guide for other projects, when expressed in terms of a single quantity such as GIFA. Reference to

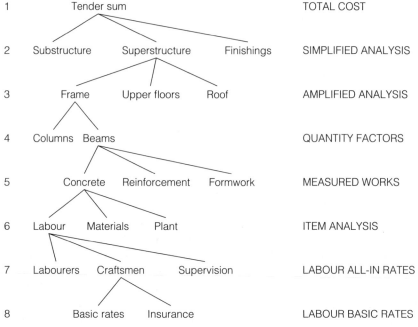

Fig. 4.2   Hierarchical structure of cost data

the *BCIS Quarterly Review of Building Prices* will emphasise this view, in the range of prices that is offered.

Opinions vary, but intuition would support the viewpoint that the more detailed the cost data, e.g. level 8, the more accurate would be the resultant estimates prepared from such data. The PSA study *Cost Planning and Computers*, however, concluded that to identify the major (100) items of work on any project and to price these accordingly would achieve a level of accuracy that could hardly be increased by more detailed pricing. This has been a method used by building contractors for some time when pricing, drawing and specifying contracts. It is in reality also the method used by contractors when pricing larger projects, since they often 'guess' the rates of the cost-insignificant items. Figure 4.3 shows that the level of estimating accuracy improves only marginally beyond the 100 items identification.

The following points are worthy of note in connection with construction cost data:

- The quantity surveyor cannot predict the error contained within the successful contractor's own estimate.
- Bills of quantities, although applicable to an actual project, include a wide variation of rates. For example, some items of a comparable nature on different projects may vary by as much as ±200%.
- Small items on bills of quantities are not priced carefully.

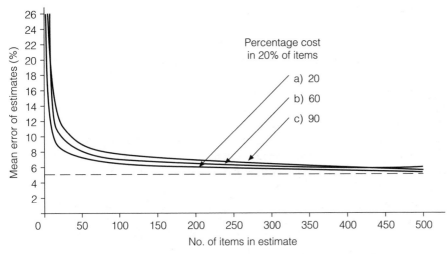

Fig. 4.3   Graph of estimating errors against the number of items in the estimate
Source: Bennett *et al.* (1981)

- The theory of feedback, as described in standard textbooks, does not occur routinely in practice.
- Although published cost data are available to the industry, quantity surveyors place greater reliance on their own generated data, on the grounds that they know and understand the important features of their own projects.
- Unless a practice is very large, the data available are generally insufficient in number or too diverse to provide a satisfactory database.
- There are many different sources of cost data, which provide different types of information to the profession.
- The optimum estimating performance is reached by the measuring and pricing of approximately 100 major items of work (see Figure 4.3).
- The provision of suitable cost data will only be possible by use of a central store of information that can be recalled by computer.
- A quantity surveyor's estimating accuracy is in the region of ±13% when attempting to forecast the contract sum.
- The use of data will be restricted when the system of operation is difficult to use or where the process of application is slow.
- The various cost data that are available show conflicting values for the work to be estimated.
- The method of measuring in finished quantities encourages inaccuracy in terms of outputs, waste and risk.
- In a bill of quantities, 80% of the costs can be attributed to 20% of the items and vice versa. The ratio is somewhat smaller on civil engineering projects, since the Civil Engineering Standard Method of Measurement (CESMM) tends to measure only the major cost items.

4.4 ACCURACY AND CONSISTENCY

None of the compilers of all types of published cost information claim that the information is accurate. They are correct to suggest that the prices quoted are nothing more than a guide. Perhaps the only exception is the builder's merchants' price lists, but even these include disclaimers that the prices are subject to change at very little notice. But how good a guide is this information? This can be measured in two different ways, through accuracy and consistency. Accuracy implies a closeness to the actual value, whatever that is. Consistency, on the other hand, is a measure of how often this accuracy can be relied on. It has been established through research that while prices overall (including bills of quantities) have levels of accuracy ±10%, the individual components (i.e. individual items) of these prices may have measures of variance ten times this percentage. In terms of overall consistency the latter aspect presents the users of cost information with substantial difficulties, comfort being taken only in the fact that the overall price will not be subject to such large extremes of variation.

Table 4.1 provides a list of typical project items together with the prices from three different sources of cost information. In Table 4.2 this information has been converted into a series of price relatives, giving the mean price of three sources a value of 100. On an unweighted basis the mean price represents a value of 100, with source C giving the highest value of 107. If it were assumed that the mean

Table 4.1   Comparison of published prices (in £): sample of items

No.	Item	Source A	Source B	Source C
1	Trench excavation/m^3	4.10	4.50	5.50
2	Earthwork support/m^2	1.10	2.00	1.60
3	Hardcore filling/m^3	18.00	27.50	25.00
4	Concrete in foundations/m^3	52.50	47.00	55.00
5	Fabric reinforcement/m^2	2.30	2.20	3.10
6	Wall in commons (HB)/m^2	20.20	21.70	26.50
7	Block partition (100 mm)/m^2	12.60	16.00	15.50
8	Hessian-based DPC/m^2	7.20	8.50	7.00
9	Floor joist (50 × 100 mm)/m	2.00	2.80	2.20
10	Clay pantiles/m^2	21.50	24.00	19.00
11	Roofing felt (3 layers)/m^2	12.50	16.00	15.00
12	Plasterboard (12.5 mm)/m^2	4.90	4.50	5.00
13	Blockboard (12 mm)/m^2	13.20	11.00	14.50
14	Standard flush door/No.	21.00	18.50	20.50
15	Float glazing (4 mm)/m^2	14.50	12.50	15.00
16	Lightweight plaster/m^2	6.10	6.50	7.60
17	Emulsion paint (2 coats)/m^2	2.20	2.80	3.50
18	KPS and 3 oils/m^2	4.50	4.90	6.00
19	UPVC rainwater pipe/m	6.80	6.00	7.50
20	Vitreous clay drain pipe/m	5.80	5.50	6.00

Table 4.2   Price relatives (based on a mean price for each item = 100)

Item	Mean price	Source A	Source B	Source C
1	4.70	87	96	117
2	1.57	70	128	102
3	23.50	77	117	106
4	51.50	102	91	107
5	2.53	91	87	122
6	22.80	89	95	116
7	14.70	86	109	105
8	7.57	95	112	93
9	2.33	86	120	94
10	21.50	100	117	88
11	14.50	86	110	103
12	4.80	102	94	104
13	12.90	102	85	112
14	20.00	105	93	103
15	14.00	104	89	107
16	6.73	91	97	113
17	2.83	78	99	124
18	5.13	88	95	117
19	6.77	100	89	111
20	5.77	101	95	104
Mean		92	101	107
Standard deviation		10.05	12.19	9.35

price of these sources was the most accurate (this of course might not be the case), then source A would provide the most accurate results for this sample, with a mean value of 92. The standard deviation represents the measure of consistency, and shows that while source C provides the highest prices, it is the most consistent in its pricing compared with the overall mean prices.

4.5 COST FEEDBACK

The traditional principle of gathering site performance data from previous projects, known as feedback, is illustrated in Figure 4.4. The estimator using standard outputs, influenced by size, complexity, quality etc., estimates the costs of the work to be performed. If the contractor is successful in submitting the accepted tender then the work is put into practice and, during construction, is monitored by site management staff. The monitoring is frequently carried out incidental to other purposes such as bonus or incentive calculations. These calculations require the quantities of work to be measured against the labour times that have been expended. The record sheets are usually copied to the estimating department in the form of

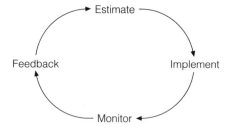

Fig. 4.4   Feedback

feedback that can be used to inform the estimates for future projects. However, in practice, such information is not routinely used by estimators in calculating or revising their outputs for the following reasons:

- It is very variable in terms of the outputs it generates.
- There is insufficient confidence, by estimators, in the site recording systems.
- The information is often not compatible with future estimating needs.
- There is a difficulty in reusing the data because of the unique circumstances under which the work has been carried out.

The traditional method used for estimating purposes has been to develop a classification system against which to record costs. The outputs achieved on similar work from previous projects should be the major source of information used in estimating. However, construction work requires a complex system against which to record this information. Research has shown that the reliability of any cost recording system substantially deteriorates when the number of cost codes exceeds 50. The cost code system used in the construction industry is a four-digit system.

The complexity of construction work and the fact that most projects are bespoke one-off designs (even projects that are considered to be 'identical' record different actual outputs and costs) make this process difficult to achieve in practice. The use of new techniques on site or improved methods of working will, of course, encourage estimators to review their own tables of standard outputs. Amending these standard outputs that estimators have recorded in their personal 'black books' on the basis of feedback from a single site will not have much influence. Evidence indicates that different sites record different feedback values for apparently similar items of work. Estimators, by their nature and through their training and experience, are conservative in their outlook. The view or hope that things will work out better on the next project has little influence on the estimator's own constants for labour and materials!

Production standards, for both labour and plant, are likely to be influenced by a whole range of project characteristics. Estimators, when adapting a standard output, need to assimilate these different factors in order to arrive at a best estimate for the work. Some of the characteristics that need to be considered include:

- Location and accessibility of the work
- Amount of repetition in the work
- Intricacy of the design
- Need for special labour skills
- Quantity of work involved
- Quality of materials used
- Standards of workmanship
- Working environment such as safety, temperature, cleanliness etc.

One the most important aspects that affects the value of labour outputs is the incentive scheme operated by the contractor. This is designed to improve the overall performance of those working on site and in other locations of a contractor's business. While there has been much debate about bonus and incentive schemes in the construction industry, their application has had a significant impact upon labour outputs. The targets set for bonus payments influence an operative's output of work perhaps more than any other single factor.

Figure 4.5 indicates how variable actual outputs can be, even on a single project. This shows the outputs per square metre of work completed for brickwork over a 21-week period, when bricklaying operation was at its greatest. The output varies from 0.2 to 1.60 m² per hour. While some of the variability can be explained, other aspects of it cannot. Week 10, for example, was during a period of bad weather, when no inside work was available. Weeks 17–19 were the result of the work coming to an end and included snagging work and completing small and intricate items. The estimator, when pricing, has usually to select a single output to represent the whole amount of work. This in itself adds a further variable that needs to be taken into account.

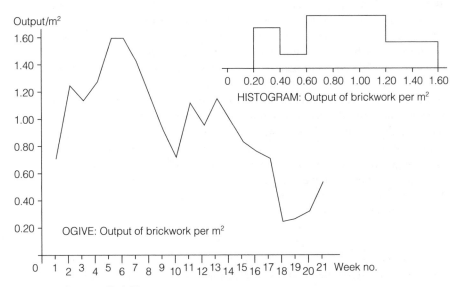

Fig. 4.5 Output of bricklayers

4.6 PRICE BOOKS

Building price books, or price guides as they are more correctly interpreted, have been in use for over 170 years. These are a traditional source of useful information. The *Laxton's Building Price Book* claims to have the longest history. Its 180th edition was published in 2008. Throughout the past two decades there has been an increase in the number of guides that are available. In addition, their contents have been enlarged, providing some indication of the increase in knowledge and information that is now accessible. Most of the current building price guides include prices for materials and measured items covering the full range of the Standard Method of Measurement (seventh edition). Other guides adopt other relevant methods of measurement. In addition the guides also present information on approximate estimating, cost limits and allowances, daywork, professional fees, memoranda. The guides may include labour outputs, wage rates, labour all-in rates, national working rule agreements and plant hire charges. All the guides have expanded the scope of their contents to keep pace with current demands. They are now more comprehensive than ever before. Guides are available for major and minor works, where the rates and prices used may vary by up to 50%. It is essential to use the correct guide for the particular size of project. Table 4.3 lists some of the more common price guides that are used in the construction industry.

A good understanding of the rules of measurement on which a guide is based is desirable prior to attempting to use the guide in practice. It is also important to identify the locality on which the price guides are based. Building costs and prices in different regions of the UK can vary considerably. The Building Cost Information Service is able to provide statistical data for the standard regions and sub-regions in the UK. The choice of the words 'price guide' is deliberate, since this is all the guides can hope to achieve.

There is frequently a hierarchy in price information with the guides only being considered as a last resort, and personal preference, priced contract documentation and third parties being more important sources of information. Despite this, it is not unusual to find an assortment of these books in the offices of firms involved in the building industry. The growth in the industry of price information for the construction industry has been considerable, indicating the need by users and the usefulness in terms of what is achieved. Most of the information is now accessible electronically as well as on hard copy.

In periods of high inflation, which was last experienced in the UK in the 1970s, the price guides suffered from becoming rapidly out of date, particularly as the information had to be prepared some time in advance of publication. Publishers attempted to overcome such problems through an updating service and in the case of one publisher a new guide was published every month.

4.6.1 Usage of cost data

It has long been recognised that the most useful type of cost data is those which have been generated from personal practice. In this situation the user can be fully aware

Table 4.3   Price guides

Building major works

Griffiths Building Price Book
Hutchins' Major Works SMM7 Price Book
Laxton's Building Price Book
Spon's Architects' and Builders' Price Book
Ti Wessex Comprehensive Building Price Book: Major Works
Ti Wessex SMM7 Building Price Book

Building minor works

BMI Building Maintenance Price Book
Hutchins' Small Works Price Book
Laxton's Trade Price Book: Small Works, Repairs and Maintenance
Ti Wessex Comprehensive Building Price Book: Major and Minor Works
Ti Wessex Alterations and Refurbishment Price Book for Small Projects
Ti Wessex Comprehensive Building Price Book: Minor Works
Ti Wessex Painting and Decorating Price Book

Building services engineering

Laxton's Mechanical and Electrical Price Book
Spon's Mechanical and Electrical Services Price Book

Civil engineering

CESMM3 Price Database
Laxton's Civil Engineering Price Book
Laxton's Highway Price Book
Spon's Civil Engineering and Highways Price Book

Landscape

Griffiths Landscape and Gardens Price Book
Spon's Landscape and External Works Price Book

Overseas

Laxton's European Building Price Book
Spon's Asia Pacific Construction Costs Handbook
Spon's European Construction Costs Handbook
Spon's Middle Eastern Construction Price Book: Volumes 1 and 2

Miscellaneous

Laxton's Industrial Premises Price Book
Laxton's Schools and Colleges Price Book: A Building Maintenance Costs Guide
Schedule of Basic Plant Charges for use in Connection with Dayworks under a Building Contract
Building Cost Information Service (BCIS)
Building Maintenance Information (BMI)
Building trade and manufacturers' price catalogues

Note: Many overseas countries produce their own building and construction price guides.
Most notable is 'Rawlinsons' throughout Australia, New Zealand and the Pacific region.

of the anomalies that might be inherent in its composition. However, due to the shortage of time, and the costs associated with operating a comprehensive system, builders and surveyors resort to one of the many forms of published cost information. Even where time and money are not significant obstacles, the retrieval of data may at best represent only a proportion of the cost information that is required.

Builders and surveyors need a ready access to cost information for the following purposes:

- Approximate estimates for proposed schemes
- Cost planning during design
- Contract estimating for tendering purposes
- Agreement of variations in final accounts
- Calculation and settlement of contractors' claims
- Loss adjustment valuations
- Maintenance management

When an estimator's or surveyor's own data are not available, the builder or surveyor will need to refer to a published source of data. On other occasions this information is used for a second opinion. In matters of urgency some items of work may be priced directly from the price book.

4.6.2 Precautions

In some circumstances the small building contractor may have to rely almost exclusively on the price book for estimating. In this case the use of the price book is being extended to a point the compilers never envisaged. Each of the price book authors is careful to suggest that the data are a price guide only. They therefore expect the user to make some comparison or adjustment to the rates prior to their application in practice. The following points should therefore be borne in mind.

Size of project   The majority of the guides base their prices on an assumed size of project. In some of the guides separate rates are provided for large and small works. Adjustments to the quoted rates may therefore be necessary where the project concerned does not fit within the price bands mentioned. Smaller-sized projects on the whole tend to cost more per unit than larger jobs of a similar type and construction.

Location   The majority of the books are based on London rates, and adjustment will therefore be necessary for projects that are intended to be constructed in the provinces. It is possible to obtain and use regional trend indices to make the appropriate allowances for this factor. Some adjustment may also need to be considered in respect of projects in city centres or country areas.

Fluctuations   The price guides take many months to prepare, and are generally published ahead of the year to which they apply. It is generally assumed, therefore, that the prices quoted are current at the date of publication. In periods of high inflation they can very quickly become out of date. Even in normal times some adjustment to the prices will need to be made to account for changes in prices due to inflation as the months elapse.

Labour costs   Each of the guides provides its own analysis of all–in labour rates. These are then used throughout in the determination of the measured rates. Some adjustment to these costs will be necessary in practice, to take into account both labour availability and the actual payments made. In addition, the output of the operatives will vary depending on the actual work being done. Some of the guides indicate the outputs used, and in these cases adjustments should be fairly easy to make.

Material costs   In some of the guides the costs of materials can be easily identified, and adjustment to the measured rate is therefore a simple process. Material prices often depend on contractor, location, quantity and discount. Contractors would therefore need to substitute their own material costs in the measured rate analysis.

Market conditions   Where a surveyor is trying to predict a future tender sum, or a builder is trying to submit a competitive price, each must be aware that market conditions can seriously undermine their forecast. During the past twelve months, although increases in costs in the basic labour rates and materials will have occurred, tender prices will not always have followed a similar pattern. The competitive state of the market is therefore influenced by the local supply of contracts and the demand for the work by available contractors. The influence of the method of tendering may also affect the prices used.

Overheads and profit   The amounts included in the published rates vary considerably. In some guides nothing has been allowed, and these expect the user to add an appropriate amount. In other information, up to 25% has been included on small works projects. Construction company analysts have quoted builders' profit margins of 3%. In addition, overhead costs have generally been reduced in order to keep costs to an absolute minimum, while seeking to maintain the required workload. Builders will therefore need to assess their own percentage and adjust this according to that quoted. Surveyors, in attempting to predict tender values, will also need to take this into account in their approximate estimate.

4.6.3 Selection

The choice of a particular price book by a builder or surveyor will depend on a large number of factors. The use of a price book over a number of years may be difficult to change, especially when the price book has provided satisfactory service during that time. The contents may not always be what is required, but tradition dies hard. The familiarity of a particular layout, and knowing just where to find the right information or whether it exists at all, are important considerations. This is why the evolution of individual price books over the years has been a gradual rather than a radical change.

The format and style are also important. They need to be clear, easy to read and easily understood. The terminology that is used should be consistent with the accepted meaning of the word or phrase, otherwise confusion will occur. It would appear that the most popular method of presentation is to gather all the relevant information under one heading. A person then using the guide, particularly the builder's estimator, can see at a glance all the relevant constituent parts of the measured rate.

A further point of importance is the relative completeness of the price data. The majority of the guides provide information on the major common items of work. They do, however, vary their contents on the minor and uncommon items, depending perhaps on the general availability and representativeness of price information. The demands of the user must also be taken into account when choosing what information should be included or omitted. There will be times, however, when required costs or prices cannot be found in the price books. In these circumstances some form of price analysis will be required.

If it is necessary to be able to analyse the measured rates in terms of profit, materials costs and labour outputs for probable amendment, then a single measured rate will be inadequate. This may be particularly important for small builders who are attempting to use a price book as a basis for their own estimating. The trend among the more recent versions has been to provide this information to an appropriate level of detail. Surveyors may be largely concerned only with the measured rate, and builders who use the books for reference purposes may also be mainly interested in this fact.

The reliability of the data provided will need close examination. There can be a wide discrepancy between identical items, some of which can be attributed solely to the subjective opinion of the author. In other circumstances errors in transcribing do occur, but fortunately not too often. The measured rates have largely been selected on the basis of contractors' current pricing, market trends and the author's own view of what constitutes a typical rate. They are therefore assumed to be average rates, for average projects carried out under average conditions. Unfortunately, they are unlikely to be pitched at a level that is consistent with the user's own prices. A previous examination of a number of published rates from various sources showed that they were very erratic when compared with each other.

A further point regarding accuracy should be noted. The publishers do not guarantee that the guides are free from either mistake or misrepresentation. The misuse of a rate will not provide any grounds for redress from the publisher. All the guides include a disclaimer in the event of this occurring. This does not imply that the use of the price books is fraught with danger, but it does presuppose in the first instance that the user has some idea of the rate to be expected. For example, errors have been found due to the misplacing of a decimal point. One assumes that the user would realise this to be the case in a half-brick wall in common bricks at £200 per m²! The possibility of misleading information being presented in price books is much less than can be expected from, say, priced bills of quantities. The latter may contain bias or distortion, whereas this is likely to be much less within the published cost information.

4.7 PUBLIC SECTOR PRICE GUIDES

The public sector of the construction industry has a long history of providing price guides for use by this sector. The *PSA Schedules of Rates* were originally prepared by the Property Services Agency (PSA) and were first published in 1966, replacing the existing *War Department (WD) Schedule of Rates*. They include seven separate schedules dealing with, for example, work as varied as building works and the preparation and maintenance of land and electrical distribution systems external to buildings. The rates in this guide are both extensive and comprehensive, being designed with government estate and PSA contracts in mind. The *Schedule of Rates for Building Works*, in addition to measured rates, incorporates preamble clauses as an indicator of the specification level expected. The schedule, in common with all the price books, is constantly under revision to take into account the changing methods of construction and price inflation. A major difference between this price guide and the price books is that the former are used in practice specifically, as a basis, to agree rates and prices on construction projects.

The *PSA Schedules of Rates* (which are now a part of Carillion, the contracting group) continue to be the key pricing documents used in Measured Term Contracts (MTCs). These contracts remain one of the most popular forms of procurement for maintenance and minor new works of the government estate. The trend towards contracting out has resulted in more clients turning from directly employed labour to MTCs. Carillion has responded to this new interest by embarking on a series of training sessions on the use of *Schedules of Rates* and MTCs.

The *PSA Schedules of Rates*, published by The Stationery Office, are one of the most comprehensive suites of schedules available. The Building Works schedule contains nearly 20 000 rates and the Electrical and Mechanical Services schedules over 10 000 items. They follow the Coordinated Project Information (CPI) Initiative common arrangement and include all the forms of labour usually deleted from standard methods of measurement, but essential for MTCs. They also contain their own measurement rules and specification notes, making them totally stand-alone documents.

Computer technology has enabled Carillion to offer a bespoke schedule service to suit individual client needs. These range from schedules based on client supplied resource costs to labour or material only schedules, or even schedules in other currencies. The following comprise the suite of price guides:

- *PSA Schedule of Rates for Property Management*
- *PSA Schedule of Rates for Decoration Work*
- *PSA Schedule of Rates for Electrical Services*
- *PSA Schedule of Rates for Mechanical Services*
- *PSA Schedule of Rates for Landscape Management*
- *PSA Schedule of Rates for Building Works*

In addition the PSA also produces *A Guide to Measured Term Contracts: Obtaining Value-for-Money in Building Maintenance and Repairs.* The MTC is an arrangement whereby a contractor undertakes to carry out a series of works orders, over a period of years, within a defined geographical area and where the work is subsequently measured and valued at rates contained in a pre-priced Schedule of Rates.

4.8 MONTHLY COST INFORMATION

Several monthly journals and commercial publications include information on material prices and measured rates. The extent of these data varies, but at best, due to the amount of space allocated to them within the journals, they can only hope to provide rather general information. In periods of high inflation they are a useful source of information, and since they tend to concentrate on the major items only they can provide a useful source for comparison. They typically contain about 300 items, based on measurement generated from SMM7. In periods where inflation is at a relatively constant level their importance and use are limited.

4.9 PRICED BILLS OF QUANTITIES

Bills of quantities and work schedules are a major source of cost information, but the information contained in them must be used with great care. Comparison of rates between two bills of quantities for the same project will show a considerable variation for many of the items. Although tenders may vary by only 10%, individual trades may differ by as much as 40% and individual items by up to 200%. Bills from different contractors for other projects can show variation in excess of these figures. When examining bill rates it is important that the user has some idea of the rate expected in order that a bill rate will not be used erroneously.

The user, in attempting to make use of rates from prices bills, should not forget their confidentiality, and should not disclose their source to third parties (Standard Form of Building Contract, 1998 edition (JCT 5.7)).

Data from bills of quantities can be summarised as follows:

- Individual rates for measured items
- Overall costs for use with the single price methods of approximate estimating, e.g. unit, square metre, cubic metre
- Elemental format analysis
- Basic price list of materials, if available

The more detailed the information presented, the more it will be subject to variability and hence the less reliable it will be.

Computing average costs from a large number of projects has less relevance than examining costs from a few well-known projects. It is important when examining bills of quantities to understand the conditions that influenced the rates and prices charged.

Variations in rates may be due to the following project factors:

- The size of the project
- The type of the project
- The regional location of the project
- The contract conditions applicable to the project
- The market conditions prevalent at the time of tender for the project
- The contract implications, particularly those affecting the contract period and the account to be taken for inflation

The following are further reasons for what have become known as the vagaries of tendering.

Distribution of preliminary items   It is important to discover the extent to which preliminary items have been priced within this section, or alternatively allocated on a proportional basis among the measured items.

Location of the site   The costs associated with projects in the countryside or in the centres of busy towns and cities will vary and may reflect the problems of access, difficulties of performing the work etc.

Deliberate distortion   The estimator may deliberately distort the bill rates either because they are anticipating variations or in order to obtain all the profit at an easy stage in an attempt to finance the remainder of the work.

Errors   It is not uncommon to find bill items priced incorrectly because of mistakes.

Lack of accurate cost data   Due to the pressure of time, an estimator may be unable to price all the items analytically. On some occasions this may be due to a lack of any available feedback or material prices.

Facilities   A firm may be able to provide a more competitive price where, for example, it has its own joiner's shop.

Site techniques   The techniques that the contractor uses to complete the works, e.g. the amount of mechanisation being used, will be reflected in the prices.

Subcontractors' and suppliers' prices   Contractors usually sublet a proportion of their work to other contractors and suppliers.

Standard of workmanship   Different standards of workmanship may be anticipated depending on the type of project, the standard of the specification and the requirements of the designer, if known.

Availability of labour   The availability of skilled labour is likely to vary at different times of the year and in different regions of the country.

Financial conditions   The financial ability of both the contractor and the client is likely to be reflected in the bill rates. For example, because of the shortage of work or the prestige of the project, the contract may have been 'bought'. It would be extremely unwise to use such rates without caution.

Special requirements   The examination of the contract documents may indicate some reasons for variability in prices, e.g. an abnormal speed of construction, work required to be undertaken at unsocial hours, phasing etc.

The above is not a comprehensive list of items affecting the reliability of rates found in bills of quantities. It must also be understood that local conditions affecting the amount of work available, degree of competition etc. may distort the overall national situation.

4.10 PRICE ANALYSIS

Occasions will inevitably arise when no suitable price information is available. In these circumstances, as in the case of new products or building details being used for the first time, no other alternative may be available but to calculate rates from first principles. There are of course advantages in obtaining rates in this manner, e.g. that local peculiarities can be built into the price. The majority of contractors will inevitably use this method, calculating their measured rates on known material costs and previously recorded labour productivity. They will do this only for the major items of work, however, minor items being priced very subjectively or even 'guessed'. The smaller contractor without a proper system of estimating may resort to using published rates, suitably adjusted, for pricing tenders.

4.11 COST ANALYSIS

The cost analysis of construction projects can provide the surveyor with a useful source of cost information. The generally accepted form of a building cost analysis is given in Table 4.4 and this is further discussed in Chapter 9. A simplified version is shown in Table 4.5. The aim of any cost analysis is to provide cost centres for the work under examination. The standard form of cost analysis, although devised on the basis of elemental subdivisions, has sought to identify the major cost items. It is to these items that project cost managers need to address themselves.

A second form of cost analysis is given in Table 4.6. This form was devised to analyse the annual user costs of building ownership. The elements listed in this case have again been identified on the basis of major cost items, although they differ considerably from those included in the elemental analysis of a new building project.

Table 4.4 Cost analysis

Detailed Elemental Analysis
BCIS *Online* analysis number: 21018

New build
BCIS code: A – 1(3) – 4,932

Job title:	Indoor Athletics Training Centre, Sutton Arena Phase 2
Location:	Carshalton, Greater London
Client:	Colin Beech
Dates:	Receipt: 21–Nov–2001 Base: 21–Nov–2001 Acceptance: 21–Nov–2001 Possession: 28–Jan–2002

Project details: Single and 2 storey indoor athletics training centre and fitness facility with sports hall and changing. External works include precast concrete and macadam paving, landscaping, services, drainage and site lighting.

Site conditions: Gently sloping green field site with moderate ground conditions. Excavation above water table. Highly restricted working space and restricted access.

Market conditions: Tender adjusted by £74 549 following expiry of tender. Design fees £142 340 excluded. Indices used to adjust costs to base price level: TPI for 4Q2001 177; location factor 1.15

Tender documents:	Employers requirements (for design and build)	**Contract:** JCT with Contractor's Design 1998
Procurement:	Design and build – competitive	**Cost fluctuations:** Fixed
Number of tenders:	Issued: 6 Received: 6	**Contract period:** Stipulated: 12 Offered: 12 Agreed: 12

Altered – see market conditions text

	Contract breakdown		Competitive tender list	
	Contract £	Analysis £	Tender £	% above lowest
Measured work	3 071 323	3 071 323	1 3 981 263	–
Provisional sums	–	–		
PC sums	26 000	26 000		
Preliminaries	693 849	693 849		
Contingencies	122 300	122 300		
Contract sum	3 913 472	3 913 472		

Table 4.4   *(cont'd)*

Accommodation and design features: Indoor athletics training centre with fitness suite on the mezzanine and 2nd floor hall. RC pad and strip foundations and bed; PCC upper floor. Steel portal frame and stairs. Rendered block walls; profiled metal cladding. Metal cladding to curved roof. Double glazed aluminium windows and curtain walling. Block partitions. Hardwood doors. Paint only to walls; vinyl carpet, rubber and Junckers flooring; suspended ceilings. Fittings. Sanitaryware. Gas HW central heating, ventilation, electrics. Lift. Fire/intruder alarms, CCTV, PA, data cables, TV, emergency lights.

Storeys as a % of gross floor area			Average storey heights	
			Below ground floor	—
			At ground floor	—
			Above ground floor	—

Areas

Basement floors	— m²
Ground floor	3 217 m²
Upper floor	1 715 m²
Gross floor area	4 932 m²
Usable area	3 515 m²
Circulation area	610 m²
Ancillary area	631 m²
Internal Divisions	176 m²
Gross floor area	4 932 m²
Area not enclosed	— m²
External wall area	2 310 m²
Wall to floor ratio	0.47
Internal cube	31 057 m³

		Functional unit		Rate	
		Total cost of element £	£ per m²	£/m² incl Preliminaries	
Element	Percentage			Tender prices	1995 constant prices
Substructure	7%	282 952	57.37	70.22	44.85
Superstructure	32%	1 248 122	253.07	309.76	197.83
Internal finishes	7%	277 365	56.24	68.84	43.96
Fittings	3%	136 599	27.70	33.90	21.65
Services	23%	891 319	180.72	221.21	141.28
Building sub-total	72%	2 836 357	575.09	703.92	449.57
External works	7%	260 966	52.91	64.77	41.36
Preliminaries	18%	693 849	140.68	—	
Contingencies	3%	122 300	24.80	24.80	15.84
Total		3 913 472	793.49	793.49	506.77

Submitted by: Sport England

Table 4.4 (cont'd)

Element		Preliminaries shown separately				Preliminaries spread	
		Total cost	Cost per m²	Element unit quantity	Element unit rate	Total cost	Cost per m²
1	Substructure	282 952	57.37	3 217 m²	87.96	346 338	70.22
2A	Frame	260 440	52.81	4 932 m²	52.81	318 783	64.64
2B	Upper floors	87 959	17.83	1 715 m²	51.29	107 663	21.83
2C	Roof	178 342	36.16	3 036 m²	58.74	218 293	44.26
2D	Stairs	147 786	29.96	7 No	21 112.29	180 892	36.68
2E	External walls	286 522	58.09	2 110 m²	135.79	350 708	71.11
2F	Windows and external doors	58 589	11.88	200 m²	292.95	71 714	14.54
2G	Internal walls and partitions	154 855	31.40	3 419 m²	45.29	189 545	38.43
2H	Internal doors	73 629	14.93	159 m²	463.08	90 123	18.27
2	Superstructure	1 248 122	253.07			1 527 721	309.76
3A	Wall finishes	33 949	6.88	5 692 m²	5.96	41 554	8.43
3B	Floor finishes	184 948	37.50	4 705 m²	39.31	226 379	45.90
3C	Ceiling finishes	58 468	11.85	1 647 m²	35.50	71 566	14.51
3	Internal finishes	277 365	56.24			339 499	68.84
4	Fittings	136 599	27.70			167 199	33.90
5A	Sanitary appliances	41 892	8.49	72 No	581.83	51 277	10.40
5B	Services equipment	–					
5C	Disposal installations	5 000	1.01			6 120	1.24

Table 4.4   (*cont'd*)

Element		Preliminaries shown separately				Preliminaries spread	
		Total cost	Cost per m²	Element unit quantity	Element unit rate	Total cost	Cost per m²
5D	Water installations	71 472	14.49			87 483	17.74
5E	Heat source	31 269	6.34			38 274	7.76
5F	Space heating and air treatment	68 121	13.81			83 381	16.91
5G	Ventilating systems	209 949	42.57			256 981	52.10
5H	Electrical installations	401 888	81.49			491 917	99.74
5I	Gas installations	included in	5F				
5J	Lift and conveyor installations	20 228	4.10			24 759	5.02
5K	Protective installations	included in	5H				
5L	Communications installations	included in	5H				
5M	Special installations	included in	5H				
5N	Builder's work in connection	41 500	8.41			50 797	10.30
5O	Builder's profit and attendance	–					
5	Services	891 319	180.72			1 090 989	221.21
	Building sub-total	2 836 357	575.09			3 471 746	703.92
6A	Site works	155 976	31.63			190 917	38.71
6B	Drainage	51 899	10.52			63 525	12.88
6C	External services	38 091	7.72			46 624	9.45
6D	Minor building works	15 000	3.04			18 360	3.72
6	External works	260 966	52.91			319 426	64.77
7	Preliminaries	693 849	140.68			–	
	Total (less Contingencies)	3 791 172	768.69			3 791 172	768.69
8	Contingencies	122 300	24.80			122 300	24.80
	Contract sum	3 913 472	793.49			3 913 472	793.49

Table 4.4 *(cont'd)*

Element		Specification
1	Substructure	Concrete pad and strip foundations and bed.
2A	Frame	Steel portal frame.
2B	Upper floors	PCC composite upper floor.
2C	Roof	Metal cladding to curved roof. Sunpipes.
2D	Stairs	Steel stairs.
2E	External walls	Rendered block cavity walls; profiled metal cladding.
2F	Windows and external doors	Double glazed aluminium window and doors with full height curtain walling.
2G	Internal walls and partitions	140 mm blockwork.
2H	Internal doors	Hardwood doorsets.
3A	Wall finishes	Fair faced block walls with paint.
3B	Floor finishes	Applied rubber flooring to athletics track, Junckers to hall, vinyl and carpet.
3C	Ceiling finishes	Mineral fibre suspended ceilings.
4	Fittings	Fittings including disabled.
5A	Sanitary appliances	Standard IPS.
5C	Disposal installations	Soil and waste pipes.
5D	Water installations	Cold water mains, hot and cold water services.
5E	Heat source	Gas boilers.
5F	Space heating and air treatment	Gas fired central heating.
5G	Ventilating systems	Local mechanical ventilation.

Table 4.4　(*cont'd*)

Element		Specification
5H	Electrical installations	Electric light and power. Emergency lighting.
5I	Gas installations	Gas supply to boilers.
5J	Lift and conveyor installations	1 No lift.
5K	Protective installations	CCTV.
5L	Communications installations	Fire and intruder alarms, PA, data cabling, TV.
5N	Builder's work in connection	General builder's work in connection with services.
6A	Site works	Site preparation. Roads and parking; landscaping; fencing.
6B	Drainage	Drainage.
6C	External services	External services and lighting.
6D	Minor building works	Undefined.
7	Preliminaries	22.40% of remainder of Contract Sum (excluding Contingencies).
8	Contingencies	3.95% of remainder of Contract Sum (excluding Preliminaries).

Credits

Submitted by:	Sport England
Client:	London Borough of Sutton
Architect/project manager:	Construction & Property Consultancy
Quantity surveyor:	John Cobb & Partners
Planning supervisor:	Construction & Property Consultancy
Services engineers:	Edward Pearce & Partners
Structural engineer:	Construction & Property Consultancy
General contractor:	R Durtnell & Sons Ltd

Source: Building Cost Information Service

Table 4.5 Simplified form of cost analysis

BCIS code C-1-45

Indices   169   230

CI/SfB
212

Local Station-1

TOTAL PROJECT DETAILS

Job title:	Reconstruction of Station Building		
Location:	Sawbridgeworth, Herts.		
Client:	British Railways Board		
Tender dates:	(1) 11th April 1972	(2)	
Contract periods:	(1) 4 months	(2) 4½ months	
Type of contract:	JCT Standard Form of Contract. L.A. Edition with quantities		
Fluctuations:	No		
No. of tenders issued: 7		received: 3	

Measured work:	£ 11 791
P.C. sums:	£ –
Prov. sums:	£ 556
Preliminaries	£ 7 222
Sub-total:	19 569
Contingencies	£ 460
Contract sums	£ 20 029

ANALYSIS OF SINGLE BUILDING

No. of storeys: 1
Gross floor area: 45 m²
Functional unit: 37 m² usable area

Type of construction:
Reinforced concrete edge beam and raft foundations, digging in existing tarmac and hardcore.
Brick load bearing external walls, felt covered roof.
No heating installation.
Extensive external works include clearing away existing building.

Element	Total cost of element	Cost per m²	Total cost of element inc. prelims	Cost per m² inc. prelims	Cost per m² inc. prelims base date 1st ¼ 1969
Substructure	2 482	55.16	4 100	91.11	66.37
Superstructure	3 579	79.53	5 911	131.36	96.10
Int. finishes	640	14.22	1 057	23.49	17.40
Fittings	1 439	31.98	2 377	52.82	39.04
Services	619	13.76	1 023	22.73	16.88
Sub-total	8 759	194.65	14 468	321.51	235.79
Ext. works	3 088	68.62	5 101	113.36	81.82
Preliminaries	7 722	171.60	–	–	–
TOTAL excl. contingencies	19 569	434.87	19 569	434.87	317.61

Issued-1974/75-128-1
For further details see Detailed Cost Analysis Section G, Local Station-1

Table 4.6   Costs-in-use analysis

CI/SfB
736

Computer centre-2-f

FINANCIAL STATEMENT FOR YEAR . . . .
Gross floor area: 566 m²

Element		Total £	Cost per 100 m² floor area £	Brief description of work
0	Improvements and adaptations	6	1.06	Sundry items only
1.	Decoration			
1.1	External decoration	–	–	
1.2	Internal decoration	–	–	
	Sub-total	£ –	£ –	
2.	Fabric			
2.1	External walls	14	2.47	
2.2	Roofs	1	0.18	
2.3	Other structural items	27	4.77	General repairs
2.4	Fittings and fixtures	84	14.84	
2.5	Internal finishes	39	6.89	
	Sub-total	£ 165	£ 29.15	
3.	Services			
3.1	Plumbing and internal drainage	42	7.42	
3.2	Heating and ventilating	116	20.49	
3.3	Lifts and escalators	–	–	Contains an element for PPM

Table 4.6 (cont'd)

					Notes
3.4	Electric power and lighting	42		7.42	
3.5	Other M and E surfaces	476		84.10	
	Sub-total	£	676	£	119.43
4.	Cleaning				
4.1	Windows	15		2.65	External only
4.2	External surfaces	–		–	
4.3	Internal	792		139.93	Contains an element for internal window cleaning
	Sub-total	£	807	£	142.58
5.	Utilities				
5.1	Gas	–		–	
5.2	Electricity	387		68.38	
5.3	Fuel oil	397		70.14	
5.4	Solid fuel	–		–	
5.5	Water rates	41		7.24	
5.6	Effluents and drainage charges	–		–	Allocations by Area/Population ratios
	Sub-total	£	825	£	145.76
6.	Administrative costs				
6.1	Services attendants	–		–	
6.2	Laundry	–		–	
6.3	Porterage	–		–	
6.4	Security	–		–	
6.5	Rubbish disposal	–		–	
6.6	Property management	218		38.52	Maintenance Officer Management only, excludes management cost for porters and cleaners
	Sub-total	£	218	£	38.52	

Note – Change in Direct Labour Force

Joiners/Labourers	11
Plumbers	6
Electrical Services	10
Mechanical Services	11
Groundsmen	19

Table 4.6 (*cont'd*)

CI/SfB
736

Computer centre-2-f

FINANCIAL STATEMENT FOR YEAR . . . .
Gross floor area: 566 m²

Element	Total £	Cost per 100 m² floor area	Brief description of work
7. Overheads			
7.1 Property insurance	$\frac{142}{649}$	$\frac{25.09}{114.66}$	
7.2 Rates			
Sub-total	£ 791	£ 139.75	
TOTAL	£ 3 482	£ 615.19	

External area . . . m²	External works Total £	Cost per 100 m² of external area	Brief description of work
8. External works			
8.1 Repairs and decoration			
8.2 External services			
8.3 Cleaning			
8.4 Gardening			
External Works Total	£	£	

Financial statement (consolidated) for years 1970/71–1973/74. Gross floor area 566 m²

FINANCIAL YEAR

Element	1970/71 Total £	Cost per 100 m² FA	1971/72 Total £	Cost per 100 m² FA	1972/73 Total £	Cost per 100 m² FA	1973/74 Total £	Cost per 100 m² FA
0. Improvements and adaptations	121	21.30	596	105.30	2	0.35	6	1.06
1. Decoration	60	10.60	535	94.52	181	31.98	–	–
2. Fabric	100	17.67	70	12.37	47	8.30	165	29.15
3. Services	503	88.87	555	98.06	412	72.79	676	119.43
4. Cleaning	488	86.22	560	98.94	659	116.43	807	142.58
5. Utilities	657	116.08	728	128.62	739	130.57	825	145.76
6. Administrative costs	191	33.75	163	28.80	176	31.10	218	38.52
7. Overheads	643	113.60	662	116.96	652	115.19	791	139.75
TOTAL	2 642	466.79	3 273	578.27	2 866	506.36	3 482	615.19
8. External works	–	–	–	–	–	–	–	–

A third form of cost analysis is one that could be developed for the renovation of existing properties. The aim is again to identify the groups of items that are of cost importance, and these will differ from those listed in Tables 4.4–4.6. In a renovation project the following might be considered as relevant elemental cost centres:

- Damp-proofing
- Rot treatment
- Roofing
- External walls
- Joinery
- Internal finishings
- Decoration
- Plumbing and engineering installations
- Electrical installations
- External works
- Preliminaries

4.12 THE BUILDING COST INFORMATION SERVICE

The Building Cost Information Service (BCIS) is the largest disseminator of construction cost information in the world. It was originally established by the RICS in 1962 as the Building Cost Advisory Service. It provides a proven and invaluable service, particularly for the chartered quantity surveyor in private practice and public service. Subscribers now include architects, engineers and contractors. The service has grown considerably since it was formed, when it originally largely operated on a reciprocal basis in that it allowed quantity surveying practices to exchange cost information between those members who were able to supply it.

The BCIS provides an extensive range of publications that are concerned with construction costs. These include the following:

Reports
Quarterly Review of Building Prices
Occupancy costs reports for different categories of buildings
Survey of Tender Prices
Review of Maintenance Costs

Forms and standard documents
Elements for Design and Build
Standard Form of Cost Analysis: Principles, Instructions and Definitions
Standardized Method of Life Cycle Costing for Construction Procurement

Indices and forecasts
Indices and Forecasts
Five Year Forecast 2008
Price Adjustment Formulae Indices
BERR Construction Price and Cost Indices

Market surveys
Asia Building and Construction Survey
RICS Construction Market Survey
RICS Commercial Market Survey
RICS Housing Market Survey
RICS Rural Market Survey

Construction Online Products and Services
Construction Duration Calculator (for teaching purposes)
Housing Online
Price Adjustment Formulae Indices
Daywork Rates – updating service
RIBA/RICS Building Cost Calculator
E-Tendering

Price books and guides
Access Audit Price Guide
Civil Engineering Price Book
Wessex SMM7 Estimating Price Book
Wessex Alterations and Refurbishment
Wessex Guide to Estimating for Small Works
Wessex Comprehensive Building Price Book
Wessex Painting and Decorating Price Book
Definition of Prime Cost of Daywork Carried Out Under a Building Contract
Housing Repair Cost Guide
Building Maintenance Price Book
Dilapidations Price Book
Guide to Daywork Rates
Life Expectancy of Building Components
Schedule of Basic Plant Charges
Review of Consultants' Fees on Construction Projects
The Greener Homes Price Guide

4.12.1 Standard Form of Cost Analysis

The BCIS *Standard Form of Cost Analysis* has recently been updated for the first time in almost 40 years. The *Standard Form of Cost Analysis* for building projects is used throughout the UK to provide data. These data allow comparisons to be made between the cost of achieving various building functions in one project with that of achieving equivalent functions in other projects. The new third edition of the *Standard Form of Cost Analysis: Principles, Instructions and Definitions* clarifies the definitions and expands the sub-elements in the light of modern construction techniques, and takes into account questions of interpretation raised by users. The *Standard Form of Cost Analysis* sets out:

- The principles of analysis
- Instructions on the information required to complete a cost analysis

- General definitions
- Definitions of the elements and sub-elements
- Element unit quantities

A sample of the various standard forms can be downloaded from the BCIS website at www.bcis.co.uk

4.12.2 Quarterly Review of Building Prices

An annual subscription to the BCIS *Quarterly Review of Building Prices* covers four publications a year (October, January, April and July). It contains a selection of BCIS data which gives guidelines on the general level of building prices. It provides average ($£/m^2$) building prices, the BCIS tender price index, the BCIS building cost index and location factors. It also includes a brief commentary on market conditions and tender prices. It provides superficial area rates for an almost exhaustive list of building project types. These data are provided by way of a simplified statistical analysis as follows:

CISfB	Building type	Mean ($£$ per m^2)	Mean ($£$ per m^2)	Range ($£$ per m^2)	Standard deviation ($£$ per m^2)	Sample size
328	Banks	560	527	285–838	140	30
270	Workshops	269	298	106–602	95	58

Source: Adapted from BCIS

It is possible, therefore, to use this review to determine both the typical price and range of prices for almost any building type anywhere in the UK. The use of the tender price indices (historic data) coupled with the economic review (forecast data) will also allow for the prediction of costs for future projects.

4.12.3 Standardized Method of Life Cycle Costing for Construction Procurement Construction Procurement

This publication brings together ISO 15686-5 and the BCIS *Standard Form of Cost Analysis* (SFCA) along with industry recognised occupancy cost codes. It provides:

- A standardised method of life cycle costing, applicable to the UK construction industry and to the key stages of the procurement process.
- An industry accepted methodology to facilitate a more accurate, consistent and robust application of life cycle costing estimation and option appraisals.
- A more effective and robust basis for life cycle cost analysis and benchmarking.
- Practical guidance, instructions and definitions, together with informative worked examples on how to undertake life cycle costing for construction.

■ Process maps to help to plan, generate, interpret and present results for a variety of different purposes and levels of life cycle cost planning.
■ Instructions on how to define a client's specific requirements for life cycle costing along with the required outputs and forms of reporting, and help in deciding which method of economic evaluation to apply.

4.12.4 BCIS Construction Duration Calculator

When planning construction projects, the more reliable the advice that is available, based on only the briefest project information, the better. The BCIS Construction Duration Calculator gives a quick yet robust estimate of how long it will take to construct a building based on a database of thousands of actual project out-turns.

Using data collected from construction industry consultants and clients as part of the annual Key Performance Indicators (KPI) survey carried out by BCIS on behalf of the Department for Business, Enterprise and Regulatory Reform (BERR), BCIS have developed sophisticated models to predict the construction duration. These models also show how quickly, or slowly, the project might progress.

■ Models are based on actual construction duration of over 4,500 new-build and refurbishment projects in the UK.
■ They calculate the possible duration of a building project.
■ They use up-to-date construction duration data. This is regularly updated in line with the BCIS Tender Price Index and Location factors.
■ The model provides an expected project duration together with a possible range of durations.

Calculations are based on the following project details:

■ Type of project (new build/refurbishment)
■ Contract value (£15 000–£100m)
■ Date of project
■ Location
■ Client type (public/private)
■ Selection of contractor (single stage tendering; two stage tendering; negotiated; partnering; other)
■ Building function (there are approximately 30 building function types including housing, offices, factories, retail, sports building and schools)

4.12.5 Life Expectancy of Building Components guide

The *Life Expectancy of Building Components* guide presents the findings from a survey of the life expectancy of common building components based on the experience of building surveyors. It helps to highlight the factors that affect the deterioration or failure of the components. A checklist of the factors to be considered when assessing life expectancy provides invaluable assistance for anyone involved in condition surveys and building design. A checklist identifying causes of early deterioration is also included to provide further advice. This comprehensive guide provides unique information on the life expectancy of building components. This will assist those who are involved in:

- Planning maintenance programmes
- Preparing whole-life cost analysis
- Undertaking value engineering exercises and decisions
- Analysing building failures or problems
- Making design decisions

4.12.6 BCIS Rebuild Online

This is a guide for surveyors, valuers and loss adjusters who carry out rebuilding cost assessments. The BCIS Rebuild Online service allows practitioners to:

- Calculate reinstatement cost assessments for houses and flats instantly
- Include all necessary specification, date and location adjustments in one calculation
- Print out tables adjusted for a specific date and property location
- Record calculations and print out reports for clients

The calculations are made from selections of the following factors: type, age, size, number of floors, quality and region. Depending on whether a property is a house or a flat, adjustments can be made for many features including roof covering, garage, conservatory, fencing, walls, paving, drainage and much, much more. Users have commented that the high quality of the data and elemental structure which the services have at their core is recognised throughout the insurance, loss adjusting and valuation professions when producing rebuilding cost allocations and advising customers. It is an excellent tool for calculating rebuilding costs for flats or houses and the real benefit is that the practitioner knows that base rates are always up to date. In any subsequent dispute there can be no question that the practitioner availed himself of the most up-to-date information available.

4.13 MARKET INTELLIGENCE

Market intelligence is the information that is relevant to a sector's markets. It is gathered and analysed specifically for the purpose of accurate and confident decision-making in determining market opportunities, market penetration strategies and market development metrics. It is the systematic process of gathering, analysing, supplying and applying information, both qualitative and quantitative, about the external market environment. It will include:

- Market and customer orientation – clarity about the current market needs
- Identification of new opportunities – new trends and practices in markets
- Early warning of changes in activity – enables countermeasures

- Minimising investment risks – detecting threats and trends as early as possible
- Quicker, more efficient and cost-effective information – avoids duplication of report acquisitions and expensive consultant work

The use of building cost data, as noted in the earlier parts of this chapter, is not simply a mechanical process of finding historic rates and prices, updating these for inflation and then applying them to new situations. It requires an understanding of how the market is changing, globally, nationally, regionally and locally. The role of an estimator is often underrated, but success in winning work and at the right price is paramount to the start of a successful project. Similarly the quantity surveyor in practice must have not just access to reliable and current cost information, but also the ability to understand the building intelligence that is published by a number of organisations and, in a fast changing market, to keep this regularly under review. Using the cost data requires a clear understanding of a client's expectations and changes that are taking place in the construction industry to help deliver maximum added value. It also includes a good understanding of the different techniques and practices, many of which are explained in this book.

Building magazine is a useful starting point, since it provides a current perspective that can be tested in the market. It provides information that is updated regularly by means of an analysis of recent past performance together with a forecast of future expectations. It provides an overview of economic activity, starting from an analysis of current data, using, for example, data from the Office of National Statistics. It combines this with economic forecasts of the country's performance, taken from a range of different sources. It then uses this information to examine the different sectors of the industry. It examines this information, where possible, on a regional basis by using regional development agencies' forecasts of workloads. The information is recorded for each quarter. BCIS, for example, provides a Commercial Market Survey report that reviews this area of activity. It has also written a Five Year Forecast in 2008.

Construction market intelligence covers the following:

- Construction market analysis and research
- Construction trend forecasting
- Resource and capacity assessment
- Labour force analysis
- Economic analysis
- Cost and price comparisons
- Construction marketing and communications development

This intelligence will cover global markets and changes in direction, since the large global players exert a considerable influence over weaker and smaller economies. Some countries are able to buck the trend because of local circumstances and government management of their economies. Construction market intelligence includes everything an organisation needs to know and understand in order to make sound decisions for the future. It seeks to understand what the industry will decide to do tomorrow and how it will do it.

4.14 INTERNATIONAL COMPARISON OF CONSTRUCTION COSTS

The construction industry has been criticised on several occasions because of its inability to produce buildings for the same costs as in other countries. When one seriously attempts to make a comparison of these costs, the task becomes very difficult and complex. To answer the question 'Where should I build my offices to be the most cost-effective?' requires the consideration of so many factors, for example land costs, professional fees, grants, taxation and costs-in-use. Even after making the rash assumption that these will be equal, the problem is still a long way from being solved. A straightforward comparison of the cost per square metre of an identical building in Germany, for example, with one in the UK is only the start of the solution. It is of course quite straightforward to convert the official exchange rate of these two countries into a common currency, but in practice this may require regular revision to account for constantly varying changes to these rates. Since, however, these rates do not necessarily reflect the real price levels ratio between countries, it is inappropriate to use them.

The correct approach is to use real price ratios between countries. A price ratio or parity is equal to the price, let us say, of sugar in the UK (in £) and of sugar in Germany (in DM), £ per DM. This gives us a parity or exchange rate for sugar between the UK and Germany. This parity is probably different from the official exchange rate and different for other products. In order to provide a realistic comparison between countries the price ratios of a number of products are aggregated to find an average price ratio. The products chosen must of course be representative in each of the countries for which the comparison is to be made. They must also be comparable in each of the countries.

Thus a selection of representative and comparable products is taken and their price ratios calculated. The average price ratio is then used to compare, for example, the costs of building walls of a similar kind. It must be further stressed that the cost comparison, to be of any value, should compare items which are the same in every respect. It may be possible to provide some indication of cost differences between three-bedroom houses in different countries. It is likely that factors such as size and quality will also differ, and this will then tend to reduce the usefulness of any such comparison. Earnings, for example, would also need to be considered.

SELF ASSESSMENT QUESTIONS

1. Identify the different sources and nature of published cost information and comment upon their usefulness in practice.

2. What precautions should be adopted before attempting to use existing cost information and applying this to new projects and new situations?

3. Select a standard BCIS cost analysis and, using the information provided, describe the physical attributes of the project to which it applies.

BIBLIOGRAPHY

Ashworth, A., 'Making sense of price book data', *Building Trades Journal*, February 1986.

Ashworth, A., 'The source, nature and comparison of published cost information', *Building Technology and Management*, 1980.

Ashworth, A. and Skitmore, M., *Accuracy in Estimating*. Chartered Institute of Building 1982.

Beeston, D., *One Statistician's View of Estimating*. PSA 1973.

Beeston, D., *Statistical Analysis of Building Price Data*. Spon 1983.

Bennett, J., 'Cost data and the QS', *Chartered Quantity Surveyor*, May 1984.

Bennett, J. *et al.*, *Cost Planning and Computers*. HMSO 1981.

Best, R., *The Development and Testing of a Purchasing Power Parity Method for Comparing Construction Costs Internationally*. Unpublished PhD Thesis, University of Technology, Sydney. 2009.

BCIS, *BCIS Quarterly Review of Building Prices*, BCIS Quarterly.

Brook, M., *Estimating and Tendering for Construction Work*. Butterworth Heinemann 2008.

Harrigan, J., Raiser, J. and Raiser, P., *CIOB Code of Estimating Practice*. Blackwell Publishing 2008.

Skitmore, M., *The Accuracy of construction price forecasts: a study of quantity surveyors' performance in early stage estimating*. University of Salford 1990.

Wood, R.D., 'Principles of Estimating', *Estates Gazette*, 1982.

CHAPTER 5

DESIGN ECONOMICS

LEARNING OUTCOMES

After reading this chapter, you should have an understanding of the economics associated with the design of buildings. You should be able to:

- Understand the nature of a client's requirements
- Evaluate the various factors of design economy
- Apply the principles of design economy to new projects
- Recognise that design economy is just one factor to consider in design
- Appreciate that evolution in design is taking place

5.1 INTRODUCTION

It used to be assumed that the only items affecting the costs of a construction project were size and quality. In the present age we are more aware that the construction costs of a particular design solution are influenced by many other factors, some of which are interrelated. Insufficient research has been undertaken to enable us to be more precise about the morphology of building and engineering structures. There is, however, some knowledge available to us that has arisen largely because of expediency in practice, which does provide us with some rule-of-thumb guidelines. It may take many years before we are able to base our judgements on a more reliable body of knowledge.

For any one project the designer will usually consider initially the several different options that are available as possible design solutions. In addition, the various attributes, other than the economic criteria, by which the client will judge the finished building or engineering structure will be considered. The designer may then decide to reject the economic choice in preference for qualities to be found in an alternative design. The examination of the following factors does not seek to restrict or to limit the architect or engineer during the design process. The aim is to identify the factors that have economic consequences in the various design options, in an attempt to select the most suitable and appropriate proposal for the promoter of the project.

5.2 CLIENT'S REQUIREMENTS

A confirmed cynic once described architecture as the design of beautiful buildings
that satisfy only the architect and not the client. In addition, the buildings often
failed to function properly and were always too expensive. As in most cynicism, the
case is vastly overstated, yet there is enough truth in the statement that it cannot be
dismissed as merely unusually harsh criticism.

The success of any construction project can be measured against several different
criteria. The promoter's requirement can be summarised as follows and as shown in
Figure 5.1. A successful combination of these factors is necessary in order to provide
a project for a satisfied client.

5.2.1 Performance

The architect must produce a basic plan concept to meet the client's requirements
in the most efficient manner. The project when completed must also have aesthetic
merit. The architecture and engineering and the construction work done on site
must be done in a manner that will protect the client against his own inexperience.
Although some of this may be in part a value judgement, there are several factors by
which these requirements can be appraised. The quality of the finished work will
have been defined in the specification and this will be a measure for assessment.
This will necessitate an adequate specification initially, the selection of an
experienced contractor and the required supervision throughout the construction
operations on site. Inadequate design and detailing and the incorrect choice of

Fig. 5.1   Client's requirements

materials are elementary problems that will cause obstacles to proper performance. The promoter will need to be satisfied that the completed structure meets the needs and requirements in terms of spatial design and the structure's function. A further important consideration that has arisen is that of the future maintenance requirements once the project is in use. Many clients have seen the advantages of providing a design in terms of the total project rather than on the basis of initial design alone.

5.2.2 Time

Once clients decide to build they are generally in a hurry for their completed projects. Although a large amount of time may be spent deliberating over a scheme, once the decision to build is made the project then becomes of the utmost urgency. The design of the project will influence the methods adopted by the contractor for construction, and these in turn will have an important influence on the length of the contract period. One method of measuring the success of a project is whether it is available for commissioning by the date promised in the contract documents. It may be necessary therefore to consider alternative methods of contracting, such as fast-tracking methods that can improve contract performance in terms of time.

5.2.3 Cost

Before clients commit themselves to the detailed design of projects, some information on the expected price is usually required. In circumstances where projects need to be carried out as a matter of urgency, cost may be of less importance, but even so it cannot easily be ignored. A budget price is generally required and this is the sum that the promoter will remember. However good the reasons may be for increasing the price, the promoter may look on such action with dismay. Eventually a tender sum will be accepted, and once the work has been completed the final account can be agreed. Research has attempted to discover the reasons for the discrepancies between these sums, but as yet this remains inconclusive. An enlightened client, however, will not measure cost solely in terms of initial values but rather on the basis of a whole-life approach.

5.3 VALUE FOR MONEY

Buildings and engineering structures are complex commodities, but in common with other goods available, clients are rightly concerned with obtaining value for money. Cheapness is in itself no virtue. It is well worthwhile to pay a little more if as a result the gain in value exceeds the extra cost. In terms of a whole-life cost it can be shown that it is often expedient to spend an extra sum initially on a construction project in an attempt to reduce future recurring costs. Although value for money may be very subjective in assessment, three factors can be examined in an attempt to evaluate it. The *appearance* of buildings and engineering structures will always be largely subjective, although the opinions of others cannot easily be disregarded in this respect. The structure can be assessed in its relation to its adequacy to support the building during its life. Architects, in particular, have developed some judgement of the aesthetics of the design based on form, shape, proportion, materials used, location etc. However, personal choice and taste are factors which will also need to be taken into account and these are extremely difficult to evaluate. The materials used may be judged in terms of their durability, appearance and freedom from future maintenance. *Function* can be judged against various criteria. These are often included in the brief given to the designer and can be measured to some extent by comparison with other similar buildings. Judgement of the adequacy of the internal space and its arrangement can be related to the extent to which it facilitates the functions to be performed in the building. The arrangement of the design, the materials used, the method of construction and the environment created or other forms of control are examples of the facets suitable for comparison. The third factor to take into account is that of *cost*. The obvious approach is to put all the measurable components on one side of an equation in the form of cost and to set these against a value judgement of form, appearance, comfort and convenience.

The need for careful assessment of a building design is all the more necessary today, when such a large and ever-increasing range of materials and techniques is available. Clearly a project design is always a compromise between the many facilities and amenities which the building is designed to provide. The determination of the most efficient design cannot be an exact science since there is so much that is unknown about the various aspects of the construction process. In addition, far too little research is undertaken to establish the relationship between design and costs.

5.4 FACTORS TO CONSIDER

The following factors have a direct influence on the cost of a project, and need to be considered during the economic evaluation of the building or engineering structure.

5.4.1 Site considerations

Each construction site has its own characteristics which have an important influence on its suitability for development. The size of the site required will generally be determined by the type of project to be constructed. The cost of the project will be affected by its location. It may be situated on a congested city site with all the problems of access, materials deliveries, close proximity of adjacent structures etc. Alternatively, it may be located in the heart of the countryside with its own peculiar problems and particularly transport costs. The availability of mains services or the costs of their provision will be an important consideration. Construction costs will also vary between different parts of the country, with costs in London currently being of the order of 13% higher than the average in the provinces (London 1.13, northwest 1.03; East Anglia 1.00; southeast 1.02). The location of the building on the site will also affect the overall cost of the scheme. Some projects, for example, may necessitate long-haul roads with the consequent costs necessary for provision and contract maintenance. Others constructed at the far end of a site will require permanent access, long drainage connections and perhaps substantial costs for landscaping. Often, because of the high costs of building land, it is essential that as much of the site be used for building purposes as the planning regulations will permit. The siting of buildings can also have recurrent implications for the user's future energy consumption.

The ground conditions of the chosen site are a factor that can substantially influence constructional costs. The increased costs of an expensive type of foundation construction necessitated by a poor ground-bearing capacity may be coupled with overall poor working conditions for men and machines as they become bogged down. These problems will be aggravated in inclement weather conditions. Water-bearing ground, and the necessity to remove obstructions or to work around them, not only slows down progress but also increases costs. The opposite conditions of running sand and hard rock create their own peculiar problems, and in cases where excavation work is necessary, foundation costs are likely to increase considerably. Steeply sloping sites can often result in large quantities of cut and fill. Should the choice be available of either sloping banks or the provision of retaining walls, then the former would always be selected for economic reasons. This may, however, need to be balanced with space considerations and land availability. The preparation of a site prior to construction operations needs careful consideration. Artificial strengthening of the ground, the redirection of watercourses or demolition can all significantly increase costs and should be avoided where possible.

5.4.2 Building size

One of the first items to be considered in connection with any construction project is its size. This is an important factor in terms of cost efficiency, because costs are not in proportion to changes in size. The designer, however, may have little influence over the size of the project as this is generally determined by clients' needs. Larger buildings have lower unit costs than smaller-sized projects offering an equivalent quality of specification. For example, a dwelling house on its own individual plot of land will cost more to construct than a similar dwelling which may be part of a large housing estate contract. Smaller factories cost more per unit than their larger counterparts. To some extent this is due to the economic theory of economies of scale. The designers' (architects and quantity surveyors) costs are also calculated on a sliding scale of charges. Smaller projects take a proportionately longer time per unit to design, and this is reflected in the design costs. Larger projects can be more efficiently managed, particularly where the size of the project warrants a resident site manager. Because of this better organisational ability and also because of the improvement in outputs of operatives, they can be completed in a disproportionate amount of time. Both of these factors will cause the unit costs to

900 m²   Plan area   8 100 m²
120 m   Perimeter length   360 m
480 m²   Wall area   1 440 m²

$\dfrac{480}{900} = 0.533$ Wall-to-floor ratio $\dfrac{1\,440}{8\,100} = 0.178$

Fig. 5.2   Variation of wall-to-floor ratio with building size

be reduced. In addition, a more intensive use of plant and a better capability of obtaining improved discounts on materials are factors that favour the larger-sized projects. A further reason for the lower unit cost is the lower wall-to-floor ratio. For a given plan shape, a larger plan area will always result in a lower ratio. For example, a building 30 m × 30 m × 4 m high has a wall-to-floor ratio of 0.533. A similar project offering nine times this area, 90 m × 90 m × 4 m high, has a wall-to-floor ratio of 0.178. This is illustrated in Figure 5.2. It might be noted at this point that just as size does not change directly with cost, neither does the wall-to-floor ratio. However, there is more correlation of cost with wall-to-floor ratio than with size. The examination of any pricing data will support the cost–size theory. Manufacturers will always give larger discounts for larger orders. Theoretical pricing data from price books will also show costs grouped in project size bands, and the examination of bill rates from actual projects of different sizes will substantiate this fact.

5.4.3 Planning efficiency

Although the outline alternative plans for a project may be similar in overall size, the way that space can be utilised within the project may vary considerably. The designer will have attempted to make the best possible use of space within each alternative design, but the ratio between usable and non-usable (circulation) space will differ. Traditional cost analysis supplied by the Building Cost Information Service only provides costs in terms of the gross internal floor area. It could be useful if the analysis also indicated an appropriate rate based on the net usable area.

Total area   = 576 m²
Usable area   = 480 m²
Efficiency factor $\frac{480}{576}$ = 83%

Total area   = 576 m²
Usable area   = 460 m²
Efficiency factor $\frac{460}{576}$ = 80%

Total area   = 576 m²
Usable area   = 455 m²
Efficiency factor $\frac{455}{576}$ = 79%

Fig. 5.3   Comparison of planning efficiency factors

One of the main aims of an economic layout will be to reduce the amount of circulation space to an acceptable minimum. The ratio of non-usable space will depend on the type and purpose of the building. This may typically represent 20% in blocks of flats or offices and as little as 13% for laboratory buildings. It may be difficult initially to analyse open-plan designs, but it is generally accepted that these incorporate a lower ratio of non-usable space. The more irregular the plan shape, the lower the amount of usable space that is likely to be available. The elimination of long lengths of corridors resulting in communication through other rooms or open-plan designs may not be acceptable, while reducing the widths of circulation areas may not be permissible under the building regulations. The planning efficiency of engineering projects such as highways will often be determined along the lines of safety considerations. There may also be examples where a client deliberately enlarges the circulation area for prestige reasons or to achieve images of grandeur. In these circumstances, a reduction in planning efficiency must be expected as a penalty for such designs.

Figure 5.3 considers the planning efficiency of three alternative layouts, each with the same gross internal floor area (GIFA). In terms of the layouts offered, the planning efficiency factors vary from 79% to 83%.

The building and planning regulations will also have other influences on the overall efficiency of the design layout, e.g. in the provision of toilet accommodation in public buildings and compliance with fire regulations, particularly those sections dealing with means of escape. The appropriate offices, shops and factories Acts will also have requirements to be complied with that can impinge not only on the design but also on the planning efficiency.

Theories that do not aim for the optimum in planning efficiency abound, such as the 'loose-fit' theory. Designs based on this idea produce buildings on the basis of maximum future adaptability. The assumption is made that the functions within almost all types of building will change, and probably more rapidly than in the past. The design should therefore include as much flexibility as possible, in order to preclude the future obsolescence of the building or the need for major structural alterations. This loose-fit approach may result in a more standardised type of structure than one particularly designed to suit the client's personal needs at that time. A loose-fit design will also make adaptation and conversion at some future date easier and therefore less expensive. Although in the long term it may be economic to be fairly generous with space, even those clients anticipating future changes or expansion generally cannot afford to waste space in the initial design.

In practice, the planning of space requirements is of prime importance to both the client and the designer. In complex building projects, studies of the various functions to be performed may be undertaken by several different disciplines, sometimes outside the construction professions.

5.4.4 Plan shape

The plan shape of any structure has an important effect on the overall cost of the project. This effect is not restricted to the external envelope costs, but also applies to the internal division elements. A square-plan-shaped structure will in the majority of cases provide the most economic solution. This is largely due to the theory known as the wall-to-floor ratio. A square shape provides the lowest amount of wall area to gross floor area (discounting of course the circular plan, which in construction terms tends to be very expensive). The more complex the shape, therefore, the higher will be the overall cost of the structure based on an agreed required floor area. The reason the irregular-shaped plan costs more can be attributed to the number of corners involved. This is known to be a factor influencing the costs of brickwork and the output of bricklayers, and it can also be shown to have some effects on, for example, roofing costs.

There have been a number of attempts to measure the cost efficiency of plan shape. The wall-to-floor area ratio is perhaps the most familiar, but it can only be used to compare buildings with a similar floor area and it does not have an optimum reference point. The length/breadth index is a mathematical concept that reduces the shape of a building to a rectangle having the same area and perimeter as the building. A development of this is the plan/shape index to allow for multi-storey construction. Other attempts have also been made to assess plan and volume compactness.

Fig. 5.4   Comparison of plan shapes

Figure 5.4 shows comparable design data in terms of plan shape, emphasising the economic compactness of the square-plan-shaped building. The lower the wall-to-floor ratio, the more economic will be the design when judged against this criterion.

The designer usually does not have complete freedom when choosing a plan shape, but has to work within a number of constraints. The purpose of the structure will often have a considerable influence on this factor. For example, it may be necessary with some industrial manufacturing processes that the correct design solution results in a long narrow building. Second, the site boundaries may be such as to restrict the design to fit within the shape of the plot of land. In some instances, particularly where land costs are high, the client or developer may wish to make the total possible use of this land by constructing the building up to the site perimeter. The topography of the site is a further reason why the designer may not choose simple square plans, and this point is discussed later. Finally, the architect or surveyor will need to balance the cost concept of a simple plan shape with the aesthetic appearance of the finished project. This does not have to imply that simple plan shapes will result in ugly designs, as this will depend very much on the elevational treatment of the project by the architect.

The square plan shape will not always form the best economic solution. On sloping sites it is generally agreed that rectangular buildings following the lines of the contours will form the correct basis of an economic solution. The reason for this is the reduction in foundation costs, particularly in terms of cut and fill. Environmental factors such as the provision of natural lighting, the reduction of noise transfer and adequate circulation space and internal division do not always favour the square shape. It is, however, favoured in respect of heating installations, in connection with both initial and recurring costs, because the volume of air to be treated is lower with the square plan shape than with other geometric forms (again ignoring the hypothetical circular building in economic terms). Buildings which require heavy loads to be transmitted at second floor level, e.g. two-storey warehouses, also do not favour the square shape, narrower buildings with shorter spans being more appropriate solutions.

Changes in the proposed plan shape of a project are likely to affect many of the major cost-important elements in the design. In order to evaluate the design solution fully, some type of cost study will be necessary to provide for comparative costs of the alternative designs. It must not be forgotten that the desire to select an appropriate plan shape will need to be balanced against the other important criteria of planning efficiency and external aesthetics. Plan shape, however, is a good measure to use in an attempt to qualify the project's overall complexity in respect of both the architectural design and the construction work envisaged on site.

5.4.5 Height

The constructional costs of tall structures are greater than those of low-rise buildings offering a similar amount of accommodation. Tall structures are thus preferred only where the land is either expensive or in scarce supply. The only exception to this rule is the addition of further storeys, to make the fullest possible use of the already expensive provisions in terms of structure or services. The following are some of the reasons why multi-storey structures are more expensive than low-rise buildings of comparable size:

- The higher constructional costs arising from building at a higher level include the provision of vertical transportation such as hoists and cranes, the problems with material storage, the delays in waiting for the construction to 'set', the increased amounts payable to operatives and safety requirements.
- The increasing costs of engineering services and their provision within the building, such as lifts, refuse disposal installations, pumping equipment for sewage disposal and protective installations such as fire fighting equipment and lightning conductors.
- The higher costs of provision for certain elements such as foundations, the necessity for a structural frame, more stringent constructional requirements for staircases, the provision of more fittings and furnishings for compactness and convenience, and the increased costs of engineering services described above.
- The improvement of fire-resistance precautions, particularly insulation between floors.
- An increase in the proportion of circulation areas required, including wider stairways, larger landing areas and areas for access.
- Less competition for the work because of the limited number of building contractors capable of undertaking the work. Although some contractors do become specialists in these types of projects, their tender prices often reflect their competitors and market conditions.
- Experimental forms of construction, used during the 1960s, tended to increase uncertainties of performance, and required large expenditure on specialised mechanical plant, which resulted in higher prices.
- Due to the complexities involved, more of the work has to be awarded to specialist subcontractors.
- Wind loading factors need to be taken into account with tall buildings, and this is likely to increase the constructional difficulty and its associated costs.

Figure 5.5 highlights some of these factors.

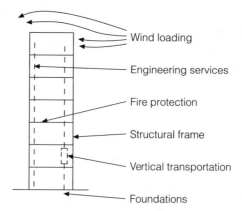

Fig. 5.5   Reasons for the relatively high construction costs of tall structures

The whole-life costs of buildings are also likely to be higher with taller buildings. This is due largely to maintenance costs associated with the improved engineering services such as lifts, plumbing and heating and electrical distribution which have all incorporated a higher cost specification than in low-rise buildings. Items such as window cleaning will need special provisions in the form of either elaborate external cradles or specially designed windows.

The psychological effects on people living or working in high-rise buildings must also be considered. There is certainly some evidence to suggest that people living in multi-storey blocks of flats do find many of the conditions intolerable under general circumstances. Tall buildings should therefore not be considered as solutions to building space problems unless no other real alternative exists.

Single-storey buildings are not a cost-effective solution either, for several reasons. Their foundations, for example, are usually capable of supporting greater loads than they actually support. A similar argument cannot be extended to tall blocks since the method of roof design and construction is likely to be considerably different. Single-storey structures are preferred, however, in buildings requiring large uninterrupted floor spans, or where huge loads are to be transmitted at floor level. They are generally also preferred for temporary structures as an economic solution.

In terms of building height, research has suggested that the cost components of a building can be divided into four categories:

1. Those which fall as the number of storeys increases (e.g. roofs, foundations)
2. Those which rise as the number of storeys increases (e.g. lift installations)
3. Those which are unaffected by height (e.g. floor finishes, internal doors)
4. Those which fall initially and then rise as the number of storeys increases
   (e.g. exterior enclosure)

The provision of required space below ground level is also an expensive alternative to the more traditional low-rise construction. However, there are circumstances where the incorporation of the basement within a structure is preferred. For example, for security reasons strong rooms of banks are generally placed below ground level. There are also examples where freezer stores have been constructed entirely below ground level, which, although it has meant an increased cost initially, has produced substantial savings in costs-in-use. Often the shortage of land is one of the major reasons for constructing below ground level. Car parking to multi-storey office buildings is often provided in this way. This may show some cost savings if the increased basement costs are offset against a more traditional foundation construction and the provision of car parking is made in some other way.

5.4.6 Storey height

The storey heights of buildings are largely determined by the needs of the user of the building. A greater storey height than normal may be necessary to accommodate large machinery or equipment, or it may be necessary to provide space within false ceilings for service ducts for cables, pipes or air conditioning ducts. In other circumstances, increased storey heights may be preferred for prestige reasons, as in the case of hotel foyers. Buildings such as churches, sports halls and theatres provide for high storey heights because of either tradition or design necessity. Excessive storey heights do have the effect of increasing the costs of the vertical circulation elements initially, and also the future maintenance costs, particularly for engineering services, such as heating and ventilation. Buildings with high storey heights will cost more per square metre of floor area than comparable accommodation with lower storey heights. Such buildings also result in higher wall-to-floor ratios. Minimum floor to ceiling heights used to be specified in the Building Regulations, and with both public and speculative housing these heights are rarely exceeded. Although storey heights are often restricted by building costs, over-generous storey heights are wasteful unless they are absolutely essential.

5.4.7 Groupings of buildings

The grouping and arrangement of buildings on a site can have an important influence on the overall cost of the project. Where it is possible to inter-link buildings or structures, a saving in costs can usually be achieved, often because of a saving in foundation and external walling costs or other common items of construction. These savings in cost will not be restricted to initial costs; there will also be a resultant saving in costs-in-use. Where it is possible to provide for some commonality in the provision of engineering services, such as district heating supplies, further savings in cost will be achieved.

Although the grouping of structures in practice generally applies only to dwellings and industrial units, any structure that can be linked to an adjoining building will generally show some cost advantage. It must not be overlooked that current fashion is towards detached units, but it must be stated that in purely economic terms these are not appropriate solutions. In very broad terms, cost studies have shown that semi-detached houses may be 6% more expensive than similar terraced versions, and detached houses more than 10% more expensive for building costs alone.

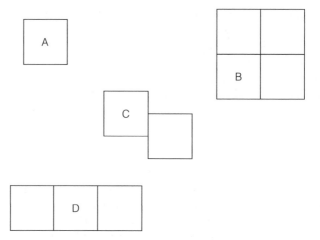

Fig. 5.6 Alternative plan groupings of buildings

Figure 5.6 shows some of the alternative plan arrangements which might be used for housing layouts. In cost terms, both initial and recurring, arrangement B offers the best solution. It provides for the use of more shared elements and can be constructed on a more confined site. Heat loss during use will also be lower. Arrangement A, although desired for other reasons, is the worst solution in terms of design economics.

5.4.8 Buildability

The buildability aspects of a construction project will also affect its economics. Buildability is defined as the extent to which the design of a building facilitates ease of construction, subject to the overall requirements for the completed building (*Buildability: an Assessment*, CIRIA 1983). Buildability is therefore largely concerned with the work on site and the practicalities of producing a structure from a design. Good buildability means that the design takes a very close account of the way in which the building will be constructed, and the conditions imposed on this process. Designs which require unnecessarily complex construction methods and procedures, or fail to take account of the mechanics of the work on site, fail in this respect. Requirements in respect of quality or aesthetics may at times be in conflict with the best buildability solutions. Wherever possible such problems should be eliminated, but it should be remembered that buildability, like many of the preceding aspects, is only one criterion against which building economics should be evaluated.

The ease with which buildings and other structures are erected on site should result in financial savings to the contractor. Contractors, during tendering, are particularly concerned with the alternative methods of construction, and this is clearly a factor which affects their final tender sum. The pre-tender method statement will plan a practical sequence of building operations. The constructional detail outlined in the tender drawings should encourage simple sequences of operations. Since good buildability allows the contractor to make such monetary savings, these, in time, where healthy competition exists, should provide some savings to the clients of the industry. The lack of buildability results in an inefficient use of the construction industry's resources.

5.4.9 Constructional details

The constructional details, materials used and methods of construction will have important cost implications for the project. These items are therefore of direct relevance to the resources input of the project in terms of labour, materials, plant and equipment, and organisation. Although it may be necessary to examine the economic consequences of each element or operation in turn, the cost influence of individual elements on each other must also be considered. Cost studies of the choices available should be undertaken in circumstances where the cost differences between alternatives may make a substantial difference to the overall cost of the scheme. However, it is not always the case that, having carried out a cost study, the least expensive solution will be selected. Other factors, such as the length of time required for construction purposes, will also need to be considered. The structural form is likely to require an early decision in the design process, and the relative merits of traditional load-bearing construction and framed construction will need to be evaluated. A choice, for example, between a steel and a concrete frame may be contemplated. Although concrete may have an initial cost advantage, the speed of steel frame erection may show a reduced contract period with a consequent saving in costs over the entire project. Concrete frames tend to be favoured in mainland Europe, whereas in the USA steel frames are preferred. The argument as to which is better depends very much on local circumstances, design and construction method, labour skills, and the costs and availability of materials. The span and bay spacing will also have a significant bearing on initial costs. Obviously, the fewer intermediate uprights the better, but careful examination and relevant experience will suggest the optimum acceptability of structural grids. It should also be noted that particular methods with a cost advantage today may show a different result at some time in the future. For example, trench fill foundations in the 1960s were generally not an economic proposition, but because of the rapidly rising labour costs in the subsequent years they became a realistic economic solution in certain situations. Constructional details that provide for *repetition* in the processes used will be able to provide economic savings on the project.

The learning curve is particularly appropriate to projects such as mass housing and pipelines contracts which require identical site operations to be repeated several times. However, there is likely to be far wider variation in costs between apparently identical items of work on civil engineering projects than on building projects. This apparent variation is due largely to the fact that civil engineering is much more *method*-orientated. The importance of this is that any cost study should seek to ascertain the appropriate construction method to be used by the contractor on site, particularly where this may affect price. The choice of *materials* depends on both the characteristics required and the price the client is willing to pay. Materials should be used to their fullest advantage, and qualities over and above those required are wasteful. Where new materials, processes or technologies are envisaged, it should be noted that contractors are likely to be rather wary and price the work accordingly. The use of *prefabricated* components or off-site techniques will generally tend to shorten the contract period, albeit often with a forfeit in terms of a higher cost. The economic comparison between traditional and system building is not easy to make. Prices will vary because of the variability in the designs offered and the constructional performance of each alternative. Prefabrication relies heavily on the necessity of a large number of units to make it viable. Inevitably, the factory process of some components has enabled cost reductions to be made and, although these are a welcome sign, the total off-site manufacture of buildings is not favoured by the majority of the industry's clients. The expected *quality* of the workmanship should be adequately described in the preamble clauses of the contract document, but those who have attempted to specify construction work soon realise that this is not an easy matter. Contractors often assess quality on the basis of project type, client and architect. The quality of the work should here be interpreted not as building quality alone, but also as the performance of the building throughout its life. The amount of *embellishment* in a design is likely to increase construction costs not only because of its provision, but also because the project is likely to be more complex to construct and the standard of finish will also probably be greater.

5.4.10 Structural morphology

There are many different ways of achieving the same structural efficacy through the use of different geometrical forms and shapes. The design economist will need to consider the relevant implications on the initial building costs and the whole-life costs. This might be through different column arrangements in attempting to achieve an optimum design solution in terms of costs. Increasing the size of structural components and thereby possibly reducing their number will have cost implications in relation to maintaining the stability of the structure (see p. 176). Smart designs will be able to properly consider the layout of the building especially in structures such as single-storey warehouses or multi-storey offices and flats.

The structural efficiency and the architectural appearance of building forms are becoming an increasingly important field of engineering, particularly because of the present widespread availability of computer analysis capabilities. The realisation of complex shapes that would not have been possible through traditional analysis becomes achievable. Structural forms can be visualised starting from a geometry based on that of regular or semi-regular polyhedra, as they form the basis of most of the building structures that are in use today. The architectural use of these forms and their influence on our man-made environment are of general importance. They can define either the overall shape of the building structure or its internal configuration. In the first case the building has a centrally symmetric appearance, consisting of a faceted envelope as in geodesic sphere subdivisions. This may be adapted or deformed to match the required space or ground floor plan. Polyhedral shapes are also often combined so that they form conglomerates, such as in space frame systems.

The arrangement of structural elements has a considerable influence on the design and the form of the building. Re-spacing and re-sizing the beams and columns will have cost consequences. What is required of the design economist is to be able to think of the smart implications in terms of the cost and design variables.

5.4.11 Standardisation, prefabrication and pre-assembly

Standardisation, prefabrication and pre-assembly are not new concepts in the construction industry. These techniques have been used increasingly within the construction industry as a means of increasing overall efficiency, reducing costs and improving quality. The Reading Construction report *Value for Money: Helping the UK Afford the Buildings it Likes* identified the following key findings in a survey of the construction industry:

- A high incidence of one-off components
- Almost half of the components on a project are unique
- The remaining components are used to a maximum of six times on average

There is a recognised desire for creativity and bespoke design solutions within the UK construction industry, and this characteristic is one that is viewed favourably by clients both in the UK and overseas and must be cherished. However, standardisation and uniqueness are not incompatible attributes. There are many benefits to be gained from greater standardisation of building components and design details. These include:

- Improved product predictability
- Improved quality
- Increased efficiency
- Improvements to systems, processes and practices
- Reductions in costs
- Ease of long-term maintenance
- Potential improvements to social and environmental considerations

Some of these points are further reiterated in Chapter 23. The maximum benefits are likely to occur where there is a careful consideration at the outset of the project and where standardisation and pre-assembly are applied together.

Standardisation can be seen in many different contexts. Some of these include the following.

Generic standardisation   An element or process is by its nature standard and is usually recognised as such worldwide, e.g. steel, concrete, plaster.

National standardisation   An item relates to country practices. The dimensions of household bricks are a good example of this.

Client standardisation   A particular client uses defined components as a trademark of their business.

Supplier standardisation   A supplier defines the extent of variability in its components. This is usually driven by market forces such as demand and costs.

Project standardisation   The project team will determine the extent of standardisation that is required.

Prefabrication or pre-assembly relate to the manufacture of building components off-site and ahead of the time that they are required, so that they can be incorporated within the building when required by the programme. These might include volume items such as doors and windows and even more extensive units such as bathrooms. They also include items where for ease of manufacture or improved quality control factory conditions are preferable. Components such as precast concrete beams or composite steel girders are typical examples.

In the case of modular construction many more of the building components are manufactured off-site. These allow them to be quickly assembled on site, often in a fraction of the time of traditional construction arrangements. This type of construction was developed in the 1960s for the construction of a large number of primary and secondary schools. These included, for example, CLASP (Consortium of Local Authorities School Programme), SCOLA (Second Consortium of Local Authorities), EDCON (Educational Construction) and ONWARD (Organisation of North West Authorities Rationalised Design). All of these systems used a type of steel frame construction clad with a combination of brick, timber or concrete. The principles were also extended to other types of public sector construction such as hospital buildings.

The main advantage of pre-assembly is rapid construction on site, providing clients with buildings that they are able to use much more quickly. This method of procurement and construction may not always be the most cost-effective solution in terms of a straightforward comparison of building costs alone. However, where a client is using the premises for production or income generation then this must also be considered, and in such cases pre-assembly has been able to demonstrate overall cost-effectiveness.

5.4.12 Refurbishment

A further factor to consider at the inception stage of a new project is the availability of an existing project that may be capable of adaptation. It is generally presumed that the refurbishment of an existing project will always be less expensive than the construction of a new building. This assumption may be true in respect of the initial building costs, but may well prove to be false when future recurring costs are taken into account. The introduction of development grants and allowances against taxation for certain parts of the building may tend to distort the real situation. Also, an appropriate existing project in the correct location may not be available for modernisation. The existing appearance and condition of the structure would be particularly important. If the structure was very dilapidated then demolition might be the only course of action. This may have occurred where the structure has been allowed to deteriorate over a number of years owing to an absence of essential maintenance. The refurbishment of an existing building would need to take into account the future running costs, and this may require insulation work to the structure, in addition to the extensive or complete replacement of the engineering services. Both the poor insulation qualities and the outmoded aspects of the design would probably require more extensive heating equipment than in a comparable new building. Older buildings do not make the best possible use of space for modern use. Extensive refurbishment may be so necessary as to make redevelopment a better alternative. Improved spatial aspects can always be achieved by a completely new design. Sometimes, where the space available is insufficient, the possibility of extending the premises may also be considered. A further factor is that of continuity of use. If demolition and rebuilding are required on the same site, then this factor will be lost and some form of temporary accommodation close by will be required. Refurbishment does, however, have the advantage that it can often be carried out on a phased basis, thus allowing some continued use of the existing premises.

5.5 WHOLE-LIFE COSTING

Once the construction works have been completed and the project is put into commission, many of the industry's clients will be responsible for the project's costs-in-use. Even clients who will eventually not be responsible for these costs will need to take into account the leasability and saleability of their completed projects. Several years ago the only cost consideration of the client was to reduce the initial construction costs to a minimum. Clients are now more enlightened and do attempt to take into account the three 'R's: running, repairs and replacement costs. These should be considered alongside the costs for the initial construction work. The emphasis, therefore, particularly of owner-use clients, is now often on an economic whole-life cost, in preference to the cheapest possible constructional design. The introduction of energy-saving measures within the design in order to reduce future fuel costs is now commonplace. There has been some attempt at introducing maintenance-free construction in an attempt to minimise this aspect of costs-in-use. Buildings are also sometimes loosely designed on the assumption that building use or usage may change within their structural lives, and thus adaptation will be simpler and less expensive. The disadvantages and disruptions caused by major repairs and maintenance can often result in costs out of all proportion to those that would have been necessary had a more durable method of construction been chosen initially. It must also be remembered that the most expensive type of construction initially does not always result in a saving in future costs. Sometimes the reverse can be true, for example where an expensive automated system can require a high allowance for future maintenance expenditure (see also Chapters 17 and 18).

5.6 SUSTAINABLE ECONOMIC DESIGN

Sustainable design and construction or sustainable design is the philosophy of designing physical objects, such as the built environment, and services to comply with the principles of economic, social and ecological sustainability. The intention of sustainable design is to eliminate, as far as is practicable and possible, the negative environmental impact completely through skilful, sensitive design. Sustainable designs require no non-renewable resources, a minimal impact on the environment and relating these criteria with people and with the natural environment.

Sustainable development is development which meets the needs of the present without compromising the ability of future generations to meet their own needs. This definition was formulated in 1987 by the World Commission on Environment and Development, known as the Brundtland Commission.

The word development in this definition incorporates two important aspects of the concept: it is omnidisciplinary, it cannot be limiteds to a number of disciplines or areas, but it is applicable to the whole world and everyone and everything in it, now and in the future. Secondly, there is no set aim, but the continuation of development is the aim of the development. The definition is based on two concepts:

- the concept of needs, comprising the conditions for maintaining an acceptable life standard for all people
- the concept of limits of the capacity of the environment to fulfil the needs of the present and the future, determined by the state of technology and of social organisation

The development of and the desire for sustainable design and construction is mostly a general reaction to global environmental crises. These are due to the rapid growth of economic activity and human population, the depletion of natural resources, the damage to ecosystems and the loss of biodiversity. The limits of sustainable design in reducing whole-earth impacts are beginning to be considered because growth in goods and services is consistently outpacing gains in efficiency. As a result, the net effect of sustainable design to date has been simply to improve the efficiency of rapidly increasing impacts. The present approach, which focuses on the efficiency of delivering individual goods and services, does not solve this problem. The basic dilemmas that are not yet well addressed include:

- the increasing complexity of efficiency improvements
- the difficulty of implementing new technologies in societies built around old ones
- that physical impacts of delivering goods and services are not localised but distributed throughout the economies
- the scale of resource uses is growing and not stabilising

Transformative technologies are hoped for, but workable options are not yet evident. Only if the scale of resource uses stabilises will the efficiency of how each resource is delivered result in reducing total impacts.

The motivation for sustainable design was articulated famously in E.F. Schumacher's 1973 book, *Small Is Beautiful*. In architectural design, sustainability is not an attachment or afterthought but is integral to the design process. This requires close cooperation of all of those who are involved in the design and construction process at all project stages. It incorporates site selection, scheme formation, choice of materials, methods of procurement and project implementation.

5.6.1 Principles of sustainable design

While the practical application of sustainable design varies amongst the different disciplines there are some common principles, which include:

- **Low-impact materials:** non-toxic, sustainably produced or recycled materials
- **Energy efficiency:** use of manufacturing processes and production of products which require less energy
- **Quality and durability:** longer-lasting and better-functioning products that will need to be replaced less frequently, reducing the impacts of producing replacements
- **Reuse and recycling:** products, processes and systems should be designed for performance in a commercial afterlife
- **Design impact measures:** for total-earth footprint and whole-life assessment
- **Sustainable design standards:** a large body of new methods emerging from the rapid development of what is becoming known as sustainability science
- **Biomimicry:** designing built environment systems along biological lines enabling the constant reuse of materials in continuous closed cycles
- **Service substitution:** shifting the mode of consumption from personal ownership of products to provision of services which provide similar functions
- **Renewability:** materials should come from the local or bioregional area, from sustainably managed renewable sources that can be composted, or fed to livestock, when their usefulness has been exhausted
- **Healthy buildings:** sustainable building design aims to create buildings that are not harmful to their occupants or to the wider environment

Green building has rapidly been gaining momentum as a design protocol and measurement standard for a building's environmental performance. Though many precepts of sustainable building were established thousands of years ago, it has only been defined and integrated into the global building industry since the late 1980s. The various green building rating systems devised in different countries provide the best definition of a green building. In many countries, green building activity was initiated in the public sector. The cost of funds for such building activity for government is low, and the average life of a public building is long.

5.7 ELEMENT EVOLUTION

A study of the evolution of building elements reveals a general aim to improve cost efficiency through design and construction. This can be observed by examining all of the building elements (see Chapter 9). This evolution has involved the adoption

Fig. 5.7   Traditional design *c*.1900

of new building techniques and the selection of modern and often artificial materials. The development of these materials is a response to the requirement for similar characteristics at a more economic cost. All of a material's characteristics should be considered and those characteristics selected that meet the demands of fitness for purpose. The development of such materials in the past can sometimes be traced to difficulties in the supply of the more traditional materials and the general aim to reduce construction costs. This has been achieved while still maintaining the client's overall short- and long-term objectives.

The roof of a building is generally an important element in terms of cost, and is therefore worthy of study in any economic appraisal. Figure 5.7 shows the details of construction which were prevalent during the early part of this century for housing. Figure 5.8 shows the comparable construction adopted over the past fifteen years or so. The point which needs to be noted is that in terms of architecture little has changed. Only the pitch and the roof covering appearance are different. In terms of aesthetics, the lower roof pitch may be more desirable, but this is largely a matter of taste. It is also possible today to create artificial materials which are comparable in performance and appearance with, and much cheaper than, the natural ones. At eye level, the unlikeness may not be noticeable anyway. The building function and performance, however, are at least as good as those outlined in Figure 5.7.

The method used to achieve the architecture has changed, but as long as the construction principles used are sound the method will be of no real interest to the client. The major differences in the modern, as opposed to the traditional, form can be summarised as follows. Each has a consequent saving in terms of cost.

- Lower roof pitch requiring a smaller area of roof covering, shorter lengths of rafters etc.
- Lighter-weight artificial materials for roof covering, offering similar appearance and the required length of life

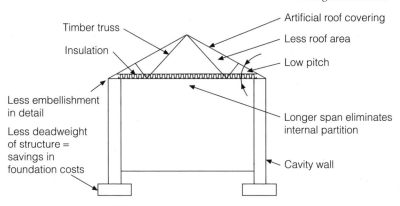

Fig. 5.8   Modern design *c*.1990

- Use of prefabricated roof trusses; smaller sectional sizes, lighter weight, ease of erection
- Improved thermal insulation showing savings in costs-in-use
- Simplified details in terms of embellishment at eaves etc.
- Overall lighter weight of roof construction resulting in a slenderer construction for the structure
- Timber impregnation guaranteeing future life
- Lighter-weight roof and structure resulting in lower deadweight, offering savings in the foundation design
- Longer roof spans resulting in the elimination of some load-bearing walls

The above have not resulted in any loss of performance or function in terms of building usage on behalf of the client.

CONCLUSIONS

Value for money in construction may be increased by improving utility with no change in cost, retaining the same utility for less cost or combining an improved utility with a decrease in cost. Value management (see Chapter 19) examines the design and is all the time searching for improvements in the project's worth. The technique of value analysis should help surveyors to achieve better results in terms of value for money when used alongside the already proven procedures such as cost planning. A possible failure of cost planning, however, stems from the fact that once the original cost objective had been achieved, no further improvement in the analysis was undertaken. A dissatisfaction with the use of cost limits, a further mechanism aimed at cost-effectiveness, was due to the adoption of unrealistic sums which did not focus the design team's mind on costs savings beyond the imposed government targets. The application of value analysis techniques should therefore allow for a better evaluation of certain aspects of the design economics of buildings and structures.

John Ruskin (1819–1900) said, 'It is not the cheaper things that we want to possess, but expensive things that cost less'. This is the particular theme of building design economics. A wide range of different techniques can be used to aid us in achieving this goal and these are described in the subsequent chapters of this book. Some of these techniques have been tried and tested over many years and are now akin to scientific methods. Other techniques, which in some cases have been borrowed from other industries, have been found to have a good application to the construction industry.

Reducing building costs, both the initial costs and the whole-life costs, or adding value as some suggest, is beneficial not only for the construction industry in general and its clients in particular, but also for society as a whole. Reducing the costs of building, without affecting value, is in everyone's interest, since not only does it encourage more buildings to be constructed but the financial implications are obvious. Reducing building costs by 30%, a recommendation of the government's former Best Practice Programme, seemed unachievable at the time it was proposed. However, the introduction of modern methods of construction and finding newer and smarter ways of working without working our cash harder has helped us to go a long way in achieving this.

SELF ASSESSMENT QUESTIONS

1. Rank in order of importance a client's requirements for each of the following construction projects:
   Speculative factory and warehouse units
   Comprehensive school
   Office building for rental to others

2. Select any construction project with which you are familiar and evaluate it in the context of its construction economy.

3. What effects does a consideration of initial design economy have on future costs-in-use? Give examples to illustrate your answer.

BIBLIOGRAPHY

Ashworth, A. and Hogg, K., *Added Value in Design and Construction*. Longmans 2000.
Banks, D.G., 'The use of a length/breadth index and a plan/shape index', *Chartered Surveyor*, B & QS Quarterly 1974.
Brandon, P.S., 'A framework for cost exploration and strategic cost planning in design', *Chartered Surveyor*, B & QS Quarterly 1978.
Du Bosky, P.S., *Building Costs – The Property Development Process*. Calus 1976.
Flanagan, R. and Norman, G., 'The relationship between construction price and height', *Chartered Surveyor*, B & QS Quarterly 1978.
Gordon, A., 'Long life, loose fit, low energy', *Chartered Surveyor*, B & QS Quarterly 1977.
Gray, C., *Value for Money: Helping the UK Afford Buildings it Likes*. Reading Construction Forum 1996.

Green, S. and Popper, P., *Value Engineering – the Search for Unnecessary Cost*. Chartered Institute of Building 1990.

Hamilton, J.C., 'Functional life as a basis for design', *Industrialization Forum 6*, USA 1975.

Kaye, S., 'Narrow or wide, high or low, profit or loss?', *Chartered Surveyor*, June 1966.

Kelly, J.R., 'Value analysis in early building design', in *Building Costs – New Directions*. Spon 1982.

Luckman, C., *Determinants of Building Costs – Creative Control of Building Costs*. McGraw-Hill 1967.

Schumacher, E.F., *Small is Beautiful: Economics as if People Mattered: 25 Years Later*. Hartley & Marks Publishers 1999.

Standardisation and Preassembly. Construction Best Practice Programme 2003.

CHAPTER 6

THE ECONOMICS OF QUALITY

LEARNING OUTCOMES

After reading this chapter, you should have an understanding about the economics of quality associated with the design of buildings. You should be able to:

- Understand what is meant by quality in buildings
- Differentiate between attributes and variables in the context of quality
- Apply techniques to evaluate the economics of quality in buildings
- Recognise that quality is just one of the factors to consider in an economic appraisal

6.1 INTRODUCTION

Defects in construction projects are a persistently worrying problem despite continually improving technology and education. The construction industry has too often in the past been discredited by bad publicity resulting from sometimes dramatic failures of both the design and the construction of its products. It must not, because of economic stringency and also because of external pressures, devote its resources to unprofitable ends by failing to achieve the desired standard of work at the first attempt.

The achievement of an acceptable standard in buildings is a combination of quality of design and quality of construction. In the former, quality is determined by the engineer or architect in terms of their skill and by promoters in what they are prepared to pay. In the latter, quality is determined by the management and operative capabilities of the constructor, and by the supervisory capabilities provided by the designer with regard to the standards required. Figure 6.1 lists some of the main points to be borne in mind when considering the implications of quality in the construction process.

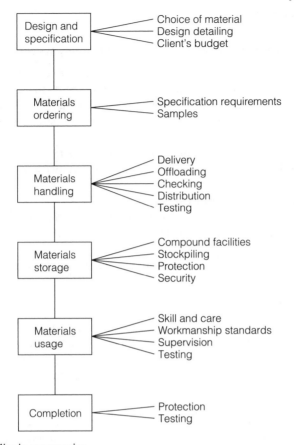

Fig. 6.1   Quality in construction

6.2 WHAT IS QUALITY?

The *Oxford English Dictionary* definition of quality includes the terms 'nature', 'character', 'kind' and 'attributes'. The Building Research Establishment (BRE), in attempting to answer this question related to buildings, defined it as:

> The totality of the attributes of a building which enable it to satisfy needs, including the way in which individual attributes are related, balanced and integrated in the whole building and its surroundings.

The BRE report considered quality in the context of three main aspects:

1. External attributes – the effect of the project on its surroundings and vice versa
2. Performance attributes – aspects of the project which make it operationally efficient and provide reasonable conditions for users
3. Aesthetics and amenity – internal and external attributes of a standard higher than is needed just to meet mandatory and performance requirements

6.3 QUALITY SYSTEMS

The following are general principles concerned with quality systems:

- A recognition of fitness for purpose based on agreed objectives and standards
- The need to set quality issues with the organisation's own strategic plans
- A recognition that quality must be planned and managed
- All aspects need to be focused on, since quality is only as strong as the weakest link
- The need for some form of continuous monitoring system
- An acknowledgement of the merits of the different quality control and assurance systems available
- An emphasis on quality enhancement
- A recognition of the importance of committed staff at all levels
- The need for accountability to the firm's customers
- A concern for value for money
- A recognition that quality and its absence both have economic consequences

6.3.1 Definitions

Quality system   Made up of the mechanisms and their documentation. A quality system is one that has the potential for producing goods to specification, although it cannot guarantee that all goods will meet it, nor can it guarantee that the specification or standards reached will be better than those provided by another organisation.

Quality control   Designed to keep a planned, systematic procedure in its planned state. It refers to using operational techniques or standardised procedures to check that products conform to specification.

Quality assurance   The setting of standards by an organisation and ensuring that these are being achieved in practice.

Quality audit   The external verification by an outside body, such as BSI, that the specified standards are being achieved.

Quality improvement   This acknowledges that the nature of something can be improved, and therefore grades of excellence can be aimed for. Continuous quality improvement, which forms the basis of total quality management (TQM), refers to the notion of 'never being satisfied' with the current degree of quality and success in meeting customers' identified needs, requirements, interests and expectations. This is why TQM encompasses the search for opportunities for improvement, rather than maintaining current performance.

Several management models seek to address issues of quality. Two of these are ISO 9000 and TQM.

6.3.2 ISO 9000

ISO 9000 is the international standard for quality systems published by the International Organisation for Standardisation (ISO) as the ISO 9000 series. This has replaced BS 5750 and is used to accredit a quality assurance system. It does not concern itself with standards for quality *per se*. They could be high or low standards, although the standard to be aimed for has to be defined and agreed, and requires consistency.

BS 5750 grew out of military quality assurance procedures and was developed by the British Standards Institute (BSI) to apply particularly to manufacturing industry in the 1960s and 1970s. The language and construction of the standard are strongly biased towards manufacture. Nevertheless, it is seen as applicable to all areas of activity, including service industries and professional practice. It provides a template for good practice in the design and implementation of quality assurance. The intention is that it should be fitted to the organisation using it, and not that the organisation should change fundamentally to fit the requirements.

The approach is to require the documentation of all activities within an organisation and evidence that procedures which should take place have indeed occurred or their omission has been justified. Internal audit and occasional external inspection are used to ensure compliance with the procedures. Thus the system would, for example, describe in detail each step in the process being applied, together with a means of ensuring that these have been complied with in the appropriate way. The documentation of the firm's procedures is central and document control is one of the important parts of the quality system. Those who are well versed see simplification of the documentation as a particular challenge.

The ISO 9000 registration kitemark is not a product quality kitemark or specification, and does not establish a level of excellence for a product. A registration does not purport to be anything more than a way of describing the capability of the system to produce goods to specification. Some of its critics therefore see the emphasis on quality systems rather than quality standards as missing the main point of quality in practice. As was recorded at the American Society for Quality Control Congress in 1990, the ISO 9000 series 'intentionally does not emphasize the ability to demonstrate continued quality improvement capability'.

There are twenty components of the standard, which cover areas from central policy to training about quality assurance. Not all the twenty components are relevant to every type of activity. An organisation is required to produce manuals which indicate how each area of its activity is covered and fits the template of the standard. Experience from manufacturing industry indicates that it is likely to take two or more years for an organisation to audit and document existing procedures, develop new procedures where there are gaps and work with staff so that there is understanding of and commitment to improved quality assurance. Once accreditation has been obtained it is maintained through inspections by an approved external agency. The inspection will verify that procedures that are supposed to take place are indeed taking place. Accreditation can be withdrawn for serious non-compliance.

6.3.3 Total quality management (TQM)

TQM is a system that seeks to realign the mission, culture and working practices of an organisation by means of pursuing continued quality improvement. This process, which is founded on individual attitude and effort in quality improvement, emphasises a commitment to satisfy the needs of the customer both inside and outside an organisation. In contrast to ISO 9000, TQM does not set out to meet a pre-defined quality goal, but rather seeks to improve quality continually by a process of research, evaluation and feedback. Once areas capable of improvement can be defined, resources can be applied to effect an improvement. TQM can be introduced only gradually and lacks the specificity of ISO 9000 in the way in which the culture change can be introduced. Indeed, an initial adoption and implementation of ISO 9000 may be a first step towards the advancement of TQM. Once TQM is adopted, an organisation is on a path of continuous quality improvement by challenging current practices and performance with a view to improvement.

TQM philosophy and practice originated through the ideas of Dr W. Edwards-Denning, an American who provided the intellectual drive behind Japan's post-war construction. Among other things he persuaded Japanese companies to involve and consult with customers in their efforts to improve products continuously. Others have developed these ideas, believing that the present performance in any function and at any level can and should be improved. The focus is towards people and towards creating an appropriate, supportive and well-disciplined climate for promoting a positive and effective commitment to improving quality.

The main features of TQM are

- An organisation-wide commitment to quality
- Creation of an appropriate climate
- A focus on satisfying customers' needs
- Management by data/facts
- People-based management
- A commitment to continuous quality improvement

TQM is implemented at the top of an organisation first. While the ultimate aim is for everyone in the company to work on improvements in quality, emphasis is placed on training senior managers first because they initiate the changes required. The leaders in the business are trained in skills to enable them to change the way they work in order to practise and promote the principles of TQM, and then to train others.

6.4 QUALITY CONTROL

In all management processes, objectives must be set by the organisation concerned in order to measure success or failure of the process. Management, therefore, carries with it the responsibility for quality and for achieving the standard required.

Perfection is a word closely related to quality. Perfection may be the standard to be aimed for, but due to tolerances in the construction process it will always be out of reach. The difference therefore between perfection and achievable quality can be measured and compared. When quality is assessed in this way it can be controlled.

The objects of quality control are:

■ To attain the quality of design and conformance which will satisfy the client, both on handover and during subsequent use.
■ To achieve this at the lowest possible cost.

6.5 THE 'M' FACTORS AFFECTING QUALITY

Markets Comparability between standards provided by different firms.

Men This is perhaps the single most important factor in achieving quality: having the right people to do the job which is required.

Money Quality costs money. If an inadequate amount of money is included in a budget, then the required quality will be difficult to obtain.

Management It is the function of management to set a company's quality policy, and this will in turn form the basis of the company's reputation in this respect.

Materials These must have been specified correctly, properly delivered to and checked on site and then stored and used in accordance with the manufacturer's instructions.

Methods The method specified must be capable of being executed in practice to the tolerance and finish required. Specifications which do not take into account these factors are unlikely to achieve their desired objectives.

Machines The correct machine for the work being carried out must be carefully selected, and to work efficiently it must be properly maintained.

6.6 QUALITY STANDARDS

Quality can be classified by variables or attributes. A *variable* is a quantity that may take any one of a specified set of values, and can thus be measured. Examples of this type of quality characteristic are dimensions and insulation properties. An *attribute* is a qualitative characteristic of an item such as appearance or colour. Variables can therefore be defined objectively and can be easily controlled by sampling. Attributes have a much more subjective definition and are difficult to control by sampling.

The purpose of setting the standard is to produce a practical, factual and measurable limit as an objective. The following are some of the factors associated with quality, and should readily be identified in the bill of quantities or specification.

6.6.1 Dimensions

These are a practical way of measuring how close the product should get to perfection. Dimensional exactness cannot be achieved under typical site conditions, but dimensions are subject to acceptable levels of tolerance. The tolerance will depend on type of material, shape, size, weight, method of assembly, expansion and contraction, and jointing techniques. The amount of tolerance is specified for the components to be used. Levels, trueness to slopes, or verticality can be specified by referring to permissible limits for the work to be carried out.

6.6.2 Appearance

This quality factor may give the client a greater amount of satisfaction than any of the other variables or attributes. It is a difficult attribute to specify, but will be of some importance to the designer. In some areas of work, such as external facing brickwork, it is of major importance. A visual standard is often required for comparison purposes. With visual examples of decoration it is especially important to define the type of light under which they are to be observed.

6.6.3 Strength

The strength of the structure is vitally important to its overall performance. This variable can be easily specified and reasonably measured in order to determine its characteristics. Representative samples of the materials concerned for the structural components, such as concrete, bricks, steel and timber, can be tested to destruction against a predetermined specification such as a British Standard.

6.6.4 Stability

This is largely a matter for the concern of the designer. It is necessary to establish in the first instance the structural stability of the members concerned, and to ensure that the constructor complies with proven practices during erection. Often structures are unstable during the course of erection, and the contractor must be fully aware of this possibility. The designer must also ensure that the construction complies fully with the construction details supplied in order that the desired solution can be achieved.

6.6.5 Materials

Materials can generally be readily identified to British Standards, and these will provide a precise indication of quality standards for construction purposes.

6.6.6 Performance

The materials and construction detailing can be specified by their relative performance, e.g. thermal insulation standards, sound insulation qualities, moisture contents. The materials and methods used can be measured against already established performance standards.

6.6.7 Finishes

This quality characteristic provides the same problem of assessment as appearance. They are both concerned with visual factors, but finishes are, in addition, also concerned with texture. As with appearance, a sample of a particular finish should preferably be obtained, and this can then be used as the appropriate standard.

6.7 CATEGORIES OF QUALITY COSTS

Quality and reliability costs can be categorised under three headings: prevention costs, appraisal costs and failure costs.

6.7.1 Prevention costs

These are the costs concerned with trying to ensure that no defective items are produced either in manufacture or construction. The premise is that money spent on prevention should be more than recouped at some future date. The word 'economic' is significant in this context. The cost of achieving conformity to a given standard is subject to the law of diminishing returns. A standard of quality less than perfection will be required, otherwise the costs in attempting to achieve it will be very high. Prevention costs arise in connection with the planning and design for quality control.

6.7.2 Appraisal costs

These are the costs expended on the measurement of specified quality characteristics to establish conformity to the specification. They are the costs involved in detecting low quality standards during construction. The construction industry relies heavily in this context on clerks of works and inspectors of works. These seek to ensure that the contractors conform to the specification and other contract documents in respect of quality standards and requirements. Quality standards will be checked at regular intervals during the construction process and also on completion of the works prior to handing over to the promoter. During construction the checking may be more of an investigative nature, with the work being objectively measured and tested. At the completion stage the quality assessment is likely to be more by appearance, apart from the usual testing of services and equipment.

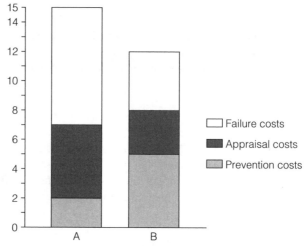

Fig. 6.2   Prevention, appraisal and failure costs

6.7.3 Failure costs

These are the costs associated with the manufacture and replacement of a defective part of the construction project. They may occur through a faulty design or because the constructor has failed to comply fully with the specification. The first type of failure is more serious and costly to the promoter. When the fault is discovered during construction, it can be remedied more easily than when it is discovered after commissioning. The costs of the latter category are complicated to evaluate, since they will have the knock-on effect of some dissatisfaction on the part of the promoter. This can result in adverse consequences in terms of goodwill for both the designer and the constructor.

Figure 6.2 illustrates the relationship between the three categories of cost. Column A provides an illustration showing an original concept before any consideration or concept of quality and cost is envisaged. The total cost of achieving the desired solution is calculated as 15 units. Column B shows the overall effect of adopting the philosophy of prevention being better than the cure. A greater proportion of the total costs is spent on prevention but this has the desired effect of reducing the total costs to 12 units. This indicates an overall cost saving.

6.8 ECONOMIC ASSESSMENT

Several techniques are available which can be used to select or optimise a course of action where alternative solutions are available. The following can be used to determine an economic basis for quality standards.

Fig. 6.3 Breakeven chart

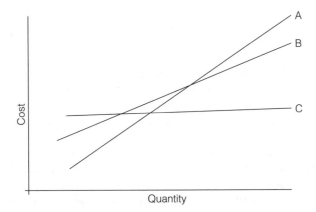

Fig. 6.4 Optimisation chart

6.8.1 Breakeven analysis

This technique is concerned with calculating the breakeven point between two or more operations. It generally involves the preparation and interpretation of a breakeven chart. The points of intersection on this chart are known as breakeven points. Figure 6.3 shows a typical example of a breakeven chart.

In the construction industry a large number of site operations comprise a fixed cost (A–B), a variable cost (B–C) and a price (A–C). The fixed cost may represent site set-up costs which are largely unaffected by quantity once they have been established. Variable costs, on the other hand, are often directly related to quantity in some way. Price is often determined in direct relationship to quantity alone over a given range of values. In order to determine this price or rate for the work, some idea of the likely breakeven point is therefore necessary.

Another example of the use of breakeven analysis is in the selection of an economic process which is known to vary depending on the quantity required. This is shown in Figure 6.4. This implies that for relatively small quantities

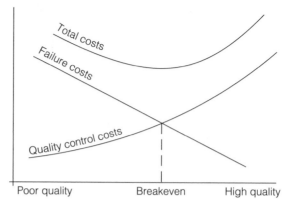

Fig. 6.5   Quality costs chart

system A should be used, but where large quantities are required then the economic option would be system C.

Figure 6.5 shows further use for this technique in the determination of the economics of quality. This implies that there is a limit, in economic terms, to improvement of the quality of construction work on site. The breakeven point indicates that methods of quality control at a high level are costing more than the failure costs. The technique must be used with care, however. The separation of costs into different categories is in practice not as straightforward as might be suggested. Also, the construction industry is not used to thinking in such terms. The linear presentation of these charts may appear to be an inaccurate and simplified reflection of how costs are incurred in practice. Research has, however, been able to show that some relationship between costs and quantity does exist within a range of values. Too much emphasis must not be placed on the breakeven 'point'; it is preferable to interpret this as the centre of a range of activities which produce this situation. In economic terms there is little point in spending $£x$ more on improving quality standards if the benefit achieved is only $£x - 1$.

6.8.2 Discounted cash flow analysis

A second technique which can be utilised is that of discounted cash flow (DCF). It has in theory long been recognised that to evaluate the costs of construction on the basis of initial costs alone is unsatisfactory. Some consideration must also be given to the future costs-in-use which are directly influenced by design decisions. The use of DCF in economic analysis therefore seeks to evaluate for the client the best-buy project, taking into account both initial and recurring costs within the context of the project as a whole. The improvement of quality standards is not therefore seen in the context of a newly commissioned project alone, but in terms of how such standards will be affected by the project in use. Do higher quality standards automatically result in a more economic solution, for example? In some cases, the provision of a higher quality initially will inevitably produce higher costs-in-use. Expensive floor finishes will often require more expensive care and maintenance than a cheaper alternative. In other instances a higher quality roofing material, although it may provide a trouble-free life, may not necessarily be the better economic solution over a cheaper alternative covering.

An objective way of determining the economics of differing processes, systems or materials is to use DCF. This technique involves the discounting of the values of future receipts and payments to a common time base. It therefore identifies the total costs of providing a particular design solution, rather than concentrating on the initial capital costs alone. It can also take into account factors such as taxation, which is appropriate to certain types of buildings, and maintenance and repair charges.

The technique of DCF does present a number of difficulties, including the scarcity and reliability of historic cost data, the interest rates to be selected, the life expectancy of components and inflationary factors (see Chapter 17).

Although some of these factors present immense problems, they can be overcome to some extent by the use of sensitivity analysis. This technique seeks to evaluate our judgements by examining alternative interest rates and the life expectancy of the components used in the DCF calculations.

Example

The following illustrates the use of DCF. Two constructional alternatives, each fulfilling the same requirements, have the following costs. Alternative A has an initial cost of £40 000, requires considerable repairs every 20 years costing £10 000, and has annual maintenance costs of £1 500. Alternative B has a lower initial cost of £30 000 but requires repairs every 10 years costing £12 000 and maintenance costs of £2 500 per annum.

The better quality of alternative A is such that it results in less attention during its lifetime for repairs and maintenance. Each alternative has a total life expectancy of 30 years and the cost of capital is 5%.

Alternative A

Initial cost	=	40 000
Repairs every 20 years £10 000 × 0.37688[a]	=	3 799
Maintenance per annum £1 500 × 15.3724[b]	=	23 059
NPV[c]	=	£66 858

Alternative B

Initial cost	=	30 000
Repairs every 10 years		
£12 000 × 0.61391 (Year 10)[a]	=	7 367
£12 000 × 0.37688 (Year 20)[a]	=	4 523
Maintenance per annum		
£2 500 × 15.3724[b]	=	38 431
NPV[c]	=	£80 321

Notes
[a] Use of the PV of £1 table (see section 13.6, p. 293, for explanation of terms).
[b] Use of the PV of £1 per annum or YP table.
[c] Comparative value of the two schemes taking into account initial and future costs of the asset.

The above example shows that, in economic terms alone, alternative A is the best-buy solution. In this hypothetical example, the initial higher quality of the project is the chief reason for the lower whole-life cost.

6.9 QUALITY CONSIDERATIONS

The economics of a construction project is only one of the factors to consider in the determination of its quality. Many other aspects of the project will also need to be examined. These include the following:

- Replacement or repairs may be inconvenient. The necessity of these two items will always cause some sort of disruption. Whether the repairs are as a result of some emergency work to a road drainage system, or whether they have been pre-planned as a programme of renewal, they will result in inconvenience.
- Replacement or repairs may be difficult and therefore expensive. Even minor repairs on highway projects require considerable protective barriers and traffic redirection, which increase the total costs of the repairs.
- The saving of money on a specific item may involve costs out of all proportion to the saving. The use of a poor-quality road foundation can necessitate the replacement of an entire carriageway even though the surface materials may have been of the best quality, and capable of a much longer life.
- Obsolescence may not be a factor to be considered. The best quality may be chosen initially on the basis that this will last for hundreds of years, and be superseded by newer technology.
- A client may be able to obtain grants against the capital expenditure, and may therefore wish to spend more initially and thereby reduce future running costs.
- High-quality construction, although more expensive, is often more durable and more pleasant to look at.
- A promoter may consider that the project is too important, for prestige reasons, to incorporate cheap or poor-quality materials and methods of construction.

CONCLUSIONS

The quality of work in any construction project will be influenced by many different considerations. High standards of quality may be chosen for aesthetic reasons, since appearance is an important facet of the majority of construction projects. The durability of the project over a long life may be another important reason for high quality. In other circumstances high quality may have been chosen for convenience reasons, to limit the amount of disruption while the project is in use. No-one generally sets out with the intention of providing a poor standard of quality. The nature of the design and the skill of the constructor, however, will largely determine the standard achieved in practice. Since quality costs money, the economic consequences of a design will need to be considered very carefully in order to achieve both a balanced project and value for money in terms of a best-buy solution overall.

There are now numerous texts on the subject of quality. The following are a number of definitions that can be applied to different situations in the construction industry:

Fitness for purpose   Joseph Juran coined the phrase, 'There is gold in the mine' to signify the potentially large cost savings to be made by identifying and solving quality problems. He believed that as little as 20% of quality problems can be attributed to those who work on the shop floor but that quality breakdowns are due to the attitudes of managers. Quality must be integral to the process. It is never achieved by accident and includes careful planning, control and improvement.

Conformance to requirements   Philip Crosby stated that quality is free, do it right the first time and achieve zero defects. Quality means conformance to requirements. These are established from a knowledge of the customer's needs. Good or bad, high or low quality are all meaningless terms to Crosby. Quality measurement is the price of non-conformance, i.e. the cost of not doing it right the first time. Quality improvement is a process that requires determination and education as well as implementation. Crosby defined five elements in a successful organisation:

- People routinely do things right the first time
- Change is anticipated and used to advantage
- Growth is consistent and possible
- New products and services appear when needed
- Everyone is happy to work there

Worker effectiveness   W. Edwards Deming is often thought to be the father of the quality movement. The core of Deming's philosophy was that in order to improve the quality of production over time, the people doing the job have to be involved in the production process. The key component in this exercise of winning worker commitment consists of teams sharing a vision of continuous improvement. Deming's philosophy is succinctly summarised in his famous 14 points which include the abolition of performance appraisals, performance related pay, productivity bonuses, etc.

Several studies have shown that poor quality can affect 10–25% of a construction project where management and accounting are properly considered. The problem is particularly acute in small- to medium-sized firms. A high proportion of payroll costs is being spent on non-value work resulting from poor quality. Some companies are prepared to spend considerable amounts of money on quality management programmes without any measurable increase in the actual quality of the processes or products that are produced.

SELF ASSESSMENT QUESTIONS

1. Describe what is meant by quality and standards in the construction of buildings, giving examples to illustrate your answer.

2. Identify the differences between quality attributes and variables giving examples to illustrate your answer.

3. Does cost versus quality always have to be a zero sum game?

BIBLIOGRAPHY

Ashworth, A., 'The economies of quality in construction' in *Quality, A Shared Commitment* (ed. R. Grover), Fifth Seminar on Construction Quality Management 1987.
Cormick, T., *Quality Management for Building Design*. Butterworth 1990.
Covington, S.A., *Quality Assurance of Building Components*, BRE Information Paper IP 24/80.
Crosby, P.B., *Quality is Free*. Penguin Books USA 1980.
Davies, W.H., *Construction Site Production*. Butterworth 1982.
Edwards-Deming, W., *Out of Crisis*. Massachusetts Institute of Technology 1986.
Fisher, W., 'A designer's view of quality control', *The Architects' Journal*, May 1980.
Forster, G., *Construction Site Studies – Production, Administration and Personnel*. Longman 1981.
Foster, R., *Quality Assurance in the Construction Industry*. Longman 1989.
Greater London Council Handbook for Clerk of Works. Architectural Press 1983.
Griffith, A., *Quality Assurance in Building*. Macmillan 1990.
Johnston, V.E., *Site Control of Materials*. Butterworth 1981.
Juran, J.M., *Juran on Quality by Design*. Macmillan, New York 1992.
Newlove, J., 'The achievement of quality in construction', *Building Technology and Management*, January 1978.

CHAPTER 7

INDICES AND TRENDS

LEARNING OUTCOMES

After reading this chapter, you should have an understanding of building cost indices and trends. You should be able to:

- Understand the purpose of index numbers
- Be aware of how indices can distort facts
- Distinguish between building cost and tender price indices
- Appreciate regional differences in costs
- Measure and forecast trends in costs and prices using different techniques

7.1 INTRODUCTION

Cost information is recorded and collected over a period of time. During this period of time, building costs, market conditions and inflation will change for a variety of reasons. The collected data therefore, if they are to be of any use to the construction industry, must be converted to a current date or appropriate future timescale. This conversion process is normally achieved both in theory and in practice by means of index numbers. These index numbers, of which there are many different types, are used in comparing price, production, employment or population changes etc. over a certain period of time. An index number will measure the change that has occurred from one period to another.

All index numbers require the selection of a base period, i.e. a date to which all of the other numbers in the series can relate. For all general purposes this is set at 100 to allow for a decrease in the value of the data as well as an increase. The base date could be set at 1, but this might result in negative values which could become cumbersome to handle.

Simple index numbers do not take into consideration the relative importance of various items concerned. These are known as unweighted index numbers and are generally meaningless. The majority of the index numbers used in the construction industry include weighted items in the order of their importance within the index. They are calculated on the principle known as 'the basket of goods'.

7.2 CONSTRUCTING AN INDEX

When constructing an index the following four factors need to be considered.

Purpose of index   The purpose must be considered carefully since it will affect all decisions relating to the other factors. The use of the index will be restricted to that purpose. Building cost and tender price indices (see p. 132) are two common indices associated with the cost study of buildings. While they have some similarities they are different and one cannot be used to predict values for the other.

Selection of items   There is a difficulty in capturing the spirit of an index by measuring a limited number of items. It is not possible to include every item but only a representative sample that hopefully will predict future values. The correct solution lies in defining the purpose of the index carefully and then selecting the items that best achieve that purpose.

Choice of weights   The balance between the importance of the individual items in the index is achieved through weighting the different values.

Choice of base year   A typical rather than a freak year should be selected since this is more likely to provide an honest and reliable index.

7.3 USES OF INDEX NUMBERS

Index numbers are used either for updating historic cost data to current pricing levels or for predicting future trends in costs and prices. The following are some of the more common applications.

7.3.1 Cost planning

The process of cost planning requires the efficient use of large quantities of historic cost data. In order for these to be used properly the data will need to be updated by use of indices. The total cost of a project, the all-in rate or the prices for an individual element can each be updated as follows:

Cost per m^2 GIFA in cost analysis = £298.31
Index for this scheme 271
Current index applicable to proposed scheme 327

The formula for updating is as follows:

$$PR = OR \times \frac{CI}{OI}$$

where PR is the proposed rate, OR is the original rate, CI is the current index and OI is the original index.

$$PR = £298.31 \times \frac{327}{271} = £359.95 \text{ per m}^2$$

The percentage difference between the two rates can be calculated as follows:

$$\text{percentage change} = \frac{CI - OI}{OI} \times 100$$

$$= \frac{327 - 271}{271} \times 100$$

$$= \underline{+20.66\%}$$

7.3.2 Forecasting

The pattern of the existing indices can be extended to a date at some time in the future. This extrapolation of existing indices must be done with great care. Some subjective allowance must also be made for the difference between the conditions prevalent in the past and the future. The projection of existing indices is a simple matter under stable conditions, but the erratic behaviour of inflation rates in recent years has made forecasting a very difficult occupation. Even experienced economic commentators have found themselves in difficulty over this matter. The methods used for forecasting trends are discussed more fully later in this chapter.

7.3.3 Variation of price clauses (fluctuations)

Indices are used to calculate the increased costs of construction under a fluctuation-type contract. It is possible to evaluate to a tolerable level of accuracy the increases in the costs of resources to the contractor. This method has distinct advantages over the traditional 'actual' cost reimbursement clauses found in many construction contracts. The method is easily understood, is quick to calculate, reduces administrative time and costs and results in earlier payment to the contractor. The single disadvantage is that it does not provide an exact reimbursement of the contractor's increased costs.

7.3.4 Comparison of cost relationships

The costs of different materials and processes do not change at the same rate. Indices can therefore be used to see the changes in the relationship between one component and another over a period of time.

7.3.5 Assessment of market conditions

In addition to the costs of building, market conditions will affect the price charged to the client. The tender price index takes this into account. A relative market condition index can be calculated by dividing the tender price index by the building cost index.

7.3.6 Pricing

Index numbers can be used for updating prices in bills of quantities or other published sources, to current or future dates. The process used is identical to that used in cost planning.

7.4 LASPEYRE'S AND PAASCHE'S INDICES

Some index numbers are known by the names of their authors. The Laspeyre index is a base year weighted index expressed by the formula

$$\frac{\text{current price} \times \text{base quantity}}{\text{base quantity} \times \text{base price}} \times 100$$

Paasche's index uses current year quantities in its computation:

$$\frac{\text{current price} \times \text{current quantity}}{\text{base price} \times \text{current quantity}} \times 100$$

Laspeyre's index is more frequently used, mainly because of the difficulties of establishing current quantities.

A third option is to use an index calculated by the formula

$$\frac{\text{current price} \times \text{current quantity}}{\text{base price} \times \text{base quantity}} \times 100$$

This will measure changes in price other than inflation factors alone.

7.5 DIFFICULTIES IN THE USE OF INDICES

- An index number can at best provide only a general indication of the changes in value of a commodity. It cannot therefore be considered to be very precise. The retail price index (RPI), for example, provides only a general indication of the changes in the costs of household goods. It is an index based on an average or typical household, few of which probably exist in practice.
- The composition of an index is based on typical commodities that should reasonably measure the appropriate change. They may in practice be totally unrepresentative of the things they hope to measure.
- Commodities which may be considered to be important may be outside the scope of the index.
- In an attempt to measure real comparisons, the same item, same quantity and same source for the commodity must be used. When any of these are altered, inaccuracies may result, from factors other than the change one is trying to measure.
- If the original commodities cease to exist then inaccuracies can occur due to the substitution of alternative items.

- The correct balance of items that were chosen initially may now prove to be false, because of the changes in fashion etc. that have occurred over a period of time. For example, a housing cost index constructed in 1900 may have included slates and stonework as predominant items. It would not have included heating, electricity, double glazing, insulation etc. The purpose of the index will influence how often the index itself needs to be revised.
- Individual weightings should be applied to reflect the importance of certain items in the index.
- Inaccuracies in the data can occur because of errors in computation or because users supplied false returns. They may do this to conceal information they do not wish to disclose.
- Although indices may attempt to measure a change in the overall pricing level, they can provide misleading results. For example, Table 7.1 gives the hypothetical costs of employing labour. The company costs, which include a bonus payment, show no change in the costs of employing labour for each of the five years. The nationally agreed wage rates show that costs have been rising by 10%. The use of the latter would therefore provide misleading information as far as this company is concerned. Table 7.2 also gives some misleading data. The indices show that costs have apparently increased by 10% this year when compared with last year. The actual costs show no change in the true values.

Table 7.1   Costs of employing labour

	Year				
	1	2	3	4	5
National rate	5.00	5.50	6.05	6.66	7.33
National index	100	110	121	133	146
Company rate	5.00	5.50	6.05	6.66	7.33
Bonus	3.00	2.50	1.95	1.34	0.77
Total rate	8.00	8.00	8.00	8.00	8.00
Index	100	100	100	100	100

Table 7.2   Misleading statistics

Costs	Last year	This year
Bricks	£12.00	£18.00
Concrete	£20.00	£14.00
	£32.00	£32.00
Indices		
Bricks	100	150
Concrete	100	70
Average	100	110

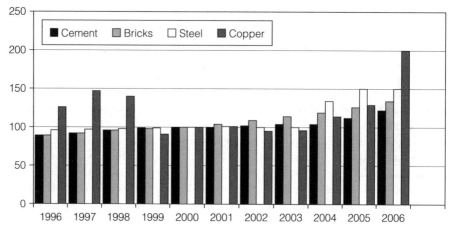

Fig. 7.1   Variability in material price indices

7.6 COST AND PRICE INDICES

7.6.1 Building cost index

The building cost index measures changes in the contractors' costs. It is constructed on a combination of actual wage rates, material costs and plant and overhead charges (see Figure 7.1). The combination of these different items is complex in attempting to measure reality. It is therefore based on the principle of a 'basket of goods' in order to represent typical ratios. The reality of the representation depends upon the type of goods selected, since not all items can be included, and their respective amounts or ratios. The task is difficult owing to the variety of different materials and methods of construction that are available.

A building cost index can be constructed for:

■ The total construction costs of a building or a type of building
■ An element (external walls) or trade (brickwork) within the building process
■ A single material, e.g. cement

The task in the first of these is extremely difficult owing to the variety of materials, construction methods and types of building available. The available published indices, of which there are several, provide little more than a general indication of cost trends. In an attempt to make such indices more useful, building types can be separately classified according to their method of construction. While this provides for more reliable information the index is nevertheless a rather blunt tool.

The selection of representative or typical items is also difficult, because of the wide variation of constructional methods and materials that can be used. Other problems include the quantity of data that needs to be collected in order to provide some statistical reliability. However, in view of the costs involved in collecting data, the amount needs to be restricted.

Table 7.3 Typical material price indices, 2000 = 100

	1996	1997	1998	1999	2000	2001	2002	2003	2004	2005	2006
Sand and gravel	95	101	105	105	100	100	108	119	127	131	137
Cement	89	92	96	99	100	100	102	104	104	112	122
Precast concrete	85	87	93	98	100	103	108	113	119	124	127
Bricks	89	92	96	98	100	104	109	114	119	126	134
Imported hardwood	101	95	90	92	100	103	104	109	113	120	127
Sawn softwood	107	110	103	101	100	98	99	102	110	114	117
Structural steel	96	97	98	99	100	101	100	100	134	150	150
Screws	119	110	108	102	100	102	101	99	103	126	142
Central heating boilers	90	94	97	99	100	101	103	102	104	107	109
Copper pipe	126	147	140	91	100	101	95	96	114	129	200
Plastic pipes	92	92	96	98	100	102	104	105	106	112	117

Source: BERR (2007)

Table 7.3 lists a range of some of the more common building materials, together with an index of their costs from 1996 to 2006 (2000 = 100). This table shows that building materials do not increase at the same rate over time. This is due to the availability of the supply of the raw materials and the demand for these products, often on a worldwide scene. This feature can be observed at any point in time. Material price speculators use such data to trade these materials, sometimes on futures markets, in the hope and expectation that prices will increase more than expected. Copper pipe increased by almost 60% over these ten years, although at the start of the 21st century it had fallen by 25% before then rising by 110% between 2002 and 2006. By comparison, the price of cement increased steadily, by almost 40%, between 1996 and 2006. The differences between these materials' costs can be attributed to many different factors such as the mining of raw materials, manufacturing and transport costs. Product development through research, development and innovation is another factor that influences changes in prices.

7.6.2 Tender price index

The tender price index is based on what the client is prepared to pay for the building. It therefore takes into account building costs, but it also makes an allowance for market conditions and profit. It may or may not include fluctuations in price, depending on the terms of the contract. It is usual to provide two separate indices which can either include or exclude fluctuations. Assuming that they have common base data, the index which includes fluctuations should display higher values.

These indices therefore take into account the tendering market, and are thus much more useful in updating the prices in a design budget. While building costs may continue to rise, a shortage of work in the industry may cause the tender price index to fall, as was experienced during the period 1980–81. Generally, there is some relationship in the trends of each index.

A tender price index can be obtained by pricing and repricing a bill of quantities at various required intervals. This will then provide the user with tender sums which can be quickly converted to index numbers. The bill items can be priced either on the basis of price book data or by using rates from bills of quantities received. The latter point does not infringe the form of contract, since the rates *and* the contractor are not disclosed to third parties. The use of a standard price book may result in some distortion, since the data contained are only an expert opinion and not necessarily a bona fide rate for a job. A more accurate index is likely to be prepared where the given data can be priced repeatedly. The larger the sample size, the greater will be the reliance that can be placed on the results.

The BCIS indices always indicate the sample size for statistical significance purposes. The BCIS emphasises that in order to achieve statistical reliability it needs to analyse 80 projects per quarter. In times of recession in the construction industry this objective may be difficult to achieve, and the users of the published indices must bear this in mind.

The majority of tender price indices do not use complete bills of quantities, but sample items from each trade usually totalling about 25% of the trade or section. A weighting which is proportionate to the value of the trade or section in the bill of quantities is then applied.

Table 7.4 and Figure 7.2 provide a broad indication of the relationship and movement of tender prices, building costs and retail prices. Building costs indicate a steady increase generally and are less susceptible than tender prices to changes in the market. Building costs therefore align themselves with the index of retail prices, but tender price trends tend to be more erratic.

At the beginning of 2008, tender prices continued to rise in Greater London but the rate of increase eased. This tended to indicate that changes in market conditions were beginning to take place. A year later tender prices had all but collapsed since the workload was drying up due to the onset of the impending recession. Some regions of the UK would fare rather better as workloads would not diminish to the same extent. Tender prices throughout the UK suffered a dramatic fall in the fourth quarter of 2008. These fell by as much as 7.5% as contractors saw a gaping hole in workload ahead of them. As private-sector building projects grind to a halt, two years of falling prices are forecast.

Table 7.4   Building costs, tender prices and retail prices

	Tender prices	Building costs	Retail prices
1995	270	400	375
2000	375	500	430
2005	460	600	450
2006	480	650	475
2007	500	675	520
2008	525	725	545

Source: Adapted from Davis Langdon (various)

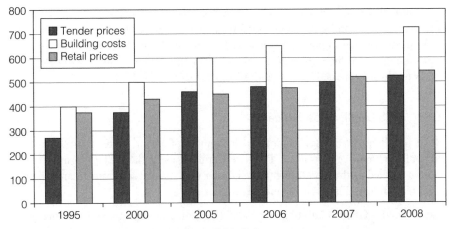

Fig. 7.2   Comparison of indices (based on Table 7.4)

Davis Langdon, the construction cost consultancy, has suggested that tender prices will fall by up to 9% in 2009 and a further 3–5% in 2010 according to a market forecast prepared for *Building*. It suggested that competition for work would intensify as projects were completed and firms chased a dwindling number of contracts. Prices fell by 1% in the first quarter of 2009 and are down by 8% compared to the same period of time in 2008.

The building cost index rose by 4.3% over the year up to the first quarter of 2008. This was the lowest recorded figure for four years. However, it was expected that incoming materials price increases and a substantial summer wage award should push inflation above 5% by the second half of the year. This was not to be.

The rate of increase in the retail prices index (RPI) slipped below 4%. The government's inflation measure, the consumer prices index, is up to 2.5%. Independent forecasters believed that the RPI would drop to below 2.9% by the end of 2008. The rate of increase retracted sharply in the fourth quarter, down to 0.9% in December after three consecutive monthly falls. Housing costs and the VAT reduction were the main drivers behind this. The index is expected to be negative territory by the end of 2009.

7.7 OTHER PUBLISHED INDICES

7.7.1 Retail prices index (RPI)

The retail prices index is the index with which most people are familiar and is used by a wide range of organisations in the UK. It is compiled by the Central Statistical Office and published in its *Monthly Digest of Statistics*. It is quoted frequently in the news and its publication, usually on a Tuesday in the middle of each month, is eagerly awaited by political and economic commentators. The general index seeks to measure the percentage changes month by month in the average level of prices of the commodities and services purchased by the average household in the UK. The commodities and services included in the index are weighted in accordance with their supposed importance. Details of the method used for computing the index are given in the *Method of Construction and Calculation of the Index of Retail Prices* published for the Department of Employment by HMSO.

7.7.2 BCIS building cost indices

BCIS building cost indices monitor the movement of labour, materials and plant costs of the building contractor. They are compiled using the NEDO formula indices applied to the work category weighting systems derived from an analysis of accepted tenders. The available indices are as follows:

Index No. 4   General building cost index (excluding mechanical and external work). This represents all construction types and covers all work except mechanical, electrical and lift installations.

Index No. 5   General building cost index; based on a weighting of indices 6–8.
Steel frame (6)   25%
Concrete frame (7)   25%
Brickwork (8)   50%

Index No. 6   Steel-framed construction cost index; represents mainly low-rise structures with both heavy and light steel frames.

Index No. 7   Concrete-framed construction cost index; represents *in situ* and precast concrete framed structures.

Index No. 8   Brick construction cost index; represents low- and medium-rise buildings with structural brick external walls.

Index No. 9   Mechanical and engineering cost index; covers mechanical and electrical work including lifts.

7.7.3 BCIS tender price indices

This is a similar type of index to the DoE tender index. It is not, however, restricted to government projects, and must therefore be presumed to cover a wider range of building types.

7.7.4 Building indicators

Building magazine provides from time to time an attempt at indicating the economic performance of the construction industry. It does this by examining certain key areas or indicators. These key areas are varied and diverse and include, for example, architect's workload, total new orders and starts and completions of dwellings, both public and private.

In addition, *Building* magazine has constructed its own housing cost index based on its own assessed weightings of a typical house. Although this is included as an indicator, it is sometimes shown separately in far greater detail.

7.7.5 Housing and Construction Statistics (DoE)

This publication is provided quarterly and is prepared by the Directorate of Statistics in the DoE. It includes indices as varied as housing finance, slum clearance and house-building performance.

7.7.6 Davis Langdon cost forecasts

The above firm of chartered quantity surveyors frequently publishes cost forecasts in both graphical and index format in *The Architects' Journal*. The indices are based on schemes in the London area and are likely to be of a smaller sample size than either government or BCIS indices. This information also appears in Spon's *Architect's and Builder's Price Book*.

7.7.7 Nationwide Building Society index

This is an index of house prices which measures the change in cost to purchasers of different house types. The data are recorded on a regional basis. The RICS collects and publishes similar data.

7.7.8 The formula method

The formula method has been devised and researched by the NEDO for the measurement of construction cost increases. It is used extensively in the UK for calculating increased costs in preference to the traditional method of using actual cost increases from time and wage sheets. The system in use is the Series 2 indices where the building work is subdivided into 48 categories, based broadly on the work sections from the SMM.

7.8 REGIONAL VARIATION

Indices of different kinds are often produced for the country as a whole, and since they measure only a typical 'basket of goods', they can at best represent a typical trend. It is possible, however, to analyse costs on a regional basis or even within a sub-region. The locality of a project produces a cost which is dependent on the market conditions such as the availability of labour and materials, workload and availability of grants. Similar buildings constructed in different localities will produce different tender sums. The BCIS produces a quarterly review of building prices which outlines the variation measured in tender sums. Over a period of time the regional variation statistics will change, depending on the amount of competition for work within a region. The recession of the early 1990s had a huge impact on the southeast region and, while actual costs declined everywhere, they did so at the greatest rate in the southeast (with the exception of Greater London). Table 7.5 compares the regional variation indices. It should be emphasised that, even within

Table 7.5   Regional variation in building

	Output proportion (%)					Difference from average costs (%)
	1975	1985	1995	2000	2005	
North	7	5	5	3	4	0.98
Yorkshire and Humberside	8	8	8	6	8	0.92
East Midlands	7	6	7	5	7	0.94
East Anglia	3	4	4	3	4	1.00
Southeast	31	39	34	49	38	1.08
Southwest	6	8	9	7	8	0.95
West Midlands	8	8	9	9	8	0.95
Northwest	10	9	10	7	10	1.03
Wales	7	4	5	3	4	0.93
Scotland	13	9	9	8	9	1.03
Total	100	100	100	100	100	

Source: BERR (2007)

the sub-regions, variations in tender levels must be expected, depending on where the projects are being constructed (see site considerations, Chapter 5).

The regional variation in building activity has, of course, some relationship to the population distribution. Wales, for example, accounts for 5% and Scotland 8% of the population of Great Britain. The population is forecast to increase by 8% in England, 5% in Wales and 3% in Scotland over the next ten years. This clearly has implications for the infrastructure and building development requirements.

7.9 MEASURING TRENDS

Time series analysis is the statistical technique used to measure the relationship between a series of data and a period of time. The name given to the graph on which such a series is plotted is a historigram, i.e. a history or record over time. The series may be plotted daily, weekly, monthly, yearly or at any interval of time. The horizontal axis is always chosen as the time axis.

Future projected trends can be assessed on the basis of the previous performance of similar recorded data. It should be emphasised, however, that past trends may not always be maintained in the pattern they suggest.

It is therefore necessary to couple with these past trends an assessment of any likely changes to the conditions of the market in general. Time series analysis actually represents a mixture of various influences; the principal ones are described below.

The long-term trend   Data are said to show a trend where the values generally increase or decrease over the whole period that they span. It is unlikely that the examination of any data, particularly from the construction industry, will indicate a smooth upturn or decline; any trend is likely to be punctuated with troughs and peaks.

The cyclical movement This is the wave-like formation generally due to the influence of booms and slumps on business activity. These trade cycles in the construction industry vary in relationship to the demand for buildings and this in turn is influenced by economic prosperity and government policy.

The seasonal variations These describe the fact that the quarterly figures follow a seasonal pattern. This pattern will vary from industry to industry and may vary within the various sectors of a single industry. For example, in the building industry productivity is at a higher level during the summer season than in the winter season.

The non-recurring influences These are random variations in a time series analysis, and since they do not occur with any statistical regularity they cannot be measured or predicted. The majority of contractors from time to time experience irregular increases or decreases in workload that cannot be attributed to any particular factor.

It is essential that we should be able to distinguish between these various influences, and to measure each one separately. This procedure is known as the analysis of a time series. Two techniques are now considered in connection with this analysis.

7.9.1 Moving average

Table 7.6 shows the data relating to expenditure on contracts for each of the four quarters between 1993 and 1997. The data have been adjusted to a common base date to remove any influences due to inflation.

Column 3 shows the actual expenditure for each quarter of each of the five years. Column 4 shows the moving average based on each of the five periods and column 5 the variation between the actual quarterly expenditure and the moving average. The question of whether to base the moving average on three years, five years or any other period is largely a matter of judgement. The purpose is to remove all the troughs and peaks in order to produce a trend line.

One disadvantage of the moving averages method is that the respective values for the end periods cannot be calculated but can only be inserted by interpolation.

The trend line, based on the data from column 4, has been plotted onto the histogram (Figure 7.3). In order to work out the second part of the problem we examine column 5 from Table 7.6. This information is first transferred to Table 7.7. These seasonal variations should add up to zero; any difference is due to rounding errors. These figures would indicate that the contractor's expenditure is considerably greater in the last two quarters of the year than at the beginning. The cyclical variation would be calculated on a similar basis over a known trade cycle taking the year's values as a whole. The difference between seasonal variation and cyclic effect is largely a matter of knowledge and experience.

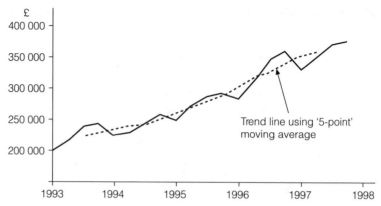

Fig. 7.3   Contractor's expenditure, 1993–97

Table 7.6   Expenditure on contracts, 1993–97 (inflation-adjusted)

1 Year	2 Quarter	3 Expenditure (£'000)	4 Moving average (£'000)	5 Variation from trend (£'000)
1993	1	200	–	–
	2	215	–	–
	3	240	225	+15
	4	245	231	+14
1994	1	225	237	−12
	2	230	241	−11
	3	245	242	3
	4	260	252	8
1995	1	250	264	−14
	2	275	274	1
	3	290	279	11
	4	295	292	3
1996	1	285	307	−22
	2	315	322	−7
	3	350	330	20
	4	365	344	21
1997	1	335	356	−21
	2	355	362	−7
	3	375	–	–
	4	380	–	–

Table 7.7 Variation from trend

	Quarter			
	1	2	3	4
1993	–	–	+15	+14
1994	–12	–11	+3	+8
1995	–14	+1	+11	+3
1996	–22	–7	+20	+21
1997	–21	–7	–	–
	–69	–24	+49	+46
Seasonal variation	–17.25	–6	+12.25	+11.5

Table 7.8 Seasonally adjusted data

Quarter	Expenditure	Seasonal variation figure	Seasonally adjusted
1	250	–17.25	267.25
2	275	–6.00	281.00
3	290	+12.25	277.75
4	295	+11.50	283.50

7.9.2 Seasonal adjustment

A popular way of presenting construction statistics is to produce seasonally adjusted data. In order to adjust a set of original figures we calculate the seasonal variations and deduct these from the original figures. For example, the seasonally adjusted expenditure for 1995 would be as given in Table 7.8.

7.9.3 Linear regression analysis

An alternative to using the above method for smoothing out data points and inserting a trend line is to use linear regression analysis. This is a statistical technique to find the formula of the best-fit line through the data.

Consider the data given in Table 7.9, which shows the number of enquiries received by a construction firm for new projects over a period of eighteen months. These data are also plotted in Figure 7.4. It can be observed that in general the trend for this particular firm is to receive more enquiries. (The type and size of enquiries in this example have been ignored.) The trend line fitted supports this viewpoint.

In order to avoid individual judgement in constructing the line through these points, it is necessary to calculate a best-fitting line. This is derived from the method of least squares analysis, i.e. the line is drawn in such a way that the sum of the

Table 7.9   Number of enquiries received for new work

Month	Enquiries x	x^2	xy	y^2
1	3	9	3	1
2	4	16	8	4
3	6	36	18	9
4	7	49	28	16
5	4	16	20	25
6	5	25	30	36
7	5	25	35	49
8	8	64	64	64
9	7	49	63	81
10	9	81	90	100
11	8	64	88	121
12	7	49	84	144
13	8	64	104	169
14	10	100	140	196
15	11	121	165	225
16	10	100	160	256
17	9	81	153	289
18	10	100	180	324
171	131	1 049	1 433	2 009

$$171 = 18a + 131b \qquad \text{(i)}$$
$$1\,433 = 131a + 1.049b \qquad \text{(ii)}$$
$$= y = -4.83 + 1.97x$$

When $x = 6$, $y = -4.83 + 6 \times 1.97 = 6.99$

Note: See pp. 388–91 for a fuller explanation of the method.

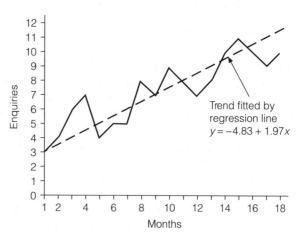

Fig. 7.4   Number of enquiries received regarding new work

squares of the vertical distances of the plotted points to the line is a minimum. The formula is written as follows (x = number of enquiries; y = month):

$$\Sigma y = an + b\Sigma x$$
$$\Sigma xy = a\Sigma x + b\Sigma x^2$$

These variables can now be replaced with the values from Table 7.9 so that on the basis of these data alone it could be predicted, for instance, that in month 24 the number of enquiries for new projects could be calculated as follows:

$$y(24 \text{ months}) = -4.83 + 1.97x = 15$$

Of course this previous trend would need to be coupled with any known prevailing conditions in the construction industry.

7.9.4 Use of time series analysis

By analysing the data and separating the seasonal variation, the cyclical variation and the trend, we hope to learn something of the behaviour of the data we are examining. This knowledge can then be used as a basis for future planning. A time series can be separated into each of these movements, and to each of these we can attribute a probable cause and a possible future course of action.

Time series analysis is widely used in business for planning on the basis that the previous performance and rhythm will to some extent be repeated.

7.10   KEY PERFORMANCE INDICATORS AND BENCHMARKING

The use of performance indicators to measure the different attributes of firms is now commonplace in different industries and companies around the world. The Egan Report (*Rethinking Construction*, 1998) concerned itself directly with improving both the effectiveness and the efficiency of the UK construction industry. It challenged the industry to meet some ambitious targets and to measure its performance over a range of activities. The creation of Construction Key Performance Indicators (KPIs) and the release of the first KPI pack in early 1999 were the first steps in the process of answering these challenges.

The KPIs are produced by a partnership of the Department for Business, Enterprise and Regulatory Reform (BERR), the Construction Industry Board and the Construction Best Practice Programme using data from BERR, the Building Cost Information service, the Construction Clients Forum, the Health and Safety Executive, Dun & Bradstreet and other third-party financial analysts.

The fourth annual Construction Industry KPIs, relating to performance in 2001, were published in May 2002. These adopted the format of an overall industry wall chart and a KPI pack containing:

- A handbook
- A number of sector-specific wall charts

- A book of additional performance indicators, containing charts for a number of industry sub-sectors
- An industry progress report
- A 'Respect for People' KPI wall chart
- An interactive CD ROM

The Construction Industry KPIs are intended for use by individual firms wishing to measure or compare their own performance.

In order to collect and use the construction industry KPI graphs, three key project stages have been identified:

1. **Commit to invest**. The point at which the client decides in principle to invest in a project, sets out the requirements in business terms and authorises the project team to proceed with the conceptual design.
2. **Commit to consult**. The point at which the client authorises the project team to start the construction of the project.
3. **Available for use**. The point at which the project is available for substantial occupancy or use. This may be in advance of the completion date.

7.10.1 Definitions of project and company KPIs

The following are the ten KPIs that have been developed as part of the benchmarking exercise.

Project KPIs	Definition
Client satisfaction – product	How satisfied the client was with the finished product, using a 1–10 scale where 10 = totally satisfied, 5–6 = neither satisfied nor dissatisfied, 1 = totally dissatisfied.
Client satisfaction – service	How satisfied the client was with the service of the consultants and the main contractor, using a 1–10 scale where 10 = totally satisfied, 5–6 = neither satisfied nor dissatisfied, 1 = totally dissatisfied.
Defects	The condition of the product with respect to defects at the time of handover, using a scale of 1–10 where 10 = defect free, 8 = defects with no significant impact, 5–6 = defects with some impact, 3 = defects with a major impact, 1 = totally defective.
Predictability – cost	There are three indicators. One is for design cost, one for construction cost and one for project cost. ■ **Design cost**. Actual cost at available for use less estimated cost at commit to invest. This is expressed as a percentage of the estimated cost at commit to invest. ■ **Construction cost**. Actual cost at available for use less estimated cost at commit to construct. This is expressed as a percentage of the estimated cost at commit to construct. ■ **Project cost**. Actual cost of the combined design and construction process at available for use less the anticipated cost of the combined design and construction process at commit to invest, expressed as a percentage.

Predictability – time	There are three indicators, one for the design phase, one for the construction phase and one for the whole project.

- **Design time.** Actual duration at commit to construct less the estimated duration at commit to invest. This is expressed as a percentage of the estimated duration at commit to invest.
- **Construction time.** The actual duration at available for use less estimated duration at commit to construct. This is expressed as a percentage of the estimated duration of commit to construct.
- **Project time.** Actual duration of the combined design and construction process at available for use less the anticipated duration of the combined design and construction process at commit to invest.

Construction cost	The normalised[1] construction cost of a project in the current year less the construction cost of a similar project one year earlier. This is expressed as a percentage of the construction cost of a similar project one year earlier.
Construction time	The normalised time to construct a project in the current year less the time to construct a similar project one year earlier. This is expressed as a percentage of the time to construct a similar project one year earlier.

Company KPIs

Profitability	Company profit before tax and interest as a percentage of sales.
Productivity	Company value added[2] per employee (#)
Safety	Reportable accidents[3] per 100 000 employed per year.[4]

# indicates that no measured data were available
Source: BERR (2007)

Notes
1. Normalisation is a statistical method for removing the effects of location, function, size and inflation.
2. Value added is turnover less all costs subcontracted to or supplied by other parties.
3. Reportable accidents are defined in the Health and Safety Statistics published by the Health and Safety Commission as: fatalities, major injuries, and over-three-day injuries to employees, self-employed people and members of the public.
4. The figures need to be compared pro rata if they cover a shorter period than a year. Such short period rates should be compared only with rates of similar periods and not the annual rates.

The statistics presented in Table 7.10 give a variable picture of the performance of the UK construction industry during 1998–2001. Whilst there is some evidence that practices are changing, there remains much scope for further improvement. In addition to the range of performance indicators shown in the table, BERR also records the Respect for People KPIs.

Table 7.10   Summary of UK industry performance

Headline KPI	Measure	Performance		
		2003	2005	2007
Client satisfaction – Product	Scoring 8/10 or better	78%	83%	82%
Client satisfaction – Services	Scoring 8/10 or better	71%	77%	75%
Defects	Scoring 8/10 or better	68%	72%	73%
Safety – Industry	Accident incidence rate	1 097	1 023	946
Safety – All contractors	% achieving zero rate	30%	50%	62%
Safety – Contractors over £10m	% achieving zero rate	1%	1%	1%
Predictability cost – Design	% on target or better	65%	63%	64%
Predictability cost – Construction	% on target or better	52%	48%	49%
Predictability cost – Project	% on target or better	52%	48%	46%
Predictability time – Design	% on target or better	53%	52%	58%
Predictability time – Construction	% on target or better	59%	62%	65%
Predictability time – Project	% on target or better	44%	46%	58%
Profitability	Median profit before interest or tax	5.4%	8.1%	8.2%
Productivity – Current values	Median added value £000	31.1	34.2	42.0
Productivity – 2000 values	Median added value £000	27.9	27.5	31.2
Construction cost	Change over previous year	5.0%	−0.8%	−3.8%
Construction time	Change over previous year	1.0%	1.3%	−0.3%

Source: Office for National Statistics (2008)

SELF ASSESSMENT QUESTIONS

1. Describe the important characteristics to be taken into account when constructing a set of indices.

2. Describe the differences between building cost indices and tender price indices, explaining where and how each would be used in practice.

3. If past performance is no guarantee of future projections then what is the point of using indices? Discuss this question.

BIBLIOGRAPHY

BCIS. *Tender Price Indices*. RICS. (Quarterly)
Building. Building Intelligence. (Weekly)
Davis Langdon. Market Intelligence, *Building*. (Various)
Department for Business, Enterprise and Regulatory Reform (BERR). *Construction Statistics Annual*. The Stationery Office 2007.
Fleming, M.C. and Tysoe, B.A. *Construction Cost and Price Indices Handbook*. Spon 1991.
Office for National Statistics. *Construction Statistics Annual No. 9*. Palgrave MacMillan 2008.

CHAPTER 8

NEW RULES OF MEASUREMENT (NRM)

Order of cost estimating and elemental cost planning

LEARNING OUTCOMES

After reading this chapter, you should have an understanding of the RICS New Rules of Measurement (NRM) for order of cost estimating and cost planning. You should be able to:

- Understand why the NRM rules have been introduced
- Appreciate the links to the measurement of building works in general
- Recognise the purpose of the NRM
- Understand the structure of the NRM
- Identify the relationship to the RIBA Plan of Work and the OGC Gateway Process and construction costs

8.1 INTRODUCTION

The Building Cost Information Service (BCIS) was established by the RICS in 1962 (see p. 80). It sets out a series of principles for the cost analysis of building works. There is no similar set of rules for the cost analysis of civil engineering works. This partially reflects the fact that civil engineering costs are related more to process than to quantity (see p. 53). These basic rules relating to the principles of analysis enable quantity surveyors to cost and compare building projects with a greater degree of reliability than was available previously, because of the consistency in the way in which the costs of the different parts or elements are now collected.

During the latter part of the twentieth century new methods of procuring buildings were introduced (see Chapter 10). These changes have in some cases diluted the need for reliable quantities that have been prepared against a given set of rules. Since changes and deviations in the methods of procurement have accelerated, the BCIS has found it more difficult than in the past to acquire detailed cost analyses for a wide range of building projects. Many of the more recent analyses are based on the simplified procedures (see pp. 68 and 166).

In 2009, the RICS introduced the RICS New Rules of Measurement (NRM). The implications of these are as follows:

- First, these rules are not prepared in collaboration with, and with the agreement of, the main (principal) contracting organisations. This change largely reflects the reality that much of the work that is now undertaken is carried out by subcontracting firms which are gathered together under the management of a principal contractor. The RICS steering group has largely produced the new set of rules in collaboration with the trade and subcontracting bodies.
- Second, the NRM now represent a suite of rules of measurement that include rules for cost estimating and cost planning as well as rules for tendering documentation.
- Third, the existing rules of measurement (SMM7) (see below) are largely UK-centric. The NRM are expected to have a much wider application and appeal around the world. This is because the profession has acquired a more international perspective whereas previously it was limited to a focus on Commonwealth countries alone, and because the world has shrunk, especially in terms of all forms of communication.
- The rules will be used by firms and organisations currently involved in cost plan tendering, and will provide greater consistency and clarity for the use of cost plans for this purpose.

8.2 STANDARD METHODS OF MEASUREMENT

The Standard Method of Measurement of Building Works (SMM) was first published in 1922, although this was preceded by a Scottish SMM, published in 1915. The SMM is currently in its seventh edition (SMM7), although a number of later editions were revised to incorporate minor changes of practice. These rules were specifically designed to advise quantity surveyors on how to measure building work items for inclusion in bills of quantities. The principal reason for measuring building works was to obtain tender prices for a building project. The consistent set of rules enabled contractors to understand precisely what a particular description was meant to include. The adoption of the rules provided transparency and avoided any misinterpretation if they were followed correctly. They also made the role of the building estimator easier, since the adoption of common standards allowed estimators to develop a system where the rates for similar items of work on different projects could be applied more easily.

The principle behind the establishment of the rules was to allow for the proper quantification of works in terms of time (labour) and materials used and to measure only those items that were of cost significance. Throughout the last century the evolution of both design and construction can be easily observed. This development was one of the main reasons for the revision of the SMM and the introduction of new rules. For example at the beginning of the twentieth century the construction of buildings was much more labour intensive than it is today. In the latter years of that century, the construction of buildings became more mechanised, with a great many items being manufactured off-site and then requiring on-site assembly. These factors obviously influenced the requirements for time and materials and the way in which the work should be quantified. Research completed in 1981 (see p. 55) illustrates the principle of measuring the cost-significant items. The revisions to the different SMMs have simplified the rules of measurement, by concentrating only on the items of the greatest cost significance. These of course vary from one trade to another.

8.3 NRM (NEW RULES OF MEASUREMENT)

The NRM for order of cost estimating and elemental cost planning were introduced in May 2009. The NRM for tender documentation will be introduced in 2010. The previous SMMs did not provide specific guidance on the measurement of building works for the purpose of producing cost estimates or cost plans. In the absence of a clear set of rules, individuals or firms simply adopt their own known practice. The NRM are informed by current best practices from the profession. This has resulted in inconsistencies in the descriptions and the calculation of costs, making it difficult for an employer and project team to understand precisely what is included in a cost estimate, cost limit or cost target provided by a quantity surveyor. This can result in doubts about the clarity and usefulness of the cost advice provided. On occasions employers have been left in doubt as to what was really included in the cost estimate or cost plan.

The RICS Quantity Surveying and Construction Faculty set up a steering group to examine the issues associated with the measurement of building works at all stages of the design and construction process. The group quickly concluded that employers and design teams required clear guidance from quantity surveyors so that a common and consistent set of rules could be produced on how to measure building works for the purpose of cost estimating and elemental cost planning.

The new measurement rules have the same status as RICS guidance notes. They provide advice to RICS members on aspects of the profession. Where procedures are recommended for specific professional tasks, these are intended to include best practice. In this context these are those procedures which in the opinion of the RICS meet a high standard of professional competence. Members are not bound to follow the rules but where an allegation of professional negligence arises against a surveyor then a court is likely to take account of relevant guidance notes that have been issued by the professional body. The courts will then decide whether a surveyor acted with reasonable competence. Following the rules may therefore provide a partial defence to the allegation of negligence but a surveyor must always act in the best interests of a client.

8.3.1 Purpose of the NRM

The NRM have been written to:

- Provide a standard set of measurement rules that are understandable by all those involved in a construction project, including the employer.
- Improve communication between project/design team and the employer.
- Provide effective and accurate cost advice to the employer and project/design team.
- Provide rules for the order of cost estimates and elemental cost plans.
- Describe and deal with costs and allowances forming part of the costs of a building, but which are not reflected in the measurable building work items.
- Provide a structured basis for measuring building works and other key components.
- Provide the essentials of good practice.

The rules do not explain estimating methods, cost planning techniques, procurement methods or contract strategies. They are based on UK practices but the requirements for a co-ordinated set of rules underlying the philosophy behind each section have worldwide application.

8.3.2 Structure of the NRM

The NRM are divided into four parts with supporting appendices:

Part 1   places the NRM for cost estimating and elemental cost planning in context with the RIBA Plan of Work and the OGC Gateway Process.

Part 2   describes the order and content of the order of cost estimate, its key constituents, explains how to prepare an order of cost estimate and sets out the rules of measurement.

Part 3   describes the purpose of an elemental cost plan, its key constituents and how to prepare a cost plan.

Part 4   comprises the tabulated rules of measurement for the preparation of formal cost plans.

The following appendices are included with the rules:

Appendix A   Core definition of gross internal floor area (GIFA)
Appendix B   Commonly used functional units and functional units of measurement
Appendix C   Core definitions of net internal area (NIA)
Appendix D   Special use definitions for shops
Appendix E   Measurement rules for elemental estimating
Appendix F   Logic and arrangement of levels 1 to 3 for elemental cost planning
Appendix G   Information requirements for formal cost plans
Appendix H   Template for elemental cost plan (based on level 1 codes)
Appendix I   Template for elemental cost plan (based on level 2 codes)

8.4 PART 1: GENERAL

8.4.1 RIBA Plan of Work and the OGC Gateway Process

This part of the NRM places order of cost estimating and elemental cost planning in context with the RIBA Plan of Work (Work stages) and the Office of Government Commerce (OGC) Gateway Process and explains the symbols and definitions that are used.

The RIBA Plan of Work is a framework recognised by the construction industry that organises the process of managing and designing building projects and administering building contracts into a number of key work stages. The full RIBA Plan of Work has 11 sequential steps as shown in Table 8.1. Whilst this appears to be a longitudinal process it should be recognised that these steps may overlap with each other or on some projects the sequence may need to be varied. Nevertheless this provides an ideal framework that illustrates the development of a building project from inception through to completion.

More detailed information about the RIBA outline Plan of Work, along with the detailed tasks to be carried out at each stage, can be found at www.pedr.co.uk.

Table 8.1 The RICS formal cost estimating and cost planning stages in context with the RIBA Plan of Work and the OGC Gateway Process

RIBA Plan of Work stages	RICS formal cost estimating and elemental cost planning stages	OGC Gateway applicable to building projects	
A Appraisal	Order of cost estimate		
		1	Business justification
B Design brief			
		2	Delivery strategy
C Concept	Formal cost plan 1	3A	Design brief and concept approval
D Design development	Formal cost plan 2		
E Technical design	Formal cost plan 3	3B	Detailed design approval
F Production information	Pre-tender estimate		
G Tender documentation			
H Tender action	Post-tender action		
		3C	Investment decision
J Mobilisation			
K Construction to practical completion			
		4	Readiness for service
L Post-practical completion			
		5	Operations review and benefits realisation

Source: Adapted from RIBA (2008)

An alternative to the RIBA Plan of Work is to adopt the UK Government's Gateway Process. This is used on projects that are initiated by central government, the health sector, defence and local government as a best practice model for managing and designing building projects. This process examines programmes and projects at key decision points in their life cycle. It looks ahead to provide assurance that an employer can progress to the next stage in the process. Project reviews are carried out under OGC Gateway Review stages 1–5. Typically a project will undergo three reviews before commitment to invest and two examining implementation and confirming the operational benefits.

The Office of Government Commerce (OGC) is an independent office of HM Treasury, established to help the government deliver best value from its spending. The OGC works with central government departments and other public sector organisations to ensure the achievement of six key goals:

- Delivery of value for money from third party spend
- Delivery of projects to time, quality and cost, realising benefits
- Getting the best from the government's £30bn estate
- Improving the sustainability of the government estate and operations, including reducing carbon emissions by 12.5% by 2010–11, through stronger performance management and guidance
- Helping achieve delivery of further government policy goals, including innovation, equality and support for small- and medium-sized enterprises (SMEs)
- Driving forward the improvement of central government capability in procurement, project and programme management, and estates management through the development of people skills, processes and tools.

OGC provides policy standards and guidance on best practice in procurement, projects and estate management, and monitors and challenges government departments' performance against these standards, grounded in an evidence base of information and assurance. It promotes and fosters collaborative procurement across the public sector to deliver better value for money and better public services; and it provides innovative ways to develop the government's commercial and procurement capability, including leadership of the Government Procurement Service.

8.4.2 Cost estimates and cost plans

Cost estimates and cost plans are prepared by the quantity surveyor or construction cost manager at the various stages of either the RIBA Plan of Work or the OGC Gateway Process. These are shown in Table 8.1. The RICS has determined a series of formal cost estimating and elemental cost planning stages. These are shown as a guide and it is recognised that some employers may require this information at different stages of the project.

The NRM include an extensive range of abbreviations and definitions appertaining to this work. These have considerably extended those that were previously provided by BCIS (see p. 165). They include a number of additions such as:

Design team   This includes architects and engineers and other technology specialists who are responsible for the conceptual design, drawings, specifications, instructions, etc. The design team is a part of the much bigger project team.

Project team   This includes the employer, design team, project manager and quantity surveyor. These members are responsible for delivering the project on time, within cost and to the required design and quality performance.

Site area   This refers to total area of the site within the site boundaries, but excluding any areas used temporarily for the building works that do not form part of the delivered building project.

8.5 PART 2: MEASUREMENT RULES FOR ORDER OF COST ESTIMATING

Part 2 of the NRM describes the purpose and content of the order of cost estimates within the context of the RIBA Plan of Work and the OGC Gateway Process. It sets out the rules for measurement of:

- Superficial floor area method
- Functional unit method
- Elemental method

Part 2 of the NRM describes the content and application of unit rates to measured quantities in order to generate the base cost of the building works. It also explains the various cost allowances to cover the main contractor's preliminaries, overheads and profit, project/design fees, risk allowances, inflation, value-added tax (VAT) and other development and project costs.

8.5.1 Order of cost estimates

The purpose of the order of cost estimate is to establish whether a client can afford a particular project. Where the answer is yes, a realistic cost limit for the building project will be ascertained. This cost limit is defined as the maximum expenditure that the employer is prepared to pay for the completed project, and it becomes the authorised budget. It excludes design team and project team fees.

The employer is required to supply the following information to allow the cost limit to be calculated:

- Location of the site
- A statement of building use
- Floor area or number of functional units

- Requirements for the refurbishment of any existing buildings on the site
- Initial design brief including a statement of quality, sustainability and fit-out requirements
- Details of any enabling works
- Indicative programme up to occupation
- Any restraints imposed by the employer, local planners or statutory undertakers
- Site conditions known or expected
- Budget or cash flow restraints
- Initial views on procurement options
- Expected life span: 10, 25, 60 years
- An indication of proposed storey heights
- Particular requirements in respect of mechanical or electrical services
- Requirements in respect of: project/design team fees, other development costs, inflation and VAT
- Methods for dealing with capital allowances, land remediation and grants

In addition other information will be required from the architect, as follows:

- Design study sketches or drawings for alternative design solutions
- Schedule of gross internal floor areas
- Minimum storey heights
- Schedule of accommodation
- Number of car parking spaces above or below ground
- Indicative specifications
- Indicative environmental and sustainability strategies
- Advice on likely site constraints
- Advice on likely planning constraints
- Definition of 'fit-out'
- Initial risk register

Other information will be required from the mechanical and electrical services engineers:

- Indicative services specifications
- Indicative environmental and sustainability strategies
- Advice on the availability and adequacy of utility service connections
- Initial risk register

Other information will be required from the structural engineers:

- Advice on probable ground conditions
- Indicative specifications
- Initial risk register

The accuracy of an order of cost estimate is dependent on the quality of the information supplied to the quantity surveyor or cost manager. The more detailed, informative and correct the information that is supplied the more reliable will be the order of cost estimates prepared.

Table 8.2 Constituents of an order of cost estimate

Base cost estimate
- Works cost estimate
  - Building works
  - Main contractor preliminaries
  - Main contractor overheads and profit
- Project and design team fees
- Other development costs

Cost limit (excluding inflation)
- Risk allowances estimate
  - Design development risk
  - Construction risk
  - Employer change risk
  - Employer other risks

Cost limit (excluding inflation)
- Tender inflation

Cost limit (excluding inflation)
- Construction inflation estimate

VAT assessment

8.5.2 Constituents of an order of cost estimate

Table 8.2 sets out the constituents of an order of cost estimate.

The base cost estimate is the total of the building works, main contractor's preliminaries, main contractor's overheads and profit, project and design team fees and other development costs. The base estimate excludes any allowances for risk or inflation.

8.5.3 Measurement rules for building works

The three methods identified above, the floor area method, the functional unit method and the elemental method, are more fully described in Chapter 14. That chapter also includes a range of other methods that have been used in practice to calculate order of cost estimates. The element analysis largely follows that developed by BCIS. Compare Table 8.3 with Table 15.3 to gain an understanding of the minor differences. The choice and number of elements used to analyse building costs is dependent on the information available. Different quantity surveyors will modify these lists to suit their own particular circumstances. Table 8.3 provides the list of elements and their respective groupings.

The calculation of rates and prices to be used in the above will be a combination of detailed analysis as demonstrated in Chapter 15 and the use of benchmark analyses that are retrieved from a practice's own database or from published analyses such as that retained by the BCIS. The updating of this information is done through the use of indices and as explained in Chapter 17.

Table 8.3   List of elements for cost analysis purposes

Group element	Element	
1 Substructure		
2 Superstructure	2.1	Frame
	2.2	Upper floors
	2.3	Roofs
	2.4	Stairs and ramps
	2.5	External walls
	2.6	Windows and external doors
	2.7	Internal walls and partitions
	2.8	Internal doors
3 Internal finishes	3.1	Wall finishes
	3.2	Floor finishes
	3.3	Ceiling finishes
4 Fittings, furnishings and equipment		
5 Services	5.1	Sanitary appliances
	5.2	Services equipment
	5.3	Disposal installations
	5.4	Water installations
	5.5	Heat source
	5.6	Space heating and air conditioning
	5.7	Ventilation systems
	5.8	Electrical installations
	5.9	Gas and other fuel installations
	5.10	Lift and conveyor installations
	5.11	Fire and lightning protection
	5.12	Communication, security and control systems
	5.13	Special installations
	5.14	Builders' work in connection with services
	5.15	Testing and commissioning of services
6 Complete buildings and building units		
7 Work to existing buildings		
8 External works		
9 Facilitating works		
10 Preliminaries (if shown separately)		

Preliminaries

The main contractor's preliminaries will usually be calculated through the addition of an appropriate and realistic percentage to the building works estimate. Alternatively benchmark data from previously completed projects can be used to assess the level of main contractor's preliminaries to be applied to new building projects. Historically cost analysis provided the alternative options of showing preliminaries either as distributed evenly amongst the different elements or as a separate item (see p. 169). If it is known at an early stage that there are costs relating to known site constraints, special construction methods, sequencing of works or other non-standard requirements, these are to be assessed and identified separately. Group element 10 (preliminaries) provides a checklist of main contractor's preliminary items under the two broad headings, Employer's requirements and Main contractor's cost items. Similar lists of typical items can also be found in the various standard methods of measurement for preliminaries.

Overheads and profit

The NRM suggest that in the order of cost estimating, overheads and profit should be shown as a separate item and calculated by applying a selected percentage to the cost of the building works estimate and the main contractor's preliminaries. The percentage addition to be applied will be derived from a properly considered assessment of the main contractor's overheads and profit found on previous building projects. This item may be included in the different elements or in the cost per GIFA.

Project/design team fees

The consultant's fees are usually added as a percentage. These may include a contractor's pre-construction fee. The NRM provides a checklist of the possible activities that are involved.

Other development/project costs

There might be costs involved that are not necessarily directly associated with the works costs, but which will nevertheless be a part of the project costs. These might include insurances, planning fees, fees in connection with party wall awards, decanting and relocation costs, marketing costs, etc. The construction of the Emirates Stadium for the Arsenal football team, for example, required not only the purchase of the site but the purchase, relocating and construction for the occupiers of the new premises.

Risk allowances

All buildings involve risks, some obvious and others less so. Proper management of such risks can save both time and money. Chapter 20 describes the alternative ways of dealing with risks. Risk exposure (the potential effect of risk) changes as the building project progresses and therefore the continual management of risk is essential. Five approaches to dealing with risk are envisaged:

- Risk avoidance
- Risk reduction
- Risk transfer
- Risk sharing
- Risk retention

Risk in order of cost estimating is usually allowed for by adding a percentage to the cost of the project. Separate allowances should be considered and added individually for:

Design development risks   This is an allowance for use during the design process to provide for the risks that are associated with:

- Design development
- Changes in estimating data
- Third party risks (planning requirements, legal agreements, covenants, environmental issues, pressure groups)
- Statutory requirements
- Procurement methodology
- Delays in tendering

Construction risks   This is an allowance for use during the construction process and provides for the risks associated with:

- Site conditions (access restrictions/limitations, existing building boundaries, existing occupants and users)
- Ground conditions
- Existing services
- Delays by statutory undertakers

Employer change risks   This risk applies to both the design and construction stages and includes:

- Changes to the scope of the works or brief
- Changes in quality
- Changes in time

Employer other risks   This is a risk allowance that might include, for example:

- Early handover
- Postponement
- Acceleration
- Availability of funds
- Liquidated damages or premiums on other contracts due to the late provision of accommodation
- Unconventional tender action
- Special contract arrangements

It is recommended that the size of the initial risk allowance is based upon the results of a formal risk assessment. It can be beneficial even at an early stage of the project to prepare a specific risk register incorporating the major risks identified and a risk management strategy. Risks should not be excluded without due consideration.
It is important not to allow the natural optimism that surrounds the early part of

a project to influence the nature of the judgements that need to be made. The risks that can influence the cost of the project change as the building project progresses through the subsequent RIBA Work Stages. The risks identified in a risk register should be reassessed at regular intervals throughout the various formal stages of cost planning which follow once the element costs have been calculated.

Inflation

An order of cost estimate is prepared using current prices at the time of its preparation. Inflation during both the design period and the construction phase will need to be considered. Inflation may be considered under two headings: tender inflation and construction inflation. It is important when using existing cost data for order of cost estimating to update this to current prices by use of indices.

Potential cost increases may be caused by tendering conditions and the effects of changes in the market. Price changes may also apply to particular materials or specialist resources. The impact of major projects in a region will need to be allowed for, since these may have the effect of sapping resources. Some of these aspects, rather than being inflationary, may also be better dealt with under the heading of risk (above).

Value-added tax (VAT)

VAT in relation to buildings remains a complex area. It is recommended that VAT is excluded from order of cost estimates. Specialist advice might need to be sought to ensure that the correct rates are applied under the correct circumstances (see p. 232).

Other considerations will include:

- Capital allowances for taxation purposes
- Land reclamation relief
- Grants

Taxation allowances, taxation relief and grants can provide valuable financial aid to an employer on building projects of particular types constructed in specific locations (see Chapter 11). This area is very complex and specialist advice should be sought from accountants or quantity surveyors who have acquired expertise in these areas.

Reporting order of cost estimates

Building cost are normally expressed as costs per m^2 GIFA or per functional unit. Other methods can be used although these remain the most popular and most used (see Chapter 14). They should cover all of the cost items described above. Items that are included in or excluded from order of cost estimates should be clearly stated to avoid any confusion between quantity surveyor and employer from arising. The following should be stated:

- Project title
- Project description
- Statement of cost and cost limit if appropriate
- Information and specification on which the cost plan was prepared
- Floor area of the project
- Assumptions used in the cost estimates
- Estimate base date
- Estimated costs of alternative proposals
- Inclusions or exclusions

8.6 PART 3: MEASUREMENT RULES FOR ELEMENTAL COST PLANNING

Part 3 of the NRM describes the purpose and content of elemental cost plans and explains the rules of measurement for the preparation of formal cost plans.

Elemental cost plans are produced as an intrinsic part of both the RIBA Work Stages and the OGC Gateway Process, as follows.

8.6.1 RIBA Work Stage

Stage C: Concept

This stage consists of the following elements:

- Implementation of design brief and preparation of additional data
- Preparation of concept design including outline proposals for structural and building services systems
- Outline specifications
- Formal cost plan (1)

Stage D: Design development

This stage consists of the following elements:

- Development of concept design to include structural and building services systems
- Updated outline specifications
- Cost plan (2)

Stage E: Technical design

This stage consists of the following elements:

- Preparation of technical design and specifications
- Co-ordination of components and elements of the project for statutory standards and safety
- Cost plan (3)

Stage F: Production information

This stage consists of the following elements:

- Preparation of production information in sufficient detail for tendering purposes
- Application of statutory approvals
- Post-tender estimate

The main purpose of elemental cost planning is to:

- Ensure that employers are provided with value for money
- Make employers and designers aware of the cost consequences of their desires and proposals
- Provide advice to designers that enable them to arrive at practical design solutions within budget
- Keep expenditure within the cost approved by the employer
- Provide robust cost information upon which the employer can make informed decisions

Cost planning is a budget distribution technique that is implemented during the design stages of a building project. It involves a critical breakdown of the cost limit into cost targets for each element. The cost plan results in a statement of how the project team proposes to distribute the available budget amongst the elements of the building. It provides a frame of reference from which to develop the design and maintain cost control. It is an iterative process, which is performed in steps of increasing detail as more design and other information becomes available. The cost targets within each of the cost plans approved by the employer are used as a baseline for future cost comparisons. Each subsequent cost plan requires reconciliation with its predecessor together with explanations of why changes in costs have occurred. In some cases this will be solely because of a re-examination of the various risk items.

Further information about costs analysis and cost planning can be found in Chapters 9 and 15.

8.7 PART 4: TABULATED RULES OF MEASUREMENT FOR ELEMENTAL COST PLANNING

The rules of measurement for building works for group elements 1–9 are detailed in Part 4 of the NRM.

The degree of detail to be measured for building works is related to the cost significance of the elements of the particular design. Where there is sufficient information available, the cost-significant items are measured by the use of approximate quantities. Composite items are measured by combining or grouping together work items to make up common forms of measurement. Non-significant cost items, such as minor items and minor labour, are ignored in the measurements but are accounted for by increasing the unit rate appropriately.

The tabulated rules of measurement follow a common pattern. Each group element in turn is selected and each of these may comprise a number of sub-elements. For example, the group element 'substructure' has four elements as shown in Table 8.4.

Table 8.4   Group element 1: substructure

Element	Sub-element
1.1 Foundations	1. Standard foundations 2. Piled foundations 3. Underpinning
1.2 Basement excavation	1. Basement excavation
1.3 Basement retaining walls	1. Basement retaining walls 2. Embedded basement retaining walls
1.4 Ground floor construction	1. Ground floor slab/bed and suspended floor construction

The tabulated measurement rules are based on four principal levels:

■ **Level 1: Group element:** This is the primary classification for the grouping of elements, e.g. substructure
■ **Level 2: Element:** Key part of a group element, e.g. foundations
■ **Level 3: Sub-element:** Part of an element. One or more sub-elements constitute an element, e.g. standard foundations
■ **Level 4: Component:** A building work item which forms part of a sub-element. One or more components are measured to ascertain the cost of a sub-element

Each of the sub-elements includes the unit of measurement, the specified measurement rules and what is included or excluded in the different items.

Students should refer to a copy of the above rules of measurement in order to understand this chapter further.

SELF ASSESSMENT QUESTIONS

1. How important is it for standard methods of measurement to be prepared in collaboration with and with the agreement of contracting organisations?

2. Will the introduction of the NRM encourage more cost plan tendering?

3. How do the order of cost estimating and elemental cost planning procedures differ depending on whether they are implemented within the framework of the RIBA Plan of Work or the OGC Gateway Process?

BIBLIOGRAPHY

RICS, *RICS New Rules of Measurement: Order of cost estimating and elemental cost planning.* RICS 2009.
RIBA, *RIBA Outline Plan of Work 2007*, (amended in 2008) RIBA 2008.

CHAPTER 9

COST ANALYSIS

LEARNING OUTCOMES

After reading this chapter, you should have an understanding of the purpose and function of cost analysis. You should be able to:

- Understand the principles of the standard form of cost analysis
- Assess elemental costs and the significant contributing factors
- Prepare a cost analysis using data from a live project
- Identify variable factors from buildings with different numbers of storeys
- Identify the cost-significant elements from different types of project
- Define the terminology associated with cost analysis

9.1 INTRODUCTION

A cost analysis dealing with construction cost data has been referred to in Chapter 4. A typical example of a cost analysis of a type prepared by the BCIS is given in Table 4.4, and reference will be made to this where appropriate.

The purpose of preparing a cost analysis of a building project is to attempt to reveal the cost relationship between the various sections of a project, and also to allow for some comparability with other schemes. A meaningful conclusion cannot always be drawn from the study of a cost analysis unless full regard is paid to the nature, quality and quantity of work involved. The nature of the project in this context is the details of the scheme as outlined on the first page of Table 4.4. This is considered later. The elemental cost analysis was developed in order to provide data for design cost planning. It is therefore presented in the form that will best meet that need. The BCIS form is referred to as *The Standard Form of Cost Analysis* (SFCA).

It needs to be understood at the outset that a cost analysis is prepared on data retrieved from an accepted tender. It may appear to be more worthwhile to analyse the *actual* costs rather than those projected in a tender sum. The reasons for analysing tenders are as follows:

- Final accounts are often complex, and the time taken to analyse the variation account correctly would be a considerable disincentive to the compilers of the service.
- The preparation of final accounts is often delayed, for a variety of reasons, and this would make the published cost analysis very outdated. Historic data for cost planning purposes can quickly become of little value, particularly in times of high inflation, even though procedures exist for updating by use of indices.
- It is difficult to allocate increased costs and contractual claims accurately to the individual elements.

The originators of building cost analysis were the then Ministry of Education who were trying to make some sense of building costs in the school building programme of the late 1950s. They needed some technique whereby they could forecast and control the costs of building during the design stage. They published their findings in *Building Bulletin No. 4 – Cost Study*, which in the early 1960s was a very forward-looking document. It provided the first serious study undertaken on building costs. It was subsequently revised and published by the Department of Education and Science in 1972.

Prior to the cost analysis of elements in projects, the only comparison which could be carried out was to consider trade totals. This at first was often difficult, since no objective account was taken of the size of the projects under examination. Realising that floor area was of significant cost importance, it was then decided to apportion the trade totals on a floor area basis. The floor area was chosen since this was something the building costs of which clients could identify. Difficulties in comparison, however, still existed since the trade totals varied not in respect of whether partitions existed at all, but of whether they were constructed in brick or timber or even concrete. The steelworker trade on one project might show a large sum indicating a steel frame, but on another project it might show only a negligible amount for that trade. One was then left to ponder whether the second project had a frame at all, and if so, what it was made of. A further difficulty also existed: if the second building did have a concrete frame then how did this compare in cost terms with the steel frame on the first building? Also, in order to get the true cost of the steel frame it would be necessary to analyse the bill to find out the relevant costs for fire protection. The analysis of building costs on the basis of trades was therefore very unsatisfactory.

In the early days of cost analysis, the Ministry of Education did consider the possibility of expressing costs in percentage terms. This was not considered to be a reliable method, since percentages could fluctuate with no apparent change in the major parts of the project. Also, percentage values were not as useful as money terms.

The first *Standard Form of Cost Analysis* (SFCA) was published by the then Building Cost Advisory Service of the RICS. It found widespread support from quantity surveyors employed in private practice and the public sector who were concerned with the forecasting of costs for building projects. The current form of cost analysis has been in use for over 50 years with only minor modifications, although it is currently being revised. In an attempt to standardise terminology and data in the construction industry, a Co-ordinating Committee for Project Information (CCPI) was established in 1979 to identify the measures necessary to ensure the use of standard conventions for drawings, specifications and measurement. Any future version of the SFCA might therefore consider the relevance of such standard conventions to cost analysis documentation. Reference should also be made to the NRM (see Chapter 8).

9.2 DEFINITIONS

Cost analysis   The systematic breakdown of cost data, generally on the basis of an agreed elemental structure, to assist in the preparation of cost plans for future schemes.

Element   One of a number of parts of a building which serve the same function. The elemental breakdown generally adopted is that developed by the BCIS.

Cost centre   Items of cost importance identified within a building project.

Element cost (EC)   The total sum of money allocated to this part of a building, in accordance with the definition prepared by the BCIS of the element components.

Element cost per square metre GIFA (EM²)   This is the element cost divided by the gross internal floor area (GIFA). It provides the elemental cost contribution to the overall rate per square metre GIFA for the project.

Element unit quantity (EUQ)   The total quantity of the element expressed in units appropriate to the element concerned.

Element unit rate (EUR)   A rate which when multiplied by the element unit quantity will give the total cost of the element.

Element ratio   The proportional relationship between an element and the GIFA. It is sometimes termed the cost factor.

Cost equations   Mathematical expressions which describe cost. In the context of cost analyses the following are described as cost equations:

$$EC = EUQ \times EUR$$ (element cost = element unit quantity × element unit rate)

$$EM^2 = \frac{EC}{GIFA}$$ (element cost per m² = element cost ÷ gross internal floor area)

$$EM^2 = \frac{ECQ \times EUP}{GIFA}$$ (element cost per m² = element unit quantity × element unit rate ÷ gross internal floor area)

Quantity factors   An expression of the quantity of a component related to a square metre of GIFA.

9.3 PREPARATION OF A COST ANALYSIS

Cost analyses can be of two types: the detailed or amplified version already referred to in Table 4.4 and the brief or concise type in Table 4.5. The amplified cost analysis is more commonly used, and the following notes refer to it. The BCIS publishes this type of information frequently for its subscribers, and such an analysis will include the following:

- Information on the total project
- Summary of element costs with preliminaries shown separately and also apportioned among the elements
- A brief specification in element form
- Sketch plans and elevations for the project
- The amplified analysis

It is preferable if in the first instance the bill of quantities can be prepared on an elemental basis. This will make the preparation of the cost analysis very straightforward, even if an addendum to the design is necessary at the tender stage. Where the bill has been prepared on either a trade or an SMM section basis the cost analysis will take much longer to prepare. The elements have already been identified, so it is a case of allocating each bill item to the correct element. In some instances, such as excavation and earthworks and work below damp-proof course (dpc) level in concrete work and brickwork and blockwork, whole pages of items will be able to be allocated to the appropriate element. In other cases, such as the brickwork and blockwork section above dpc level, it may be necessary to separate individual bill items, perhaps between three or four different elements. The prime cost sums will also need to be allocated to their respective elements. Unusual and difficult items should be included with the most appropriate element. If the analysis is done correctly, the total cost should equal the total sum in the bill of quantities. One item omitted from both is contingencies, since it would be impossible to allocate this to any of the elements, and anyway it bears only a slight resemblance to the project design. In some instances it may be necessary to refer back to the dimension sheets, in order to allocate an item correctly to its elements. Once all the items have been correctly identified, the totals, subtotals and details are transferred to sheets in the form of amplified analysis.

It also needs to be stated that elemental bills are very unpopular with contractors and their estimators. In theory they prefer to see the work in a site operations context, since they claim that this helps them to price the work more realistically and correctly. In practice, since they are geared up to pricing traditional bills of quantities, they are not too keen when the surveyor diverges from this format. A bill called the master bill has been devised but only little used, which seems to suit everyone (not to be confused with the computerised bill system of this name).

It can be prepared in elemental format and then shuffled to a trade presentation and back again quite easily. Operational bills, although supposedly preferred by contractors, have hardly been used at all in practice. This type of bill does not make elemental analysis easy to carry out. It should also be noted that most of the manual cut and shuffle systems are element-coded so that an elemental cost analysis can easily be produced.

9.4 PRINCIPLES OF ANALYSIS

The basic principles for the analysis of the cost of building work are described in the SFCA. This is prepared in four parts: (1) principles of analysis; (2) instructions; (3) definitions; (4) guidance notes. Readers should refer to that publication, but the following points should also be noted.

- The analysis should bring out those features in different buildings which bear most heavily on cost.
- An element should be easily definable and capable of having the appropriate costs allocated against it with a minimum of effort on the part of the user.
- An element should be of cost significance and thus a cost centre for building projects.
- The qualitative aspects of the project should be expressed by reference to accompanying specification notes.
- Other information relevant to the costs in the analysis should be provided as background to these data.

9.4.1 Information on the total project

This information is generally provided at the beginning of the cost analysis. It contains vital information about the project which often cannot be quantified. For example, a negotiated tender is generally supposed to cost more than a tender awarded by selective competition. There is sufficient evidence available to support this view. The user of a cost analysis, however, would like to know how much more expensive a negotiated tender is, so that an appropriate adjustment can be made to costs and prices in the proposed cost plan. Commentators suggest about 5%, but this is inconclusive. The following information is required to be provided on the total project.

Job title Identifies the building type, since for future analytical purposes a project of a similar type will be required. Projects are usually grouped under their type for filing purposes.

Client Contractors' prices are often influenced by both the ability and willingness to pay and agree the final account.

Location   Required so that regional variation can be taken into account by use of appropriate indices.

Tender date   A date for indexing purposes so that the cost analysis can be updated for current use.

Project details   A brief description of the construction project, which should be read in conjunction with the specification notes.

Contract   Details of the contract conditions used and the contract period. A short contract period may indicate that a rapid construction time was required, and this will probably have been reflected in the contract prices.

Market conditions   Details of any political or economic conditions prevailing at the date of tender.

Contract particulars   Information on how the tender was procured, who stipulated the contract period, the number of tenders invited and returned and whether increased costs were included.

Competitive tender list   This will show whether the successful tender was in line with the other prices or whether it was underestimated. The list will also show whether these prices were from local or national companies.

Design/shape information   This section attempts to quantify the design in terms of cost significance. The information provided includes floor areas at different levels, the number of functional units in the building, the wall-to-floor ratio and storey heights, etc.

Brief cost information   The final section under this heading provides for any analysis of the contract sum, identifying the amounts of prime cost and provisional sums. It should be noted that the higher these sums are, the less real will be the cost analysis. This is due to the fact that such items are only an approximation to set against possible future expenditure. They are in effect quantity surveyors' amounts, and not based on a contractor's estimate of cost.

9.5 ELEMENTAL COST STUDIES

Cost studies of the various parts of a building project will need to be undertaken from time to time. It may be found that under a given set of conditions a method of construction is the economically correct choice in one case, but in other circumstances an alternative method of construction would be preferred. There are underlying principles of analysis, but the individual circumstances of a project will also need to be taken into account. The merits of a particular cost solution may also vary with time. For example, a study of trench-fill in foundations has shown that at certain chronological times it has been the best economic proposition. The differentials between the costs of labour, material and plant will vary and these should identify themselves in a cost study. Also, changes in technology and manufacturing, and modern methods of construction, will have an input to a cost study. It is therefore necessary to be aware of the underlying trends which do affect building costs and to ensure that these are correctly represented in a cost study.

The following examine the cost considerations of the various elements in a building. Further, more detailed studies are to be found in the bibliography at the end of this chapter.

9.6 PRELIMINARIES

The costs associated with these items may be presented in an analysis in one of two ways. They may be shown separately as an individual element or they may be apportioned and distributed among the other elements. The BCIS SFCA makes provision for showing them in both ways. The analysis may also wish to incorporate insurances with the preliminaries – these are often added as a final item to the summary page of the bill of quantities. Opinions differ on which is the best method of presentation for the preliminaries, remembering that the chief reason for the analysis is to allow for reuse of the information provided.

The sums calculated by the contractor for the preliminary items in a construction project may also be shown in various ways in the contract documents. The Standard Method of Measurement of Building Works and the Civil Engineering Standard Method of Measurement both provide details of the items of work which the contractor may need to provide. These will of course be identified separately for each project concerned. Although the typical number of pages in a bill of quantities for this section is about 25, it is unusual to find more than a handful of items (about 20) which have been priced. The remainder of the unpriced items are there for information only, or their appropriate costs have been allowed for elsewhere. There are generally three ways in which these costs may be shown:

1. The relevant items have each been priced independently, and may also show a breakdown of fixed and time-related charges. This is perhaps the best method of presentation, but this amount of detail is rarely encountered in practice.
2. The costs of preliminaries have been calculated and shown in the documents as a lump sum. The costs may have been analysed or they may have been assessed on the basis of a percentage calculation.
3. Preliminaries have been left apparently unpriced. In these circumstances their costs will doubtless have been added to the measured rates. In only a few isolated circumstances could a project be completed without the necessity for any preliminary items.

The problems for the cost analyst are aggravated by the fact that at the last minute the contractor may decide to adjust the tender sum by either adding to or reducing the amount of preliminaries. The reality of the sum included in the bills may therefore be questionable. This supports the argument of those who do not prefer to see the value of preliminaries shown separately in a cost analysis.

The true amounts allocated to the preliminaries element will be influenced by the following. These should therefore be identified in the analyses being examined, and in the new project to which they may be applied.

Location   This will identify whether the project is in the countryside or a town centre and the expected difficulties associated with this, such as travelling and subsistence payments, access to and egress from the site and buildings, distance from road networks and the necessity for temporary roads.

Space on site   Required for storage of materials, plant and temporary accommodation. Considerations for working in confined areas or on a project being occupied throughout the contract period.

Security   The necessity of temporary fencing, hoardings, gantries, public safety and protection from vandalism and pilfering.

Contract period   A large number of items included within this section are assessed on the basis of a time analysis in conjunction with the contract period. Contracts which overrun often involve the preparation of a contract claim based on a time analysis of the preliminary items. A short contract period may necessitate overtime and weekend working. Long contract periods often require a provision for increased costs, which would otherwise be included within the contract sum.

Construction method   Preliminary costs can be greatly affected by the choice of construction method used by the contractor. Plant-orientated construction or the use of innovative techniques often have special costs allocated within the preliminaries bill.

The value of preliminaries often represents approximately 8%–15% of the contract sum. The actual amount will be influenced by type, size and length of the contract period, and this section should be priced to reflect the varying costs on site associated with the particular project concerned. In addition the contractor will need to determine his method of working, such as the use of tower cranes and the amount of prefabrication off-site, etc. Some consideration should also be given to the method of tendering procedure adopted, since this can have an overall effect on the rates used in the bill of quantities.

The value of preliminaries to be inserted into a cost plan is therefore a somewhat hit-and-miss affair. Not knowing who the contractor will be (seldom known at the cost planning stage) or what methods are proposed for construction (generally left to be determined by the contractor) leaves the client's cost adviser in a bit of a quandary. This situation could be considerably improved if the designer were also the constructor. Given the present situation of design by one party and construction by another, and that this is unlikely to change considerably in the immediate future, great care needs to be exercised when examining preliminary costs in cost analysis.

9.7 SUBSTRUCTURE

This element and the allocated costs have to include all the factors necessary to strengthen the ground prior to the erection of the structure. This may involve nothing more than a straightforward strip foundation. It may, however, necessitate extensive cut and fill operations, a system of piling, a method of dewatering, a form of vibrocompaction, the use of explosives for breaking up hard rock or the difficulties encountered in digging through running sand. The costs of providing a basement may also be included, although the costs of the enclosing walls to a basement are included under external walls.

The type and size of foundations required for a building are influenced by:

- The type of construction to be used for the superstructure
- The bearing capacity of the ground
- Subsequent usage and loads applied to the structure
- Proximity of adjacent buildings
- Method of construction such as the type of plant to be used

Detailed information on the subsoil conditions is a prerequisite to choosing the correct and most economical type of foundation. Small buildings may justify only a visual examination of the site conditions and the digging of trial holes. Large projects on awkward sites will always necessitate the costs of a proper site investigation. The investigation costs are generally well spent, both by providing peace of mind and by avoiding the selection of an inappropriate foundation type. The costs of a thorough site investigation may be carried out prior to a contractor being appointed and the costs included and borne by the consultant's fees.

A two- to three-storey building will generally require only a traditional type of foundation to a minimum depth. In bad ground or in areas of shrinkable clay, short-bored piling can be competitive with the traditional foundation type. In good virgin ground the trench-fill method can also be an economical option (see p. 174). This particularly depends on the relationship between materials, plant and labour cost. Where it is used, special care needs to be exercised to avoid over-digging, which then begins to make it less economical, since backfilling with concrete will then be required. Raft foundations are introduced for lightly loaded structures where settlement may occur. Piling and ground beams are the most expensive choice for foundations and are therefore used only as a last resort either in bad ground or where heavy loads will need to be transmitted. Basements or a form of split-level construction are used on sloping sites in order to reduce expensive filling. It is generally more economical to excavate and cart away than to provide costly hardcore filling.

An optimum balance, of course, needs to be struck to minimise costs. Basements are also sometimes provided to tall buildings to reduce the pressure on the subsoil below the foundations, or to act as a counterbalance to the superstructure.

The cost implications of foundations usually involve an assessment of the following:

■ The alternative costs of different foundations where a choice in the method of construction is available
■ The adjustment of formation levels providing for a correct economic balance between cut and fill
■ The provision of a basement, or a split-level design, on steeply sloping sites

Foundation costs in the cost analysis will generally be expressed in terms of cost per square metre of the GIFA. This will vary with the shape (Figure 9.1) and size (Figure 9.2) of the building, and also with the number of storeys. For cost analysis purposes, it is preferable to separate the costs of the lineal items for walls and the superficial items for slabs. Where there are a number of column bases or piles these should have their costs separately calculated.

Assuming that a similar method of construction is to be used for the foundation construction in both of the buildings in Figure 9.1, their approximate costs can be calculated as follows:

	Building X			*Building Y*	
Ground slab 500 m² @ £80	=	40 000	500 m² @ £80	=	40 000
Foundation 90 m @ £170	=	15 300	120 m @ £170	=	20 400
	=	£55 300		=	£60 400
Costs per m² GIFA	=	£110.60		=	£120.80

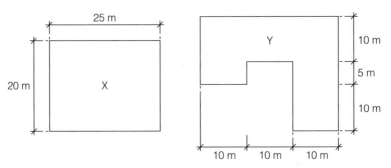

Fig. 9.1   Influence of shape on foundation costs

Fig. 9.2   Influence of size on foundation costs

The principle of the wall-to-floor area ratio holds true, with building Y being about 9% more expensive in foundation costs alone. It must be remembered that the costs per square metre GIFA are often the only indicator in the cost analysis. It cannot therefore be overstressed that only similar buildings should be used for analysis purposes, and where it is necessary to use widely differing designs then full account must be taken of the differences.

The building size in Figure 9.2 has been increased by a factor of 10 on external dimensions. It should be noted that this results in the GIFA increasing by a factor of 100. The same method of foundation construction is used in building Z as in building X (Figure 9.1). Thus, for building Z,

Ground slab 50 000 m² at £80	=	4 000 000
Foundation 900 m at £170	=	153 000
		£4 153 000
Costs per m² GIFA	=	£83.06

This emphasises the second principle of building economic theory, that the larger the building the lower will be the costs per square metre of GIFA. This does not make any allowance for the fact that building Z may well be constructed at more advantageous rates, since larger projects do not take a proportionately longer time to construct. The cost analysis for this project shows that the cost per square metre GIFA for this element is 25% less expensive than for building X.

9.7.1 Storeys

Because the costs in a cost analysis can be expressed per square metre GIFA, the greater the area of upper floors the lower these costs will be. If building X (Figure 9.1) represented single-storey construction, then the substructure cost per square metre GIFA would be £110.60. However, if building X were two-storey the cost per square metre GIFA would represent half of this amount (£55.30), and for three storeys it would be one-third (£36.87). This is assuming of course that the total cost of the foundations does not change, and this is a point to be noted. The information can be tabulated as shown in Table 9.1. It will be noted that the costs decrease rapidly up to three storeys and then begin to level off. This is further illustrated in Figure 9.3.

Table 9.1 Substructure cost factors

No. of storeys	Substructure costs at constant rate (£)	Cost ratio	Rate per m² GIFA, included in analysis (£)
1	110.60	1.00	110.60
2	110.60	0.50	55.30
3	110.60	0.33	36.87
4	110.60	0.25	27.85
5	110.60	0.20	22.12
6	110.60	0.17	18.80

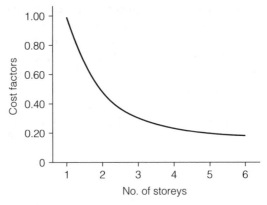

Fig. 9.3   Substructure cost factors

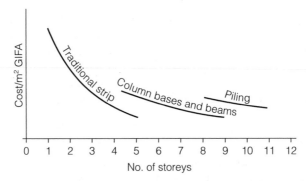

Fig. 9.4   Foundation costs versus number of storeys

Traditional strip foundations may be satisfactory up to four or five storeys, but beyond this it is likely that the method of construction will need to change to something more substantial. Reinforced concrete bases and ground beams may be appropriate up to nine or ten storeys, but beyond that there is little choice other than to select a type of piled foundation. Specialist piling companies must then be sought for their advice and relevant cost information. Figure 9.4 gives an indication of the way that costs per square metre GIFA may fare for the three most popular types of foundation in relation to the number of storeys in the building.

9.7.2 Trench-fill foundations

Figure 9.5 shows a traditional strip foundation and a trench-fill foundation. In some respects the trench-fill foundation is similar to a deep strip foundation except for the fact that the trench is deeper and requires more concrete. With the trench-fill foundation the ground is excavated and then filled with concrete. It is recommended that the concrete should be more than 500 mm deep and finish 150 mm below ground level. Where the ground is relatively solid then there will be no need to support the sides of the trench, since no-one will be working in it. One other

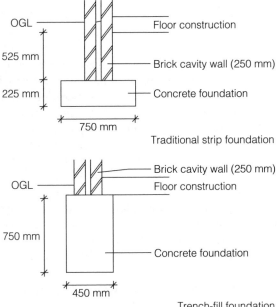

Fig. 9.5 Traditional versus trench-fill foundations

advantage is that trench-fill foundations are quicker to prepare than traditional strip foundations. This means that the project can progress more quickly. This provides for further cost reductions in terms of the contract programme.

The cost comparison, which can be used to compare any differences in construction methods or site operations, only includes those items that differ. In the example, it is only concerned with items below the formation level at the underside of floor construction. The actual quantities are based on a 10 m length of wall.

The simple cost study in Table 9.2 indicates that the trench-fill, using the above assumptions and rates, is about 50% less expensive than the comparable strip foundation.

Table 9.2 Comparison of costs of traditional and trench-fill foundations

	Rate	Traditional		Trench-fill	
		Quantity	Cost	Quantity	Cost
Excavate trench	£50	5.63 m³	281.25	3.38 m³	168.75
Backfill excavation	£20	2.65 m³	53.00	0.00 m³	0.00
Remove from site	£30	2.98 m³	89.25	3.38 m³	101.25
Earthwork support	£8	15.00 m²	120.00	0.00 m²	0.00
Concrete foundation	£125	1.73 m³	215.63	3.38 m³	421.88
Cavity wall construction	£110	5.30 m²	583.00	0.00 m²	0.00
			1 342.13		691.88
Per linear metre			£134.12		£69.19

9.8 SUPERSTRUCTURE

9.8.1 Frames

The frame of a building, where one is required, is an important element in a cost analysis. Frames are generally not required for simple domestic buildings, although they have been used in steel-framed housing developments, and some countries overseas use this a standard form of construction. Framed structures are invariably used, however, for single-storey warehousing, factory units and some farm buildings. They are extremely effective for shed-type structures. In these situations the choice will be a composite frame and roof structure (portal frame) of either prefabricated steel or precast concrete. Many of these structures are erected either on a speculative basis or through a design and build arrangement. They are invariably clad with less expensive non-structural cladding. The choice between steel or concrete depends almost solely on common practices and costs, since the construction time and overall performance are about the same. Evidence suggests that steel portal frames are generally more economic where fire resistance requirements are less critical.

On buildings above three storeys in height the provision of some form of structural frame becomes very much a necessity.

Until the seventeenth century, few buildings other than churches required any form of structural frame (e.g. internal columns). Buildings either had solid supporting walls or were of a form which integrated walls, floors and roof in one unified construction system. The advent of the industrial revolution led to the demand for large workshop buildings with maximum free floor space. The use of timber was restrictive for these designs and new forms of construction incorporating cast iron pillars were developed. Bage's Mill in Shrewsbury, Shropshire, built in 1796, is acknowledged as the world's first iron-framed building. The manufacturing processes at the time were dangerous, with a significant risk from fire in the timber structures. Concrete was developed in the late eighteenth century, with the French playing a leading role in the development of reinforced concrete (Joseph Monier, 1823–1906) and prestressed concrete (François Hennibique, 1843–1924). Although steel was available in the eighteenth century, it was not until the invention of the Bessemer process that it could be produced in large quantities at an economical price. The first multi-storey, wholly steel-framed building was built in Chicago in 1890: the Manhattan building. In the future there is likely to be a structural use for hi-tech materials such as carbon fibres.

Materials

The four main materials in use today are as follows, and their respective market shares are given in Table 9.3. Indicative costs of structural frames are given in Table 9.4.

Structural masonry   Generally suitable only for low-rise buildings. The last major multi-storey structure (16 storeys) was the Monadnock building in Chicago, which

Table 9.3   The market for structures in multi-storey buildings

	Market share for frame structures (%)		
Material	2000	1990	1980
Steel	65	51	33
In situ concrete	17	27	49
Precast concrete	14	17	14
Masonry	4	5	4
Total	100	100	100

Source: Construction Markets for Corus (formerly British Steel)

Table 9.4   Structural frame indicative costs

	Costs in £ per m² GIFA		
Materials	Normal range	Higher range including increased spans and storey heights	Extreme range
Precast concrete	100–150	140–190	190–240
In situ concrete	85–130	130–160	170–215
Steel	70–140	130–150	150–165

Source: Monk Dunstone Associates (1992)

had 1.5 m thick walls at ground floor level. This type of construction is no longer economically competitive.

Steel   High strength-to-weight ratio. Poor behaviour in fires. High tensile strength.

In situ concrete   Can be easily formed into complicated shapes. Has low tensile strength and requires reinforcement. Aspects of quality control on site are sometimes less than ideal. Loads cannot be applied immediately.

Precast concrete   Gives benefits of production and quality control.

Tables 9.5 and 9.6 illustrate the various advantages of steel frames over concrete frames.

Table 9.7 provides a cost study comparison between steel and concrete frames for two separate building briefs and using a range of different design solutions.

The necessity of having to provide a frame for a multi-storey structure will usually result in a choice between steel and *in situ* concrete. In terms of economics the choice between the two is not clear-cut, and depends on a number of specific project factors. *In situ* concrete generally has a small cost advantage (Table 9.4) but this is offset by a marginally shorter construction time for steel (Table 9.5).

Table 9.5   Structural frame indicative durations

| Material | Approximate periods (weeks) for structures with a GIFA of 10 000 m^2 | |
	Lead-in	Erection
Steel	8–14	20–40
In situ concrete	2–8	35–50
Precast concrete	8–16	25–45

The lead-in and erection times will partly overlap. Experience shows that the overall project times for steel- and concrete-framed buildings are very similar.
Source: Monk Dunstone Associates (1992)

Table 9.6   Value benefits of steel construction

Speed and predictability	Off-site prefabrication increases speed, quality and safety
	Improved predictability of cost and programme
Flexibility and adaptability	Long spans allow column-free areas
	Increased flexibility of floor layouts
Sustainability	Steel is 100% recyclable without any loss of quality
	Typically 86% of steel sections are recycled in the UK
	13% are reused in their existing form
	Recycling of bar reinforcement, by comparison, is negligible
Continuous development	Long history of development in the UK
	Efficiency improvements in manufacturing and fabrication
	Improved structural systems in areas such as fire engineering, durability, vibration control and acoustic performance
Technical support	Through Corus and the British Constructional Steelwork Association (BCSA)

Source: Corus and British Constructional Steelwork Association

Table 9.7   Steel versus concrete cost comparison

Building A is based on a developer's specification for a 2 600 m^2 office in Manchester

	Composite steel beam and slab	Slimdeck (steel shallow floor)	Long span
Frame and floor cost ($£$/m^2)	71	90	91
Frame construction time (weeks)	7	6	6
Overall building cost ($£$/m^2)	916	925	947
Overall construction time	42	40	41

Table 9.7 *(cont'd)*

	Reinforced concrete flat slab	*In situ* concrete frame with precast floors
Frame and floor cost ($£/m^2$)	118	101
Frame construction time (weeks)	8	8
Overall building cost ($£/m^2$)	979	962
Overall construction time	43	43

Building B is a prestige office of 18 000 m^2 in Central London

	Composite steel beam and slab	Slimdeck (steel shallow floor)	Long span
Frame and floor cost ($£/m^2$)	83	100	106
Frame construction time (weeks)	13	13	13
Overall building cost ($£/m^2$)	1 530	1 525	1 543
Overall construction time	67	67	66

	Reinforced concrete flat slab	*In situ* concrete frame with precast floors
Frame and floor cost ($£/m^2$)	144	170
Frame construction time (weeks)	18	19
Overall building cost ($£/m^2$)	1 636	1 638
Overall construction time	76	77

Source: Corus and British Constructional Steelwork Association

Figures 9.6 and 9.7 illustrate the continued growth in the use of steel frames over concrete frames in multi-storey buildings. The reason for this is further emphasised in Figure 9.8 in terms of their comparative costs.

The costs of the frame element are influenced by:

- Size of project
- Height of project
- Loadings
- Column spacings
- Frame sizes

It may be worthwhile, for example, to increase the size and span of roof beams and the size of column members in order to reduce the number of column members. There are obviously limitations to this idea, but the use of prestressed concrete has extended the possibilities. In normal circumstances the most economical solution is likely to be achieved with column spacings of between 2 m and 5 m. The use of deep beams will increase the overall building height in order to maintain ceiling heights. The frame element will in turn have an important influence on the following elements:

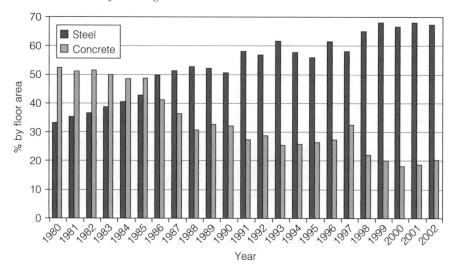

Fig. 9.6   Growth of steel in construction: steel versus concrete in multi-storey buildings
Source: Construction Markets for Corus (formerly British Steel)

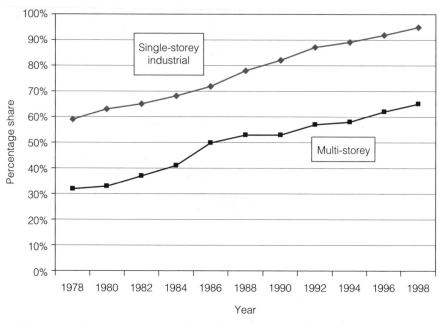

Fig. 9.7   Growth of steel in construction: UK share of steelwork in single- and multi-storey buildings
Source: Construction Markets for Corus (formerly British Steel)

- Substructure: number and size of bases and materials used for construction
- Upper floors: type of construction to be used and size of structural members
- Roof: similar considerations to the floors
- External walls: type of cladding to be used, and whether as load-bearing or infill

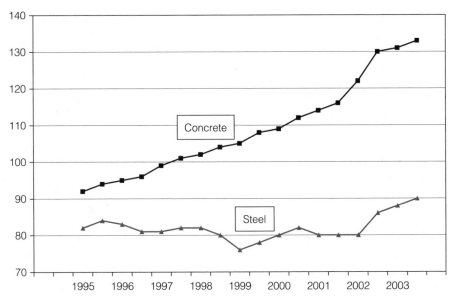

Fig 9.8 Comparison of steel and concrete frame and floor costs ($£/m^2$)
Source: Corus and British Constructional Steelwork Association

In addition, the method of construction of the frame can have an important
influence on the factors which affect the preliminary items' costs. Since a steel
frame can be used immediately after erection, this will offer cost savings against
concrete on some of the preliminary items. This reduction could be considerable
on large contracts, and the apparent cost and time savings are the reasons for the
shift towards steel-framed structures (Table 9.3). The smaller contractor who does
not have much expertise in concrete technology will also indicate a preference for
steel-frame construction.

The necessity of providing fire protection to structural steel members reduces
any cost advantage these may have over the already protected reinforced concrete.
The protection of the steel is often afforded by the *in situ* concrete casings, although
less expensive and lightweight, but adequate, protection has been provided using
plaster and asbestos-substitute boards. Structural steel frames generally offer a size
reduction compared with reinforced concrete of comparable strength.

Structural steelwork is more advantageous when the future adaptation of the
building is to be considered. Since reinforced concrete structures are monolithic in
design, the removal of one member may create stresses in other members that could
affect the overall structural stability of the building. Its relative ease of modification
by cutting, welding and bolting favours the use of structural steelwork in this context.
The architect Alex Gordon, in expounding his 'three Ls' concept of long life, loose fit
and low energy design presumed that all projects should be constructed on the basis
that, at some time in their life, adaptations on some scale would need to be carried out.

The location of the project may also influence the method of construction.
A congested city site, for example, will present its own problems throughout the
construction operations. It may not be possible, due to site limitations, to set up a

batching plant on site, necessitating the use of ready-mixed concrete at a higher cost. It may be impossible to use large cranes for steel erection purposes owing to site restrictions. The availability of raw materials locally or an adequate supply of steel erectors may tip the balance in favour of one method or the other.

During construction, greater care will be required with *in situ* frames in order to maintain structural stability, and it may therefore be necessary for the designer to provide for a higher factor of safety. The reduction in member sizes on each floor may not be cost-advantageous owing to the labour and material costs required to modify the formwork casings. The repetitive use of the same shape and size of forms can offset any possible cost savings which might be achieved through lower quantities of concreting materials.

In addition to the overall economics of the problem, it will be necessary to consider the aesthetics where these might vary, future adaptability, speed of construction and overall general suitability. The selection of a concrete or steel frame to a building needs to be considered in the light of conflicting information. The advice prepared by the different manufacturers and trade associations is biased in favour of their own products, is supported by illustrations or information or circumstances which support their case and, at times perhaps, may be economical with the truth. Care always needs to be taken in using such information. Rather, the general principles involved should be applied to each individual building project.

9.8.2 Upper floors

The upper floors can be an important element in the design of a building. The greater the number of storeys, the more important this element will be as far as cost-sensitivity is concerned. For example, assume that the upper floors to a building are prestressed precast concrete hollow units, 150 mm thick and costing £75.00 per m². Then the cost of this element per square metre GIFA in a two-storey building will be £37.50, and in a three-storey building £49.50. There is obviously no elemental cost in a single-storey building! These amounts are taken from Table 9.8. The cost factors are derived as follows. Assuming a building of four storeys high with plan dimensions of 10 × 10 m, the GIFA will equal 4 × 10 × 10 = 400 m². The total area of upper floors will equal 3 × 10 × 10 = 300 m². The cost factor is obtained by dividing the upper floor area by GIFA = 300/400 = 0.75.

The upper floors also provide an interesting and straightforward element for cost analysis purposes, and one worthy of study because of its cost importance within a

Table 9.8 Cost factors for upper floors

No. of storeys	Cost factor	×	Floor units per m²	=	Cost per m² GIFA
1	–		–		–
2	0.50		£75.00		£37.50
3	0.66		£75.00		£49.50
4	0.75		£75.00		£56.25
5	0.80		£75.00		£60.00

Table 9.9 Relative costs of floors

	Floor type	Relative costs
Timber:	25 mm softwood and joists	165
	22 mm chipboard and joists	115
Concrete:	150 mm reinforced concrete	200
	125 mm precast solid units	110
	150 mm precast solid units	120
	125 mm prestressed hollow units	110
	150 mm prestressed hollow units	120

building. The availability of different types of construction is considerable. The structural part of the floor may be timber, concrete or steel and these materials can also be used for the floor covering. The type of building and its use may somewhat limit the type of construction that might be provided. Timber floors, for example, are uncommon in today's commercial and industrial buildings. In addition it is necessary to consider the following:

■ Type of structure
■ Size of project
■ Erection time
■ Loading – live, dead and superimposed loads
■ Span
■ Fire protection
■ Sound insulation
■ Type of finish

Table 9.9 is a cost index of the more usual types of upper-floor construction. This index will of course vary depending on the span required. A concrete floor construction will become the only economical choice beyond a maximum span of around 5 m. Prestressed units become essential on larger spans. It should also be noted from the above that all the concrete floors will require to be finished off with a screed of cement and sand prior to the laying of the actual floor finish. This will add approximately 25 points to the index.

In domestic construction it is generally accepted that chipboard flooring provides the most economical initial solution compared with the other types of flooring available. Unlike on a roof, the possibility of this becoming saturated and thus deteriorating is remote, other than in bathrooms and kitchens. On a high-quality project, however, we would expect to find the traditional tongued and grooved boards, since it has more recently become fashionable to leave these exposed. Precast concrete hollow units have provided some competition, and although they are marginally more expensive than chipboard they are less expensive than softwood boarding. They also provide for speed in erection, and on some large housing projects have been found to be economically competitive with timber construction. Table 9.7 represents the element costs only, and it has already been suggested that the costs of other elements will be influenced indirectly. In addition, any reduction in the time required for construction will show some savings in the preliminary costs. The introduction of cross-beams will reduce the floor thickness and hence the overall element costs of upper floors. On balance, therefore, under the right conditions precast hollow units may be the correct choice for the construction of upper floors. A further point must be borne in mind. When the plan shape is irregular this tends to have an adverse effect on the costs of precast concrete units. It is also relevant for a cost guide to be prepared showing the effects caused by changes in spans and floor loadings.

9.8.3 Roofs

The cost-sensitivity of an element in a building depends on its total cost compared with the total cost of the building. Its quantity relative to the gross floor area is therefore significant. Its quality and performance are cost-sensitive only where the quantity factor of the element is high, and the level of this sensitivity depends on the combined costs of other elements. For an element to be cost-sensitive, any substantial change in its cost must significantly affect the total initial building cost.

Roofs are generally considered to be an element worthy of a cost study because of their relatively high cost. However, the importance of this cost diminishes as the number of storeys increases. In a single-storey building, for example, the element cost would be very important since the ratio of its quantity factor and the gross internal floor area is about 1 : 1. It needs to be remembered that when comparing analyses for different projects, the quantity factor of this element is the flat plan area and not the area of the roof slope. In the case of a twenty-storey structure the ratio of roof unit quantity to GIFA would be about 1 : 20, and many other elements would be examined as a cost/value priority in preference to this element. These ratios are similar to those of the substructure element outlined in Table 9.1.

Roof costs are generally analysed under the following headings:

- Roof structure, which covers the construction
- Roof coverings, which include the finishings and flashings
- Roof drainage, which includes gutters and downpipes
- Roof lights and associated work

The alternative materials and methods of construction are many and varied. There is never a single correct solution, with different options often fulfilling the client's and designer's needs. Each building must be considered on its own merits. Poorly constructed roofs can be troublesome throughout their life, and while designs need not be extravagant, this element is not one on which to cut corners. Also, it can produce an overall good or poor effect on the architectural aesthetics generally. The plan shape of the building will have some effect on the configuration of the roof lines, particularly where the roof eaves generally adopt the same outline. The prime objective in designing a roof is to keep out the rain and other inclement aspects. The following factors are important:

- Insulation
- Life expectancy of building and roof

Table 9.10 Relative roof costs by building type, based on costs (£) per m² of roof area

Building type	Typical cost	Cost range
Advanced factories	65	60–120
Warehouses	75	40–100
Factories	75	50–100
Housing	85	70–120
Old people's homes	110	90–140
Flats	125	80–160
Sheltered housing	125	100–140
Primary schools	140	100–180
Health centres	145	100–180
Offices	180	80–220
Hospitals	180	130–240
Churches	275	220–320

- Thermal gain
- Maintenance (ease and costs)
- Initial and recurring costs
- Aesthetics
- Wind and snow loadings
- Purpose of building (see Table 9.10) – for example, the roof may be used for car parking
- Type, shape and size of building
- Clear span requirements
- Classification of roof – pitched, flat, etc.
- Type of construction used in the building's structure
- Materials used for coverings
- Restrictions imposed by planning or building regulations

Table 9.10 provides an indication of roof element unit costs that are related to the type of project. The mean and range of possible costs are given as guidelines. It should be noted that the average figure generally falls below the centre of the range, showing that the prices for building elements often follow a skewed distribution rather than that of the median. There is a lower limit below which it is almost physically impossible to build. The upper limit will always reflect what a client is prepared to pay. For aesthetic reasons some clients are prepared to spend much more than Table 9.10 indicates. While care should be taken with all forms of cost information, because buildings used for the same purpose often employ similar construction methods the building costs given in this table may be of some relevance.

Pitched roofs

The natural shape for a roof in wetter climates, such as in the UK, is that of a pitched roof. The traditional form of pitched roof construction has changed from one that required load-bearing internal wall construction and trussed rafters which were manufactured on site to off-site manufactured low-rise trusses (see Figure 5.8). This has resulted in the use of slender and lighter structural sections, achieving economy in design throughout the structure. This type of construction, which generally incorporates a low pitched roof covering, has reduced the ability to provide future loft conversions. Refinement in the design of domestic roofs has provided considerable cost savings on a national basis (see Chapter 19). Further cost improvements are possible, particularly in respect of more inexpensive roof coverings, methods of fixing and overall longevity. Other cost savings have been achieved through design and the elimination of hips and valleys and detailing at the eaves. Although most of the less expensive domestic housing is now designed with gables rather than hipped ends, the latter are often more attractive, less expensive and less troublesome in the longer term. On lower-cost structures, roofs have now become very utilitarian, with few design features.

The additional costs of hipped ends include the hip rafters (the equivalent of one extra rafter), hip tiles, eaves course, eaves boarding and gutters. The savings are achieved by having fewer ridge tiles and verge tiles, less verge boardings and no gable end. It is in this last item that the bulk of the cost saving is achieved. One must presume that the time factor difference is the main reason for constructing dwellings with gables rather than hipped ends.

The range of coverings available for pitched roofs is considerable, varying from natural to man-made materials. Some of the more common types are given in Table 9.11. This index provides for an initial cost comparison alone, and it needs to be stressed particularly in respect of roofing materials that the long-term whole-life costs should always be borne in mind. However, even on this basis, it is doubtful that natural slating would ever now be selected. Factors other than economics must also be considered, such as appearance, durability and local planning regulations.

The type of roof covering selected will also determine the pitch of the roof and will affect the costs of the roof structure. Low-pitched roofs are preferred on account of this. These roofs also require a smaller quantity of roof coverings for the same plan shape. In addition, the weight of the coverings will affect the structural nature of the roof and the remainder of the building. In some cases it may be possible to increase the distance between the rafters, and although this may

Table 9.11   Relative costs for pitched roof coverings

Covering	Relative costs
Natural slating	290
Preformed cement slating	155
Plain tiling (machine-made)	195
Plain tiling (handmade)	320
Concrete plain tiling	220
Concrete interlocking tiling	110
Galvanised steel sheeting	210

require larger tiling battens, it may be cost-effective overall. Low-pitched roofs require more consideration in respect of possible snow loading, and this may have the effect of increasing the sizes of the structural members. Insulation should be placed immediately above the ceiling joists to give maximum benefits to the users. The thickness of the insulation must comply with legal minimum requirements, but there are long-term advantages of exceeding these requirements to reduce the expenditure on heating fuels.

The main alternatives to timber as a structural component are either steel trusses or a steel or concrete portal frame. These are widely used on industrial premises where large uninterrupted spans are required. Similar structures are used competitively for warehouse and farm buildings. They are not generally economically advantageous for domestic dwellings. The shape of the roofs on industrial premises can vary from mono-pitch to a north light design. The coverings in either case are often a mixture of glass and some type of troughed sheeting. The cost differences between coloured corrugated steel sheeting and fibre cement roofing can be as much as 70% at the extremes, but are comparable at different quality levels.

Generally, in terms of ease of construction and trouble-free maintenance, parapet and valley gutters, perforations in roofs for dormers, and stacks should be avoided wherever possible.

Flat roofs

Flat roofs are an unpopular choice and are often selected on the basis of initial cost as the main criterion. They have a bad image among the public, are aesthetically uninteresting, are out of fashion and can be troublesome and expensive throughout their life. They require a careful design, good choice of materials, the use of good workmanship and proper supervision if these disadvantages are to be kept to a minimum. However, under these circumstances they need not automatically be ignored on a whole-life cost basis.

The coverings available for flat roofs include bituminous felts of different grades and types, which are the least expensive, followed by asphalt, lead, zinc, copper and other man-made materials. Some of the more expensive materials are more vulnerable to theft than to the weather, and although this is more a matter for insurance, companies are known to incorporate lead clauses into their policies. There has been considerable striving through applied research by manufacturers of flat-roof coverings to improve the quality and image of these products. As a result, various high-performance single-ply and built-up felt flat roofing specifications have evolved which attempt to reduce the risk of future early failure. The considerations of eaves details, gutters, flashings and work around roof lights have a minimal effect on overall cost, unless one is dealing with small roof areas. Specialist care by skilled craftsmen is essential if the best results are to be achieved. The lack of this, work carried out in unsatisfactory conditions and the poor reputation of some unaccredited roofing companies are disincentives to the selection of flat roofs. Some comparative costs are given in Table 9.12.

Table 9.12   Relative costs of flat roofs

	Relative costs
Roof structure	
Reinforced concrete 125 mm thick including reinforcement, formwork, insulating screed and vapour barrier	450
Prestressed concrete beams, insulating screed and vapour barrier	440
Hollow tile construction, insulating screed and vapour barrier	460
175 × 50 mm softwood joists, firring pieces, insulation, vapour barrier and woodwool slabs	300
Ditto with channel reinforced slabs	360
Ditto with 18 mm plywood	400
Ditto with 18 mm chipboard	290
Roof finishings	
Three layer bituminous felt	110–120
Sheet lead code 3–5	330–530
Sheet zinc 12–14 gauge	330–370
Sheet copper 0.55–0.70 mm	550–660
Asphalt	110–150

The costs of removing dilapidated coverings make felt marginally less expensive than asphalt. Table 9.12 indicates that the use of felt roofing on chipboard roof boarding on timber joists will be the most economical in terms of initial costs, followed by asphalt on *in situ* concrete. The latter may in addition require a more substantial structure to carry the deadweight of the roof. Using expensive copper sheeting could increase the overall costs of the roof element by a factor of almost two and a half. The relative insulation values of these materials must also be taken into account. The avoidance of interstitial condensation must be considered in the alternative designs for the roof. For ease of reference, these are known as cold roofs, warm roofs and inverted or upside-down roofs.

The troublesome image of flat roofs makes them unpopular. Both chipboard and woodwool have the disadvantage that once they become wet, their damage is irrecoverable, they lose their strength and need to be replaced. A minor leak in the roof covering, unnoticed or left unattended, can result in expensive repairs, often out of all proportion to the initial cost saving. In addition, there is inconvenience caused to the building's users due to the need to make repairs.

The main factors to be considered in connection with coverings for all types of roof are as follows:

- Weather protection
- Appearance – colour, texture, form
- Durability
- Initial and recurring costs (assessed through whole-life costing)
- Longevity (see Table 9.13)
- Ease of maintenance and repair

Table 9.13   Life expectancy of roofing materials

Covering	Life expectancy (years)	Notes
Slate	100	Corrosion more likely from fastenings than from delaminating of slates
Clay or concrete tiles	50–60	
Fibre cement slates	40	
Industrial sheeting	40	Some vulnerability to physical damage
Lead	100	
Other metals	30–70	
Asphalt	40	
Standard-grade felt	15–20	
High-performance felt	20–30	

The exclusion of water is the main purpose of any roof, and pitched roof coverings, in which an impermeable membrane is not essential for weather-tightness, have longer lives than most flat roof coverings.

Roof lights

These are sometimes required for the natural lighting of corridors, or in the centre of large square buildings where it is not possible to provide windows and the only other option is artificial lighting. Roof lights include dome lights, lantern lights and sky lights. They are relatively quite expensive, but even taking into account the costs associated with forming the opening they are not considered to be a worthwhile subject for a cost study unless there are large numbers on many different schemes. When compared with the overall roof cost they are an insignificant cost item.

Roof drainage and rainwater disposal

The necessity of removing rainwater requires the provision of gutters and downpipes. The gutter in a flat roof may be constructed as an integral part of the roof, and although this may be more expensive than providing separate gutters, it is less unsightly. Pitched roofs generally need some form of plumbing goods, or alternatively a parapet gutter could be formed. The costs of roof drainage are not a significant item within the overall roofs element. However, factors which should be considered are durability, ease of laying, self-coloured materials and the long-term availability of replacement items. The specifying of a system to British Standards should ensure the necessary performance requirements. Comparative indices are given in Table 9.14.

In addition, it is necessary to compare the costs of fittings, painting, the provision of pipe casings and the availability of the sizes calculated. The designer must consider four aspects: cost, appearance, function and performance. A simple

Table 9.14   Relative costs of rainwater goods

Material	Relative costs	
	Gutters (100 mm)	Downpipes (60 mm)
Cast iron	180	210
Aluminium	220	165
UPVC	110	110

comparison substituting different materials on a given plan layout can easily be undertaken and can also incorporate the whole-life costs. It may be preferable for the sake of appearance to paint these items, even though this will increase the initial and recurring costs, rather than to leave them in their natural colour, although fashions change in this respect. In terms of cost alone, PVC will always be preferred, although there is a wide range of quality in this material. This preference will apply both initially and to costs-in-use. Rainwater disposal comprises three components: collection, such as by gutters and roof outlets; distribution through the pipework; and disposal, which includes the connection to the drainage system. The latter will also affect the costs of the drainage element. The study of different arrangements on large and repetitive projects can show worthwhile reductions in overall costs of these elements.

9.8.4 Staircases

This element comprises three components: the structure, the finishings, and the balustrades and handrails. It represents a minor cost element in a building, even when its form may be elaborate. Building regulations exist to determine the rise, going and width, the angle of ascent and minimum headrooms. The number of staircases in a public building is carefully controlled and is determined in conjunction with the fire officer. Because they are largely of a functional nature, their structure costs are comparable with each other and the differences in the elemental analysis are therefore determined by the finishes which might be applied. Table 9.15 provides some indication of the appropriate costs.

9.8.5 External walls

This is a very important element as far as cost is concerned. Its initial cost will be influenced by the following:

- Type of construction used
- Plan shape of the building
- Building height
- Building size
- Type, size and number of openings

Table 9.15 Relative costs of staircases (per storey height)

Description	Relative costs
Concrete – granolithic finish	340
Concrete – PVC finish	385
Concrete – terrazzo finish	480
Softwood	110
Steel	550
Fire escape	700

These elements will affect the rate per square metre GIFA found in the cost analysis. A further important factor which needs consideration at this stage is scaffolding. Although the costs of this are normally assigned to the preliminaries section, it is the external walls element that will influence the type and quantity of scaffolding and the period of time for which it is required on site. The costs are calculated by multiplying the girth of the scaffolding required by the height. The costs of provision and removal, and maintenance and use, are analysed separately.

The cost of the external walls element, in addition to the costs of the external enclosing walls themselves, includes the costs for basement walls, chimney stacks up to eaves level, curtain walling, tanking, insulation and external finishes.

A variety of different types of construction can be used, from traditional cavity walling, infill panels, lightweight cladding to curtain walling. The costs of external facing bricks can vary enormously, depending upon the quality and manufacturing process as well as the volume of particular brick that is produced. Specials and handmade varieties are the most expensive. The thermal insulation characteristics of the wall must be taken into account in accordance with the current building regulations. The choice of external walling will have an influence on many other elements. For example, the use of a lightweight construction will enable a slimmer frame to be considered, which in turn will reduce foundation loading and substructure costs. Curtain walling will eliminate the necessity of some internal finishings, but because of its low thermal values it may result in higher sums being expended on the heating installation and the respective costs-in-use.

The use of traditional brick construction is likely to provide the most economic solution for this element. With a reasonable external facing brick and a load-bearing block inner skin this should not exceed £180 per m² of walling at 2009 prices. The incorporation of additional insulation into the cavity will add an extra few pounds to this rate. Even with the use of high-quality facing brick, the cost per square metre is unlikely to exceed about £225/m². The life expectancy of this method of construction should be sufficient for all the building's needs. Repointing, however, may be necessary every 30 years, although few buildings seem to require this. These costs are considerably less than those of anodised aluminium curtain

Table 9.16   Relative costs of external walling

Type	Relative costs
Cavity wall facings £225 per 1 000	110
Cavity wall facings £275 per 1 000	140
Shiplap cedar weatherboarding on 190 mm blockwork	135
Plastic weatherboarding on ditto	125
Plain Portland stone, ashlar and cavity wall	550
Steel curtain walling, single-glazed	365
Anodised aluminium curtain walling, double-glazed	525

walling at an initial cost of around £350 per m^2 if single-glazed and £400 where double glazing is used. Even the possible enormous saving on the costs of the structural frame, foundation and internal finishings will not make this a competitive option where construction economics are important. Fire-resistant structural double glazing with one toughened skin can cost in excess of £1,500/m^2. Security features push these costs even higher. Environmental and aesthetic considerations will also need to be taken into account. Although curtain walling may be virtually maintenance-free, it will need some cleaning from time to time. On multi-storey structures this will normally require the provision of cradles and the associated control gear. The approximate costs of other types of external walling are given in Table 9.16.

The costs of a minimum internal finish to all but the curtain walling will increase the cost index by about 20 points. It will be noted from the relative costs that stone cladding is likely to be the most expensive solution. More expensive masonry materials and the incorporation of enrichments in the design will tend to increase this sum still further.

Profiled steel or asbestos sheeting has been used to a large extent only on industrial premises and warehouse buildings. Designers and owners have considered these materials to be unsuitable for other types of property. Reinforced polyester panels have, however, been used successfully, but only in a limited way. Timber-framed construction, which is now used on some housing projects, appears to be a good idea if installed correctly, but may not always be suitable for the UK climate. It requires meticulous installation and supervision during construction.

The cost per square metre GIFA of this element is not a good indicator, unless we are comparing similarly sized and shaped buildings. For example, plans R and F (Figure 9.9) represent two widely differing designs, with an identical type of construction for the external walling. Building R is a single-storey structure with external walls 2.7 m high, and 30% of the walling is taken up by windows. Building F is a three-storey structure with a total height of external walls of 9 m, and only 20% of the area is windows. The rate for the external walling is identical in both buildings, at £120 per m^2.

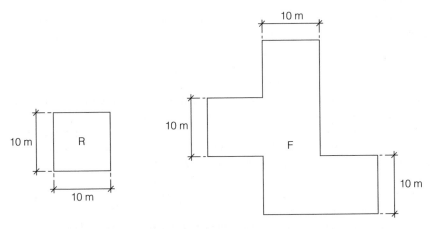

Fig. 9.9 Plans to illustrate external walling cost differentials

Building R

GIFA = 10 × 10 = 100 m²
Area of external walls = 10.00 × 4 × 2.70 × 70% = 76 m²
Cost of external walls = 76 m² × £120 = £9 120
Element cost per m² GIFA = £9 120 ÷ 100 = £91.20

Building F

GIFA = 10 × 10 × 5 × 3 = 1 500 m²
Area of external walls = 10 × 12 × 9 × 80% = 864 m²
Cost of external walls = 864 m² × £120 = £103 680
Element cost per m² GIFA = £103 680 ÷ 1 500 = £69.12

The rates per square metre multiplied by the GIFA and divided by the quantity factor will give our element unit rates.

9.8.6 Windows and external doors

This element, using the BCIS form of analysis, can be subdivided between windows and external doors. The two sections are so similar, however, that they may be considered together. The costs of this element must allow for:

■ Window/door/sidelight construction
■ Ironmongery
■ Glazing
■ Decoration
■ Treatment around the opening such as lintels, sills, cavity dpcs and plaster

The cost significance of this element will depend on the number, size and quality of the units concerned. High-quality windows used in prestigious buildings may initially be five times as expensive as the standard metal or wood windows used on mass housing schemes. This element is unlikely to be cost-sensitive, however, except in circumstances where the design is almost verging on curtain walling. General cost comparisons can be misleading, since there is an extensive range of types and qualities available. The following factors are worthy of note.

Materials used   A large variety of materials can be used for the framework, such as softwood, hardwood, steel, aluminium and plastics. Frames made of a good-quality hardwood, such as afrormosia, may be two and a half times as expensive as a similar frame made in softwood. The difference in cost between a softwood casement window and a standard metal window is only marginal.

Manufacture   An important factor to consider is whether to select a standard stock pattern unit or a purpose-made design. Often in the case of refurbishment work the architect has no alternative but to select the latter. Using a non-standard size of softwood window may result in an extra 50% being added to the catalogue price. The size of the section framing will also affect the price.

Performance   Windows and external doors described as high performance in respect of weather-proofing, ventilation requirements and manufacturing quality can be 35% more expensive than a typical dwelling-type window.

Size   Small windows have a higher cost per square metre than larger ones of a similar construction. This is due both to the costs of manufacturing the window and to forming the opening in the wall. Cost is much more correlated with window perimeter than window area.

Glazing   The type of glazing selected or required because of the size of the window or aperture in an external door can have an important influence on cost. Glazing in small panes might show a cost increase of 20% over the optimum size. Very large units in single panes are also expensive because of the type of glass required. Sheet glass is the least expensive, followed closely by cast and float glass. Georgian wired polished plate glass is almost three times as expensive as ordinary glazing-quality sheet glass. The introduction of double glazing increases the cost even more. The provision of special types of glazing for safety, security or thermal comfort are at the top end of the range.

Opening lights   This is one of the most important cost factors in window design, particularly in respect of high-performance windows. Opening lights show a considerable price difference compared with areas of fixed glazing. This is due to the extra costs of manufacturing the casement, angle jointing and ironmongery.

Window type   The selection of the window design, casement, sliding sash, pivot, etc., will also be a factor to consider. The style of the window, such as Georgian, or whether the window has curved members, resulting in curved glazing, will also need to be considered.

Fixing   The method of fixing and any special requirements regarding fixing will need to be included with the analysis. This may require a timber sub-frame for steel or aluminium windows. The BCIS rules for cost analysis require allowances to be made for the costs of lintels and the treatment of jambs and sills, including finishings, in order to make a realistic comparison. This is particularly the case where an elemental cost is required.

Ironmongery   There are enormous variations in the quality, type and hence cost of available ironmongery. The amount of ironmongery may also vary, particularly on external doors, and this can cause a wide variation in the total cost of the element.

Security   Security in openings in external walls is of vital importance. Inexpensive locks, particularly the rim type, are totally unsuitable as a deterrent to would-be thieves. A good-quality five-lever mortice deadlock and suitable bolts are becoming essential in an attempt to prevent wrongful entry into premises. In many circumstances it has now become necessary to fit intruder alarms, many which will now alert the owner to wrongful entry. Opening lights in windows must also be fitted with approved locking devices, which have now become standard.

Special requirements   In certain circumstances special considerations may be necessary when determining the costs of windows. For example, in multi-storey buildings the particular problems of cleaning, re-glazing, general maintenance and safety need to be taken into account. Cradles hung from the roof can deal with many of these factors. In offices, safety aspects have been taken care of by providing window ventilators and fixed lights. Alternatively, windows have been designed which can be reversed for cleaning and other purposes.

Costs-in-use   The cost of provision of double glazing and the elimination of draughts by efficient weather stripping can be partially offset by future savings resulting from reduced heat loss. Whole-life cost comparisons between various window types have not always been conclusive, particularly where new improved techniques of manufacture are constantly becoming available. The majority of steel, aluminium and plastic windows are self-finished and require no immediate treatment or future decoration. This provides real savings in respect of both future disturbance and cost, although how such windows might appear in 25 years' time is questioned. Timber windows have the disadvantage of requiring immediate decoration and the repeating of this process at regular intervals. They are also susceptible to rotting, and their replacement often involves costs not directly attributable to this factor. The remainder of costs-in-use associated with windows is attributable to panes being damaged either accidentally or through vandalism. These occurrences are, however, generally insurable.

Table 9.17 provides a general summary of the relative cost range expected for windows for different building types and varieties of specification. An important aspect to consider is the cost of non-standard windows, which can add as much as 35% to the overall supply price. The relationship between this element and those

Table 9.17   Relative costs of window (installed)

Description	Residential	Industrial	Retail	Offices	Hotels
Standard softwood					
Single-glazed	165–200	160–190	150–190	165–200	160–210
Double-glazed	200–130	185–230	195–230	200–240	200–240
Purpose-made softwood					
Single-glazed	200–240	200–240	200–230	205–245	205–255
Double-glazed	240–285	240–270	240–270	240–265	245–280
Steel					
Single-glazed	170–210	160–195	160–200	170–200	170–215
Double-glazed	205–240	195–235	195–235	200–240	210–250
Standard hardwood					
Single-glazed	210–275	195–235	195–235	200–240	220–250
Double-glazed	205–240	250–300	255–300	255–300	255–300
UPVC					
Double-glazed	180–450	330–420	350–420	350–420	350–420
Aluminium					
Single-glazed	–	195–230	195–230	195–230	195–240

concerned with heating and ventilation must not be overlooked. Reports suggest that environmental policies are likely to cause ever-increasing changes to the building regulations, with ventilation and U-values receiving greater attention. Due to the special nature of the product, quotations for the types of window to be used need to be obtained at the cost-planning stage of the project. The quotations can then be used to identify the high-cost aspects such as opening lights, size of panels and curved members. A performance specification would need to address U-values, special glazing, finishes to the frame, elevation proposals, maintenance requirements and delivery and site installation factors. About 6%–10% of the total contract sum is attributable to this element, with actual tenders differing by as much as 25%. Planned maintenance associated with this element includes cleaning, inspection, redecoration, repairs to ironmongery and general overall repairs. The breakdown of double-glazing units has become troublesome in recent years.

9.8.7 Internal walls and partitions

The costs of this element can vary immensely as there are so many different types of project, including open-plan office blocks and hotels which may be constructed on a cellular system. The type of project will often determine the need for this element, although what is required is susceptible to changes in fashion. Partitions, where they occur, will be affected by the plan shape and the storey height. A change in plan shape is likely to have a similar effect to that occurring in the external walls element, but it is more difficult to evaluate, since it depends on the building type and the spatial layout. The storey height will have a direct influence on the total

Table 9.18   Relative costs of internal walls and partitions

Description	Relative costs
Half-brick wall plastered both sides	230
One-brick wall plastered both sides	350
100 mm lightweight blocks plastered both sides	190
100 × 50 mm timber studding, plasterboard both sides	200
48 mm demountable partition including battens	110

costs of the internal walls and partitions. On large schemes even a small reduction in the storey height can produce considerable savings. The element total cost can be conveniently broken down into four sub-elements:

1. Structural walls
2. Non-load-bearing partitions
3. Screens and borrowed lights
4. Proprietary demountable partitioning

It is necessary during cost comparisons to ensure that an allowance for finishings is included in all the items since this may be an integral part of some of them. The comparison in Table 9.18 gives some indication of the variation in cost of some walls and partitions. Although the costs outlined are a good indicator of the cost-effectiveness of partitions, other factors also need to be considered to make the study more comprehensive.

The insulation qualities, for example, will need to be taken into account. An important factor in all types of building is the transmission of sound around the building and from room to room. The heavy solid partitioning used in old buildings provided the type of confidentiality that people grew to expect. The dry partitions used throughout the 1960s and 1970s, particularly in housing, served as little more than a dividing screen. They were introduced in the first place to provide speed in erection and for a lower price than the more traditional type of partition. They contributed to the elimination of wet trades on site, since the latter always have an adverse effect on on-site production and speed of completion. In practice the erection procedures for dry partitions could be a little finicky because of the necessity to have timber packing pieces in all sorts of positions. Care also had to be exercised to ensure that the panels did not become wet, and often it was found necessary to provide a skim coat of plaster to even up a wall. Timber studding can be filled internally with a suitable insulating quilt to reduce sound transmission. This does, however, have the adverse effect of increasing the cost, making timber studding an uneconomic choice when this noise insulation needs to be considered. Generally the thermal insulation qualities of this element, except in extreme situations, are of no real importance.

A further factor to consider is the control of the spread of flame, or fire protection; for example, timber studding would not usually be considered a satisfactory solution unless a double layer of plasterboard or a fire protection board was incorporated in the partition. Where this is an important consideration it is usually recommended that one of the solid masonry-type partitions be used.

A consideration that can affect cost is the influence that the choice of partition may have on other elements. Demountable partitions often arrive at the site, for example, with their own doors and frames complete. Because of the method of construction used these partitions often prove to be more costly than the traditional type of construction. The necessity of having doors in dry partitioning often requires some form of strengthening battens and fixing pieces for door frames. A major effect of using lightweight partitions is to reduce weight on the other parts of the structure. It is possible to use a less expensive and lighter upper floor construction, and this in turn will transmit a smaller load to the foundation below. Studies have shown that this can provide worthwhile savings in the overall costs.

It should also be noted that any simplification in the construction method alone can have a bearing on the reduction of the contract period and hence provide a saving in the preliminaries cost. A point worth noting here is that prefabrication of components should always result in a reduction in the time necessary on site. Overall cost savings, however, are likely to be negative, since the costs of prefabrication and the transport to site will often outweigh the site cost advantages. It does depend to some extent on the mass production facilities available.

Screens and borrowed lights do not have a substantial effect on total cost, and are thus not worthy of detailed cost study. Some of the comments applicable to windows are relevant here.

The cost-effectiveness of this element must be balanced with the following generally accepted criteria:

- Fire resistance
- Acoustic properties
- Weight
- Ease of construction
- Appearance
- Demountability
- Effects on other elements
- Durability
- Flexibility

9.8.8 Internal doors

The cost relationship of some of the more common types of door is shown in Table 9.19. On building projects such as hotels or mass housing projects the precise selection of a door type could perhaps save the contract a few thousand pounds. Even so, this is likely to be only a small amount compared with the contract sum, and therefore this element would not be considered to be particularly cost-sensitive. In an attempt to reduce the costs of these components, both prefabrication by way of door sets and bulk purchase agreements have been used by some employers who have large building programmes. The constituent items of the elemental analysis are the same as those for windows and external doors.

Table 9.19 Relative costs of doors

Door type (each 762 × 1 981 mm)	Relative cost
44 mm softwood, ledged and braced door with 19 mm matchboarding	195
Ditto in afrormosia	935
44 mm softwood, framed, ledged and braced door with 19 mm matchboarding	290
35 mm internal flush door with skeleton core, lipped two long edges, hardboard faced	110
Ditto plywood faced	145
Ditto with glazing panel	180
44 mm external flush door with skeleton core, lipped two long edges, plywood faced	225
44 mm internal half-hour fire-check door, lipped two long edges, plywood faced	260
33 mm softwood internal panelled door, type 2G	255

9.9 INTERNAL FINISHES

The basic design concept behind all forms of internal finishings is to disguise the structural elements to provide an aesthetic and functional appearance for the occupier. Most clients are now aware that the environment within a building dictates how occupants and visitors perceive the premises. This might relate to productivity in a factory, sales in retail shops or contentment in housing. Table 9.20 lists some of the criteria to be considered.

The emphasis placed on the internal finishes element in a building has increased during the past twenty years, but varies depending on the intended use of the premises and the individual needs of the client. Costs may vary from as little as £40 per m² GIFA up to in excess of £300 per m² GIFA, depending on the type of project (Table 9.21). Higher costs should also be expected for the more exorbitant finishes. They rarely account for less than 5% or more than 15% of the total project costs (Table 9.22). Expenditure on internal finishes is extremely sensitive to variables specific to a given project, especially the following:

- Overall quantity
- Ease of installation
- Extent of on-site cutting and detailing
- Degree of finish or component standardisation
- Installation rate

Residential developers are aware that purchasers' assessments are greatly influenced by their observation of the internal finishes, and as a consequence they spend about 10%–14% of their budget on these items. Industrial clients, conversely, are prepared to spend very little on this element, most of which goes on floor finishes. Their building motivation is largely directed towards function and profit,

Table 9.20   Internal finishes: principal considerations

Environmental	Performance	Procurement	Financial

1. Abrasion
    Foot traffic
    Wheeled traffic
	1. Aesthetic appeal
	2. Hygiene
	3. Fire rating
		1. Lead-in/delivery
		2. Installation
		3. Guarantees
			1. Initial costs
			2. Maintenance costs
			3. Capital tax
			    allowances

2. Distortion from
    loading
    Furniture
    Machinery
    Plant
	4. Vibration
	    transmission
	5. Sound transmission
	6. Impact resistance
	7. Frequency of repair
	    and maintenance
			4. Value for money

3. Soiling
    Foot-borne
    Spillage
    Air-borne
	8. Lifespan		

4. Climatic conditions
    Sunlight
    Humidity
    Pollutants			

Source: Dearle & Henderson (1992)

Table 9.21   Relative costs on finishes by building type ($£/m^2$ GIFA)

Building type	Floor	Wall	Ceiling	Total
Residential	30–50	30–60	20–50	80–160
Industrial	25–40	20–40	10–25	55–105
Commercial	40–80	30–60	30–60	100–200
Retail	45–90	40–65	45–70	130–225

Table 9.22   Proportional expenditure on internal finishes (%) by building type

Building type	Floor	Wall	Ceiling	Overall expenditure
Residential	30–40	40–45	20–30	10–14
Industrial	40–50	30–35	20–25	4–6
Commercial	40–50	15–25	25–40	10–14
Retail	35–45	25–30	35–45	7–9

Source: Dearle & Henderson (1992)

with an emphasis on trouble-free finishes which will not require too frequent
future temporary shutdowns of industrial operations for redecoration or repair.
Commercial functions are centred more on people than on plant, with an awareness
of providing more stimulating working conditions and projecting an image and
ambience to clients. There has been increased expenditure in recent years on raised
floors in commercial premises to allow for the easy installation of and access to
communication networks.

9.9.1 Floor finishes

A wide range of different sorts of floor finishes are available, some of which are
listed in Table 9.23, together with their advantages and disadvantages. Table 9.24
gives a simple whole-life cost to illustrate how floor finishes can be compared

Table 9.23 Characteristics of some floor finishes

Finish	Advantages	Disadvantages
Carpet	Speed of installation Aesthetic appearance Low acoustic transmission Immediate use	Lifespan Soiling characteristics
Vinyl, rubber, linoleum	Relatively cheap Speed of installation Soiling characteristics Lifespan Non-slip, non-conductive	Aesthetic appeal Difficult to clean
Hard surface and stone	Aesthetic appeal Lifespan Low maintenance	Expensive Extended lead-in period Rate of installation High acoustic transmission Setting period prior to use
Timber floors	Aesthetic appeal Material characteristics	Low impact resistance Additional surface treatment High maintenance Sound transmission
Liquid and seamless floors	Low cost Rapid installation Rapid drying time Durability Low maintenance Impact resistance	Aesthetic appearance Sound transmission
Raised floors	Accessibility Dry construction Enhanced flexibility	Expensive Noise generation (in cheaper systems) Potential fire hazard

Source: Dearle & Henderson (1992)

Table 9.24   Whole-life cost comparison of floor finishes (in £): quantity, 100 m²; life expectancy, 25 years; discount rate, 6%

Description	Option 1: fitted carpet	Option 2: vinyl tiles
1. Capital costs	2 000	1 300
2. Annual maintenance costs	2 434	7 943
3. Replacement costs		
Year 10	1 117	
Year 15		542
Year 20	622	
Net present value	6 173	9 785
Annual equivalent	483	765

Source: Dearle & Henderson (1992)

over the building's entire life. Chapters 17 and 18 explain in detail the methods used in whole-life costing calculations and the principles involved. Floor finishes costs include the costs of screeds, skirtings and the floor coverings themselves. The finishes to staircases are included with the stairs element (BCIS 2D). Some discrepancy between the GIFA and this element unit quantity occurs where items such as floorboarding are used and are included in the upper floors element in the cost analysis.

9.9.2 Wall finishes

The internal wall finishes include the appropriate preparatory work. The element unit quantity will generally be something less than the external wall area plus twice the internal wall area. The wide range of finishes available means that the basic cost analysis without a specification is of only limited use. The use of wet trades continues to predominate, particularly in the domestic and commercial sectors. The four common materials used are plaster, tiling, dry lining and demountable partitioning for internal walls. A decorative finish of paint or wall covering is applied to the base material.

9.9.3 Ceiling finishes

This element includes the full costs of suspended ceilings in addition to the more usual forms of directly applied ceiling finishings. Table 9.25 gives the comparative breakdown of suspended ceiling costs. These are based on quantities ranging between 1,000 m² and 2,000 m². Suspended ceilings are used in buildings to provide a high standard of ceiling finish, to conceal aspects of the structure and to allow for the unobtrusive distribution of the building services network. Demountability and accessibility are therefore important criteria when selecting a particular specification. The key performance requirements include dimensional requirements, finishes, structural classification, spread of fire requirements,

Table 9.25   Relative costs (%) of suspended ceilings components, 1993

Component	Low cost (up to £50/m²)	Medium cost (£50–£100/m²)	High cost (over £100/m²)
Tiles	40	25	30
Hangers and framework	40	25	20
Accessories	10	20	30
Edge trim	10	15	10
Fire barriers	–	15	10

Source: Cyril Sweet & Partners

acoustics, thermal insulation, vapour resistance, accessories, electrical requirements, durability, and supply and installation.

9.10 FITTINGS AND FURNISHINGS

This element includes the items which are to be provided under the terms of the building contract. It will therefore lean towards including the more permanent fixtures rather than loose furniture, which will be brought to the project on occupation by the promoter. The sorts of items included are cupboards, shelving, curtain tracks, pin-boards, etc. In addition, on some contracts there may be a requirement to provide loose furnishing materials, free-standing equipment and works of art such as sculptures. The cost guidelines are thus very broad and the element cost is likely to be related to the project type rather than any other factor. There is not a great deal to be saved, in either time or money, in attempting to refine a single fitting in an office block. However, where this fitting is repeated several times throughout the scheme, the relative cost may of course be important.

9.11 ENGINEERING SERVICES

The ever-increasing importance of engineering services in buildings has meant that greater care and attention need to be paid to their cost implications. On large projects, at least, the importance of these services has necessitated the role of a specialist quantity surveyor. One does not have to go far back in history to find dwelling houses that incorporated no engineering services whatsoever. Today they may represent 10%–15% of the initial capital cost and a majority slice of the costs-in-use. In some buildings, such as laboratories, the services cost content can be above 50% of the initial cost. A point that needs to be noted is that the greater the capital cost expenditure on services, the greater the costs-in-use of the property are likely to be. The costs of services often cannot be examined in isolation from each other, since they are often very much interrelated. It also needs to be emphasised that to cost plan only the 'building' work will result in a severe loss of the overall cost control function.

Cost studies may include a comparison of the alternative material costs, taking into account their life expectancy, their efficiency and sums required for their replacement. The costs of engineering services will vary directly as a result of changes in the building's morphology. Generally, the more complex the plan shape, the more costly will be these groups of elements. As with so many theories, there are exceptions to the rule. For example, large square buildings are not favoured in connection with lighting costs since there will be less reliance on the provision of natural lighting, thus increasing the costs of artificial lighting. It must also be remembered that, in many types of building, although there is a demand for cost-effectiveness this must be balanced and tempered with the need to make the best possible use of the internal layout of the accommodation. This will often be in opposition to the simple plan shape. Cost studies of this element will seek to achieve the most economic way of installing engineering services in a building in order to achieve a given standard of performance when in use.

9.11.1 Plumbing installation

It is possible to subdivide a study of this element into its various component parts such as sanitary fittings, waste disposal and hot and cold water services. Rainwater disposal is normally considered to be an integral part of the roof element. The significant variable under this heading is the number and type of sanitary appliances. The costs of the pipework and ancillary equipment, while not directly in proportion to these items, do show some general cost relationship. Although the quality of the fittings can vary considerably, the total costs of an installation are unlikely to vary by more than about 50%, as between high- and low-quality fittings.

An important and practical way of achieving cost savings is to consider properly the design layout of the installation within that building. In respect of the design of the building, regulations will often determine the minimum number of appliances to be provided in public buildings, offices, shops and factories. There has been a recent trend in housing to increase the number of fittings provided. The minimum has been to provide a three-piece bathroom suite, and a sink unit in the kitchen. In many modern dwellings the number of fittings has been increased to about sixteen, by the introduction of laundry rooms, additional bathrooms and the provision of more fittings generally. This trend has perhaps now reached its maximum. If possible the sanitary accommodation should be grouped closely together in a part of the building, and in the same plan positions on the various floors of a building. This will ensure reduced lengths of pipework and more efficient operating costs of the building in use. The choice of material for the distribution pipework will depend on the use and life expectancy of the building, but in every case attention should be paid to the whole-life cost factors.

The provision of hot and cold water supplies to the sanitary fittings can allocate costs to three items: first, the source items, which include bringing the supply onto the site, the storage tanks and appropriate overflows and valves – these usually account for about half of the costs of the hot and cold water installation; second, the pipework with its insulation and fittings; and third, the connections to the sanitary appliances, although these may represent only a relatively minor cost. The introduction of pumped systems necessary in high-rise buildings can almost double the source costs.

9.11.2 Heating

The consideration of this element can be logically subdivided into heat source and choice of fuel, distribution pipework and outlets or emitters. Cost studies of this element are intrinsically tied up with insulation costs and costs-in-use. Regarding choice of fuel, this is largely a contest between solid fuel, gas, oil and electricity. Although it is generally presumed that these are in an ascending order of cost, this remains unclear since the various energy suppliers have complex tariff systems and also some bartering of price does occur between the private suppliers. Fuel efficiency must also be taken into account when comparing these rates. Only gas and electricity can be considered to be fully automatic, since in the case of the other two fuels delivery to the building must be considered. Some solid fuel systems may also necessitate the use of an attendant at periodic times. The provision of district heating schemes or plants which burn refuse, although good in theory and possibly more economical in practice, is not favoured by the majority of users. Sectional cast-iron boilers for gas-fired systems are about 40% less expensive than comparable oil-fired boilers and about 80% cheaper than boilers for a solid fuel system. Distribution and emitters are comparable in every case. The gas-fired system is also able to provide capital cost savings due to the absence of any fuel storage requirement. The storage and feed requirement for solid fuel systems, and provision for the removal of ash, are further disadvantages of this alternative.

The comparative costs for central heating systems in a typical domestic house are given in Table 9.26.

The cost of heating depends largely on two factors:

1. The quantity of heat required, which depends chiefly on the architect's design skill
2. The cost per unit, which depends on the engineer's design skill

The amount of heat necessary to maintain comfortable living conditions is influenced by the shape and size of the building, the thermal transmittance of the structure, the orientation and degree of exposure and the amount of ventilation provided.

9.11.3 Ventilation

Mechanical ventilation systems are designed for three basic purposes:

1. To provide a continuous supply of clean air for breathing
2. To extract waste products from the air
3. To eliminate contaminants produced by particular manufacturing processes

Table 9.26   Capital costs of heating

	Gas	Oil	Solid fuel
60 000 BTU per hour boiler	1 025	1 400	1 500
Fuel storage	–	450	350
Pump	100	100	100
Hot water tank	200	200	200
12 No. radiators @ £150	1 800	1 800	1 800
Pipework 140 LM @ £9	1 260	1 260	1 260
Total	4 585	5 210	5 210

The typical systems comprise ducting and extractor fans, and these are designed to remove stale air from lavatories, kitchens, car parks, etc. They may in some instances work in conjunction with the heating system or be operated on the same circuit as an electric light. Kitchens, for example, may require 20–40 air changes per hour. The capital costs of such systems are not generally too high, although provision may need to be made for concealing the ducts within the building fabric.

9.11.4 Air conditioning

Air conditioning is an attempt to produce and maintain a certain atmospheric condition within a space automatically, irrespective of the conditions surrounding the space. Five factors need to be considered:

1. Temperature of the air
2. Humidity of the air
3. Density
4. Air movement and distribution
5. Control required

The above will influence the cost of the installation both initially and in use. In general, systems which have a ceiling distribution and extract will have the highest distribution cost. Perimeter systems are generally the cheapest, although they may be less efficient than their more expensive counterparts.

Air conditioning tends to be an expensive, some would say a luxury, element in a building and for this reason it is not commonplace in the majority of buildings. It is, however, necessary to provide city-centre offices with this attribute; otherwise experience has shown that difficulties may occur in letting or leasing the property. It is also expected in higher-class hotels, and essential in tropical climates for the average European. Since the process is one of controlling the condition of the air it is preferable to keep this volume to a minimum, and hence the square-shaped plan will provide the most economic solution. Long thin buildings make both the provision of air conditioning and its maintenance much more expensive.

Table 9.27 Relative costs of air conditioning (based on 10 000 m²)

System	Initial costs (£/m² GIFA)	Running costs (£/annum)
Two-pipe fan coil system with supply air and ventilation system	160–200	10 000
Four-pipe fan coil system with supply air and ventilation system	170–230	12 000
Constant air volume system	190–240	11 000
Variable refrigerant volume system with inverter and supply air and ventilation system	210–240	9 500
Variable air volume system	220–270	13 000

Source: Silk and Fraser

When a client asks for air conditioning in a building, this may not really be what is intended. Often what is perceived as air conditioning is in fact comfort cooling, which is a much simpler and cheaper option. Installation costs will vary from project to project, depending on the individual requirements. Table 9.27 indicates a comparative price range for a medium-sized office building.

'Sick building syndrome' causes complaints of malaise from the building's occupiers. It is often identified with the use of air conditioning, although little has been established about its specific causes. Areas identified as potential causes of the complaint are:

- Sealed airtight buildings
- Air conditioning installations
- Use of materials giving off irritating fumes or dust
- Flicker or glare from fluorescent lighting
- Energy conservation measures
- Lack of individual control over environmental conditions

The medical conditions associated with sick building syndrome are:

- Dryness of skin, eyes, throat and nose
- Allergic symptoms such as watery eyes or runny nose
- Asthmatic conditions such as a tight chest
- A general feeling of lethargy, headache or malaise

In order to reduce future problems, the following should be taken into account during the detailed design of engineering services:

- Allowance should be made for the adequate maintenance of all equipment, concealed device spaces and duct spaces
- A good supply of clean air should be provided
- Materials which emit toxic chemicals should be avoided
- Dedicated ventilation systems should be provided for smoking areas, avoiding the recirculation of smoking odours

9.11.5 Electrical installations

The costs of this element can be analysed in several ways. The most useful method is to consider the incoming mains, power, lighting and dedicated supplies for computers, lifts, alarms, etc. Costs can often be related to the number of sockets and switches which are provided. For example, lighting points cost about £120 and socket outlets £145. Electrical installations in domestic property composed of the basic PVC insulated and sheathed cables were costing about £55 per m² GIFA (2009), compared with heavy-duty conduit at about £75 per m² GIFA. The typical square metre rate for commercial buildings was £115. The cost of a full installation in a one-bedroom flat varied between £950 and £1,400, and that of a four-bedroom house between £1,550 and £3,250. The theoretical cost unit reduction which might be achieved on larger projects is often offset by the trend to provide a greater degree of sophistication and flexibility. Larger projects also often incorporate a wider variation of electrical equipment.

9.11.6 Lifts installation

The major factors influencing the costs of this element are the height, floor plan layout and quality of the project concerned. The costs of the element can be subdivided into shaft structure, motor room, control gear and of course the lift compartment. A passenger lift is usually provided in buildings of four storeys or more, while in eight-storey buildings it is common practice to provide for a second lift. The specialist contractor's quotation for the provision of his work will include the lift motor and hoisting gear, the car, dual safety doors, indicator panels and the necessary electrical work. The high-cost items are the car and the lift motor. Lifts to lower-rise buildings are thus more expensive per floor than lifts to multi-storey buildings. The typical cost of an electric lift with steel cage, collapsible gate and push-button control to service four floors is about £35,000. This cost will vary depending on the performance in terms of speed and load, the control arrangements and the quality of the specification. Increasing the size of the car capacity can increase the above cost by up to £6,000. The building work in terms of the pit, shaft and meter room may amount to around £5,000. For commercial developments in London with a minimum GIFA of 10,000 m², a medium-quality speculative lift installation will cost between £50 and £65 per m² GIFA, a high-quality speculative installation up to £80 per m² GIFA, and an owner-occupier installation £110 per m² GIFA. These rates will increase for smaller-sized developments and decrease for projects in the provinces. Typically, lift installations cost between 2% and 4% of the construction budget.

An approximate analysis of passenger lift costs is shown in Table 9.28.

A lift installation has to:

■ Provide adequate handling of people
■ Keep travel time to a reasonable minimum
■ Ensure that passengers are not kept waiting

Table 9.28 Passenger lift analysis

Motor control and equipment	35%
Cars and landings	35%
Controls	20%
Safety and shaft preparation	10%

The criteria for lift selection therefore include:

- The number of floors to be served
- The size of the building
- The shape of the building
- The composition of users

In premises where heavy concentrations of pedestrian traffic occur, an escalator may be the most convenient mechanical conveyance. This is the situation in large stores and below-ground railway transit systems. Indeed, in circumstances like these, where lifts only are provided, inconvenient waiting times may be expected. An escalator cost is dependant on the distance to travel, the amount of rise and the various appearance options. They range from £80,000 to £120,000.

9.11.7 Fire fighting

The range of alternatives under this heading is so wide as to make comparison extremely difficult. The majority of public buildings now provide at least some equipment, which is often required by statute. The Fire Officers Committee publications provide detailed information, and it is now generally necessary to consult a fire officer during the design of major building projects. Table 9.29 provides a general comparison of the alternatives available for a fire protection installation.

9.11.8 Communication installations

An increasing amount of expenditure on modern buildings is allocated to the various forms of communication systems. These include security control, closed

Table 9.29 Fire fighting cost index

	Index
Dry riser	110
Wet riser	280
Hose reels	155
Sprinklers	350
Fire alarms	165

circuit televisions, computers, telephones, data cabling, public address systems and master clocks. Invariably the largest amount of the initial cost is spent on the provision of the central item, e.g. the computer in a computer system. The provision of the above can also add considerable sums to costs-in-use. It is expected that in the near future more of the capital expenditure will be attributed to these items, and therefore while their cost significance may be relatively minor it will increase substantially. Some attention therefore needs to be given to the provision of ducting to accommodate such service supplies.

9.12 EXTERNAL WORKS AND DRAINAGE

This element, although expressed in terms of cost per square metre GIFA for conformity, often bears no relationship to the building size. Some private housing, for example, is situated in acres of grounds, whereas there are many city-centre office buildings which literally contain no items of cost significance as external works. This group of elements includes siteworks, drainage, external services and minor ancillary building work. The last two elements will represent only minor cost items, and what has already been said will apply to the last element. One way of reducing the costs of external services is to eliminate the provision of one or other supply facility. This may not be possible, and anyway it is likely that its elimination would have a knock-on effect on other elements and thereby negate such savings.

The costs of drainage can largely be summed up as the costs of pipework and inspection chambers. The biggest cost implication comes from the choice of a separate or a combined system. Inspection chambers should be kept to a minimum and constructed from the most cost-effective materials. Precast concrete rings are cheap and quick to install, but they may be less convenient for coping with branch pipes than other construction methods. The typical cost today of a brick manhole is about £1,000, and £1,500 for a similar-sized concrete manhole. Plastic inspection chambers, which are now much more common than in the past, cost about £300, compared with £700 for a brick or precast concrete chamber. Long drain runs cost less per linear metre than short branch connections. The latter therefore need to be kept to a minimum. A wide variety of materials is available for pipework. The pitchfibre pipe was at one time beginning to challenge the traditional vitrified clay pipe for long runs of drains, but was never competitive on the short branches. Overall vitrified clay still seems to have the edge on price, and becomes even more competitive in pipes with larger diameters. Socketed drainpipes are still marginally cheaper than flexible jointed pipe in all situations. However, for the extra few pence involved as against the ease and time saved, the latter have become more popular in recent years.

There is a huge variation in the materials available for siteworks and fencing. Soiling and grass seeding are the most economical method of dealing with large areas initially, but will require regular maintenance throughout the project's life.

The construction details of site paving are now being looked at much more closely by architects. They are beginning to move away from the more conventional concrete, precast concrete and tarmacadam coverings, to expensive types of brick paviors. It is a fairly straightforward process to advise on the cost implications of the choices made by the architect. The recommendations for fencing need to take into account the initial costs, durability, maintenance requirement and length of life.

9.13 WHOLE-BUILDING COST ANALYSES

The BCIS *Standard Form of Cost Analysis* shows a breakdown of the costs of a project distributed amongst a number of standard elements. Whilst this form has largely remained unchanged, a small number of modifications have been made and more may be required if it is to become consistent with NRM Cost Analysis (see Chapter 8). There is no real substitute for examining costs in less detail than the elemental model, since the guide prices per square metre of floor area can show a considerable range from the lowest to the highest price. Much skill and understanding are required to select appropriate rates.

Building magazine in collaboration with Davis Langdon provides a regular feature examining the cost models for a wide variety of different kinds of buildings. The following information is based on an analysis of large retail developments.

9.13.1 Retail developments

The UK has about 530 million ft^2 (53 million m^2) of retail space, 30% of which is located in shopping centres. There remains a potential for growth, as the UK has only half the retail space per capita of that of the USA. The retail sector led the UK economy out of the 1990s recession, partly through the development of large regional shopping centres (Ashworth, 2008). However, the development of retail premises in city centres is not currently expected to resume until 2011 at the earliest. Retailers are in a constant state of reinvention, and are continually changing their formats and ranges to increase their market share or to respond to changing customer expectations. Retailers can adapt at a far faster rate than the buildings they occupy, so flexibility in unit design and leasing arrangements is important.

One consequence of these charges has been the emergence of alternative models of retail development, characterised either by enclosed malls or by the development of more traditional streetscapes with a mix of covered and open space. Enclosed malls have many advantages which include the creation of a high-quality retail environment, but they can be difficult to integrate into an existing town centre. In contrast, the absence of weather protection found in open schemes is seen as a disadvantage by some developers.

9.13.2 Town-centre retail premises

In 2008, nearly five million square feet of space in large retail developments opened in the UK, in projects ranging from mixed-use schemes to enclosed malls. The next generation of centres could be developed to very different value criteria, involving the review of many aspects of investment including:

- Public realm
- Logistics
- Sustainability
- Links to public transport

The revival of retail development has also coincided with the work of the Urban Task Group and CABE (Commission for Architecture and the Built Environment). These have championed the cause of mixed-use urban regeneration. The wider development challenge involves the reintroduction of town-centre housing and the creation of a vibrant, mixed-use economy. The current retail development cycle has been focused on town centres, largely as a result of planning guidance that has encouraged these in preference to out-of-town shops. Town-centre developments are complex and time-consuming projects requiring land acquisition, design and construction management. Such projects copy many of the positive aspects of out-of-town development, i.e.:

- Large unit size
- Convenience
- Development quality
- Shelter

Brownfield sites   In addition to land acquisition, demolition and decontamination costs, town-centre schemes may be required to retain and refurbish existing buildings. Other issues include engineering services diversions.

Design quality   Many of these city-centre schemes are of multi-storey construction, resulting in higher costs than their out-of-town counterparts. The quality of facades and public realm elements will also be higher. Schemes may also have to be developed to match existing street patterns.

Construction   The logistics of working in tight town-centre sites without affecting existing traders inevitably makes programming and sequencing of operations more complex. Programmes are typically longer, driven by phasing, and can also be affected by noise limits and restrictions on deliveries.

Mixed use   Mixed-use development has been a key feature of most in-town schemes that bring together retail, leisure and residential use. In some mixed-use solutions, developers may be expected to provide community facilities as part of the contribution they must make to comply with section 106 of the Town and

Country Planning Act 1990. Mixed-use buildings may also be difficult to adapt or redevelop in the future, owing to different lease structures.

Development density   Common facilities such as parking, transport interchanges and goods yards have to be developed to greater density in town-centre schemes where land is more valuable and where the site footprint is more restricted. Underground servicing and car parking are less efficient from a space-use perspective than above-ground, are more expensive and will extend the construction programme.

Public realm   Investment in a high-quality public realm is a key part of the long-term success of an in-town scheme, creating distinct urban spaces and enhancing the quality of a visitor's experience. All shopping destinations have increased the quality of finishes and furniture, fixtures and equipment. Products used in open malls need to be more robust than those in enclosed malls, to withstand the elements and general wear and tear in a 24-hour environment.

Design management and procurement   Design management is a crucial discipline, to make sure not only that the programme is met, but also that production information is fully co-ordinated and ready for construction. Phasing the work to enable early trading by key retailers may be complex, with later phases being delivered in a trading environment. There is a greater use of modern methods of construction, including prefabricated wall panels and service pods, which are aimed at reducing programme durations, on-site labour requirements and vehicle movement.

The importance of risk management, the certainty of delivery and the involvement of banks in the choice of procurement strategy has resulted in the adoption of approaches that focus responsibility for design management, project delivery and commercial risk on the contractor. Two-stage design and build has been the commonest approach, with executive design team roles novated to the contractor.

The principal contractor then appoints a range of subcontractors and specialist firms. The advantages of the two-stage, single contractor route include programme compression, single-point responsibility, central co-ordination of design and construction and volume purchasing.

9.13.3 Cost breakdown

The indicative costs shown in Table 9.30 are based on large retail outlets. These represent city-centre developments. They are typically a more expensive solution than out-of-town developments, where land costs will be lower, but where obtaining planning permission has become difficult because of government priority being given to helping to revitalise city centres.

Table 9.30   Retail buildings: indicative construction costs

	$£/m^2$ (2008)
Shopping centre mall	3 000–4 300
Factory outlet	500–800
Retail shells	700–1 200
Centre management	1 800–2 600
WCs	2 000–3 000
Service-level yard	800–1 000
Rear of house areas	1 400–1 900
Shopping centre refurbishment	1 000–1 800
Leisure shell	800–1 200
Multiplex cinema shell	1 000–1 400
Car parking: Surface	1 500–2 200 per space
Car parking: Multi-storey	10 000–17 000 per space
Car parking: Basement	25 000–35 000 per space

Source: Adapted from Davis Langdon (2008)

CONCLUSIONS

The current analysis of the costs of previous building projects provides us with the best guide for determining optimum economic efficiency in the future. There is a considerable way to go, however, before we shall be able to achieve a correct understanding of how costs are really determined. There is always a danger in giving any sort of advice without the correct facts, or in an opinion based only on assumptions and intuition. Cost studies are therefore an important aspect of the design of buildings. Since the time and money associated with these analyses are scarce, however, in the first instance it is important to concentrate on the elements which are of cost significance.

SELF ASSESSMENT QUESTIONS

1. Why are cost analyses prepared using tender information rather than the actual costs of constructing buildings?

2. Explain the importance of qualitative and quantitative data in a cost analysis, giving examples to indicate how these data are used in practice.

3. Describe what is meant by cost sensitivity and explain whether the upper floors element would be considered cost-sensitive in:

   ■ A high-rise block of luxury flats in a city centre
   ■ A speculative single-storey warehouse
   ■ A multi-storey car park
   ■ A home for the elderly in a country setting

BIBLIOGRAPHY

Ashworth, A. *Precontract Studies*. Blackwell Publishing 2008.

Davis Langdon, *Building*. United Business Media October 2008.

Davis Langdon, *Spon's Architects' and Builders' Price Book*. Spon Press.

Dearle and Henderson, 'The finishing touch', *Chartered Quantity Surveyor* October 1992.

Department of Education and Science, *Building Bulletin No. 4 – Cost Study*. HMSO 1972.

Gordon, A., 'The three L's concept – long life, loose fit, low energy', *Chartered Surveyor* B & QS Quarterly 1976.

Gray, C., 'Analysis of the preliminary element of building production costs', in *Building Cost Techniques: New Directions*. Spon 1982.

Hook, A., 'Air conditioning', *Chartered Quantity Surveyor*, September 1992.

Lockie, S., Butterss, I. and Surgenor, A., 'Cost Analysis: Sustainable schools', *Building*, September 2008.

Monk Dunstone Associates, 'In the frame', *Chartered Quantity Surveyor*, May 1992.

Moore, P., 'Roofing', *Chartered Quantity Surveyor*, July/August 1992.

Orchard, G.F.A. and Hill, P., 'Concrete trench fill for house foundations', *Building*, April 1972.

Rowntree, B., 'Cladding', *Chartered Quantity Surveyor*, November 1992.

Williams, B. *et al.*, *Design Economics for Building Services*. Building Economics Bureau 1984.

CHAPTER 10

PROCUREMENT OF CONSTRUCTION PROJECTS

LEARNING OUTCOMES

After reading this chapter, you should have an understanding about how construction works are procured. You should be able to:

- Understand the general evolution of procurement in the construction industry
- Recognise the issues and tensions that are involved
- Realise the importance of these different routes in respect of the cost studies of buildings
- Appreciate that there is no panacea solution and that the recommendation of a particular procurement route depends upon client and project circumstances

10.1 HISTORICAL PERSPECTIVES

The role of the independent architect as the designer of buildings was not established until the end of the eighteenth century. This is barely 200 years ago. At the start of the nineteenth century the general contractor came into existence. Prior to this time the architect would work with and commission the individual trades. To these two parties were added quantity surveyors and the way in which these evolved has been described by Ashworth and Hogg (2000). The quantity surveyor's bills of quantities provided a detailed analysis of all the resources that were required for a construction project. These could be priced by a number of contractors to formulate their bids for the work and the lowest-price tender would usually be awarded the contract on the basis of a lump sum.

The system of traditional single stage selective tendering is based on the rigid separation of design and construction activities. The client appoints a design team of consultants who prepare drawings, specifications and bills of quantities. These are then issued to a selected number of contractors to assist them in preparing their bid prices. Whilst there is much debate now about considering factors other than price in awarding the contract, such other factors are very difficult to evaluate consistently, fairly and objectively. For example, a contractor's price is based against a fixed period of time for construction purposes. Increasing or decreasing the time period will affect the price to be charged. If the successful tenderer is to be assessed against both time and cost then contractors will need to know precisely just how much the employer values time. Equally, if other attributes are also a part of this equation then competing contractors need to be able to assess their own attributes in line with the employer's expectations.

10.1.1 Evolution of procurement systems

This chapter recognises the limitations in the existing systems for the procurement of buildings and the need for a more strategic approach to be implemented. This is needed to meet the increasing demands for improved building performance and added value. Whilst traditional methods of tendering provide a clear and auditable approach, criticisms are expressed at a lack of contractor input during the design phase. The traditional method of procurement fails to integrate *what* is to be built (the design) with *how* it is to be built (the construction).

The immediate solution to the procurement problem has been to introduce a variety of design and build solutions. In their early days there was much criticism of these since it was felt that the design and build method failed to consider the design attributes and aesthetics properly and that these factors were limited and constrained by the build aspects. Statistics from Contracts in Use surveys by, for example, the RICS Construction Faculty (Davis Langdon 2002) have indicated a considerable growth in this type of procurement arrangement over the past 25 years. In 1985, design and build represented just 8.0% per value of contracts and 3.6% by number of contracts. By 2001, these figures had increased to 42.7% and 13.9% respectively. These statistics suggest that it is the larger projects that have seen the shift to this form of procurement.

Whilst design and build was seen as a possible solution to the procurement problem, the adversarial nature of construction procurement still remained a major problem. Different ways of settling differences were introduced, such as adjudication, and these went some way to resolving issues. Projects were also becoming more complex and clients were more demanding and a more co-ordinated approach was required.

The emergence and introduction of any new idea is often heralded as the solution to resolve most of the identified problems. Both design and build and management contracting suffered from this praise and then eventual dissatisfaction, especially in respect of the latter. Partnering is the current response to building procurement, although it is by no means the final solution. Readers need to recognise this fact. Reports are beginning to emerge to indicate that this is by no means the hoped-for panacea.

Under a given set of criteria, a system can be devised that meets as fully as possible the demands placed upon it. All systems have their advantages and disadvantages, but clearly some systems are more suitable than others.

The fact that the industry in many other respects is continuing to evolve and change and that procurement methods need to take into account these other factors

Table 10.1   Factors to consider in the procurement of buildings

- Relationships amongst employers, consultants and contractors
- Type of client
- Size of project
- Type of project
- Risk allocation
- Form of contract to be used
- Major objectives of the client
- Status of the designer
- Relationships with contractors/consultants/subcontractors
- Type of contract documentation required

in building design and construction needs to be noted. For example, most buildings today include a much larger amount of off-site manufacture. This will have an effect on the systems used for the better procurement of buildings. The New Rules of Measurement (NRM) (Chapter 8) have involved trade subcontractors in the development of the principles of measurement rather than the main or principal contractor, since in many cases the latter has become the managing agent.

The choice of procurement method depends upon a combination of, and the relative importance of, a range of characteristics. These will assist in the selection of the most appropriate procurement arrangement to be adopted. It further needs to be noted that the examination of procurement practices yields a wide degree of variation, even for similar kinds of projects carried out under similar conditions. Some of the factors to consider are shown in Table 10.1.

10.2 EMPLOYER'S ESSENTIAL REQUIREMENTS

Employers when buying a particular service seek to ensure that it fully meets their needs. The consultant or contractor employed needs to identify with the employer's objectives within the context in which the employer has to operate and particularly to recognise any constraints which may be present. *A Study of Quantity Surveying Practice and Client Demand* (RICS 1984) identified some of the following criteria as important requirements for the majority of employers:

- Impartial and independent advice
- Trust and fairness in all dealings
- Timely information ahead of possible events
- Implications of time, cost and quality and the interaction between them
- Options from which the employer can select the best possible route
- Recommendations for action
- Good value for any fees charged
- Advice based upon a skilled consideration of the project as a whole
- Sound ability and general competence
- Reliability of advice
- Enterprise and innovation

Table 10.2 Procurement routes

Selection criteria	Traditional lump sum	Modern methods of procurement
Price certainty before commitment to build	5	2
Lowest construction cost	4	2
Programme commitment from contractor	5	4
Shortest overall programme (inception to completion)	2	4
Control over design and materials	5	5
Control over trade subcontractors	3	4
Input from contractor during design	2	4
Control over site programming	2	2
Single point responsibility for client	1	1
Contractual relationship between client and designers	5	5
Contractual relationship between client and contractor	5	5
Damages for late completion	5	5
Ability to introduce changes to the contract	4	4
Suitability for complex design	4	3

(5 = achieves very well)
Source: Adapted from Institute of Directors (2001)

Table 10.2 identifies a range of criteria that compare traditional lump sum tendering practices with modern methods of procurement. This comparison provides a broad overview, since within these two alternative systems there are many options that seek to address some of the shortfalls. Whilst there is a real desire to increase the levels of satisfaction of clients, a solution that meets the industry needs of contractors and designers may not always achieve those same objectives for clients. It is recognised, however, that happy and contented clients are more likely to want to build again than clients who have had a poor building experience. It is therefore in the interests of the industry to provide high levels of satisfaction wherever possible and not to be wedded to their preferred practices that clients may dislike. Table 10.2 has been adapted from a report produced by Davis Langdon for the Institute of Directors.

10.3 PROCUREMENT SYSTEMS

A wide range of procurement systems is available (Ashworth 2006). Some of these systems have been in use for a considerable time, others for a much shorter time. There are different ways of representing these procurement systems. There are four essential questions that need to be answered at the outset:

- Consultants or contractors?
- Competition or negotiation?
- Measurement or reimbursement?
- Traditional or alternative procurement?

10.3.1 Procurement selection and options

The two ways of selecting a contractor (or consultant) are either through some form of competition or by negotiation with a single organisation. Selective competition is the traditional and most frequently used method for awarding construction contracts. Some believe that it is one of the fairest methods. In essence a number of firms of known reputation and capability are invited to bid against each other for a project. In preparing such a short list of tenderers the following factors are considered:

- The firm's financial standing and record
- Recent experience of building over similar contract periods
- The general experience and reputation of the firm for similar building types
- Adequacy of management
- Adequacy of capacity

The Construction Industry Board's *Code of Procedure for the Selection of Main Contractors* (1997) is a useful document to refer to for guidance and good practice about the awarding of construction contracts.

A wide range of different options can be used to address the client's objectives in relation to the time, cost and quality of construction. These objectives are not mutually exclusive. Ashworth (2006) provides a selection of the possible alternative methods.

10.4 PARTNERING

Many of the problems that exist in the construction industry are attributed to the barriers that exist between employers, designers and contractors. In essence, partnering is about breaking these barriers down by establishing a working environment that is based upon the mutual objectives of teamwork, trust and sharing in risks and rewards.

Within the UK construction industry, partnering activity is a relatively recent phenomenon, being given significant impetus by the Latham report (1994) and many subsequent publications and positive action. The University of Reading report for the RICS, *Improving Value for Money in Construction* (1995), was prepared in recognition of the findings of the Latham report, *Constructing the Team* (Latham 1994). This report highlights the benefits of partnering as one factor that is critical to the success of construction projects in providing the basis for improving value for money in construction. The report stated that:

> 'An attitude of co-operation amongst the project team members and, not least, with the client must be created. Partnering can help enormously without running the risk of being uncompetitive or introducing complacency'.

Partnering also featured prominently in the report of the Construction Task Force, *Rethinking Construction* (Egan 1998). This report was regarded as central to improving the performance of the construction industry. The prominent recognition given to partnering in the last few years of the twentieth century is a clear indication of the strong belief that the partnering approach can make a major contribution to improving or adding value within the construction industry.

10.5 PUBLIC PRIVATE PARTNERSHIPS

A Public Private Partnership (PPP) refers to any alliance between public bodies, local authorities or central government, and private companies to deliver a public project or service. The Private Finance Initiative (PFI) is a more formal approach to PPP that has been adopted in the UK.

In PFI, the public sector contracts to purchase quality services on a long-term basis so as to take advantage of private sector management skills, with the incentive of having private finance at risk. The private sector partner takes on responsibility for providing a public service, including maintaining, enhancing or constructing the necessary infrastructure, and the public sector specifies a level of service in return for an annual payment, called a unitary charge. In choosing whether or not to use PFI the government ensures:

- The choice of procurement route is based on an objective assessment of value for money
- There is no bias between procurement options
- Value for money does not come at the expense of employee terms and conditions

The PFI unitary charge is an annual payment made throughout the lifetime of the contract. This covers the cost of capital expenditure, private finance and the services needed to operate and maintain that asset. The total of all such PFI payments currently is less than 2% of the total annual resource budgets across government departments. In 2005–06 this represented £6 billion from a budget of £304 billion. Its payment is conditional on the private sector reaching the required service levels set by the public sector. Payment only commences when the asset has been satisfactorily completed. Around 50% of PFI projects by capital value are reported on departmental balance sheets. The accounting treatment of a PFI project follows rules set and audited by a series of independent national and international organisations.

Partnerships UK published a report into operational PFI/PPP projects in 2006. The report, which commented on the largest survey of PFI projects ever undertaken, contains a comprehensive review of the performance of PFI projects during their operational phase. The findings show that public sector managers and users are content with their PFI/PPP projects. Specifically:

- 96% of projects are performing at least satisfactorily
- 66% of projects are performing either to a very good or a good standard
- 89% of projects are achieving the contract service levels either always or almost always
- 80% of all users of PFI projects are always or almost always satisfied with the service being provided

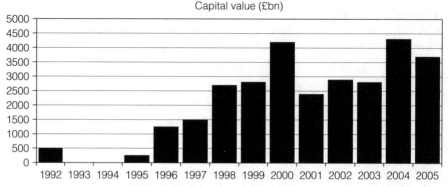

Fig. 10.1   Value of PFI Projects since 1992
Source: HM Treasury (2006)

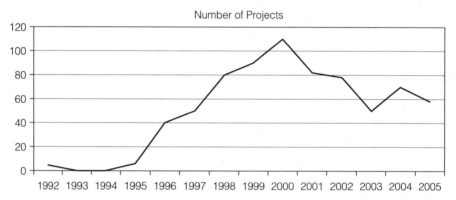

Fig. 10.2   Number of PFI projects since 1992
Source: HM Treasury (2006)

Public sector managers believe that they have developed an effective partnership with the private sector to deliver services. Over 97% believe that their relationship with their private sector partners is satisfactory or better and around 80% of public sector managers agree that the payment mechanism supports the effective contract management of the project.

PFI's record of delivery means that the government is committed to using PFI as a procurement option to deliver future infrastructure investment, wherever it is value for money to do so. In 2007, £26 billion of PFI investment across 200 projects was being proposed. This is one of the largest programmes worldwide. It includes the provision of over 60 new health facilities and 104 new schools. Figures 10.1 and 10.2 show the amount invested through PFI projects and the number of such projects since 1992.

PFI is a small but important part of the UK Government's strategy, its approach being based on its commitment to efficiency, equity and accountability and on the principles of public sector reform. PFI is only used where it can meet these requirements and deliver clear value for money without sacrificing the terms and conditions of staff. Where these conditions are met, PFI delivers a number of important benefits. It requires the private sector to invest its own capital at risk and to deliver defined levels of service to the public over the long term. PFI also helps to deliver high quality public services and ensure that public assets are delivered on time and to budget.

PFI is used to fund major new public building projects, including hospitals, schools, prisons and roads. The Conservative Party first introduced the scheme in the early 1990s. Under New Labour, PFI has become a major, and sometimes controversial, element of private sector involvement in Britain's public services. Since 1997, about £50bn worth of public-private deals have been signed, despite opposition from Labour backbenchers and public sector unions.

Private consortia, usually involving large construction firms, raise the capital to design and build a public sector project. They are also contracted to maintain the buildings while a public authority, such as a council or NHS trust, uses them. This means the private sector is responsible for providing cleaning, catering and security services. Once construction is complete, the public authority begins to pay back the private consortium for the cost of the buildings and their maintenance, plus interest. The contracts typically last for 30 years, after which time the ownership of the buildings passes to the public authority.

It is likely that expanded capital building programmes to replace out-of-date schools, hospitals and housing could not be achieved over the same period of time if the funding was being provided only by the public purse. In the short term, PFI allows the government to keep its promise of improving public services without actually raising income tax to pay for it.

According to the healthcare think-tank, the King's Fund, the physical condition of most hospitals is now vastly improved. By 2004, 30 major PFI schemes, with a capital value of around £3.36bn, had been opened or were being built in the NHS. Only four publicly funded schemes of a similar scale were completed or had begun construction during the same period.

Some critics suggest that PFI contracts are poor value for money and commit the public sector to accepting expensive financial responsibilities for decades to come. Some NHS trusts have also found it too expensive to pay the annual charges to PFI contractors for building and servicing new hospitals. The Audit Commission warned that the deficit at Queen Elizabeth Hospital NHS Trust in Woolwich, southeast London, would accumulate to £100m by 2008–09 unless the government restructured its PFI debt. The deal added about £9m a year to the costs met by an equivalent hospital built with money borrowed from the public purse. Unions, such as Unison, claim that PFI leads to poorer services because private companies maintain the buildings as cheaply as possible. The pay of cleaning, catering and security staff in PFI buildings is typically lower than, and conditions are inferior to, those of their counterparts in the public sector. A more recent Treasury report has suggested that PFI deals failed to provide value for money in the provision of many of these services. Future deals would be unlikely to automatically bundle cleaning, catering and security as part of new contracts. Some consortia appear to have benefited considerably from PFI deals. Some have made excessive profits. Others have been plunged into financial crisis.

PFI has been further criticised as being too inflexible because it ties public services into 20- to 30-year deals despite the fact that it is difficult to plan how these services will be delivered even a few years ahead. For example, the government's latest health and social care white paper has proposed moving 5% of the work of large general hospitals into the community. This shift in resources calls into question the plans for PFI investment in building large hospitals. The government's design watchdog has also attacked the cheap appearance of many PFI hospitals and schools.

10.6 BEST VALUE

The concept of best value (DETR 2001) emerged from the Local Government Act 1999, which specifies what is required. The key statement in the Act is 'A Best Value authority must make arrangements to secure continuous improvement in the way in which its functions are exercised, having regard to a combination of economy, efficiency and effectiveness'. The concept of best value applies as well in the private as in the public sector. Best value aims to achieve a cost-effective service, ensuring competitiveness and keeping up with the best that others have to offer. It embraces a cyclical review process with regular monitoring as an essential part of its ethos.

Best value extends the concepts of value for money that have been identified for a long time within both construction and property. Egan (1998), for example, defines value in terms of zero defects, delivery on time, to budget and with a maximum elimination of waste. In order to show that best value and added value are being achieved, it becomes essential to benchmark performance, including costs. It is also necessary to benchmark the overall cost of the scheme so that improved performance in the design can be assessed against its cost. The sharing of information underpins the whole best practice process. Even the leaders in an industry need to benchmark against their competitors in order to maintain that leading edge. Whilst the aspiration of best value is both admirable and essential, its demonstration in practice presents the challenge.

The best value concept for local authorities is managed for the Department for Local Government, Transport and the Regions by the Audit Commission. The performance framework placed on local authorities by central government is a challenging one and, since April 2000, all local authorities in England and Wales have had a duty to plan to provide their services under the principles of best value. Each local authority's service review must show that the authority has applied the 'four Cs' of best value to the service, and show that it is:

- **Challenging**: explaining why and how the service is provided
- **Comparing**: the performance with that of others, including non-local government providers
- **Competing**: showing that it has embraced the principles of fair competition in deciding who should deliver the service
- **Consulting**: local service users and residents on their expectations about the service.

Local authorities are required to show that they are continuously improving the ways in which their services are delivered.

10.7 AUDIT COMMISSION

The Audit Commission has set up a new inspection service to guide the work of best value. In common with other forms of inspection services and benchmarking, best value seeks out best practices and uses them to help all local authorities to improve their general levels of performance. Best practices today are unlikely to be best practices tomorrow, since the achievement of improvements in quality is always a journey and never a destination. The enhancement of quality remains the long-term goal. Best value inspectors use a simple framework of six questions to make sure that they collect the right information and evidence to support their judgements. These are:

- Are the authority's aims clear and challenging?
- Does the service meet its aims?
- How does its performance compare?
- Does the best value review (BVR) drive improvement?
- How good is the improvement plan?
- Will the authority deliver the improvements?

Best value is a concept that is important not only to the public sector but also within private sector organisations, however it is achieved.

10.8 CONSTRUCTING EXCELLENCE

Constructing Excellence is a movement that is based in the UK. It has wide support from its various stakeholders in the construction industry. Its aims are to achieve a significant improvement in construction productivity by tackling the market failures in the sector and selling the business case for continuous improvement. Constructing Excellence has developed a clear strategy to deliver the process, product and cultural changes that are needed to drive major productivity improvements in the sector. There are focused programmes in:

- Innovation
- Best practice knowledge
- Productivity and engagement

Constructing Excellence was formed in 2004 through the amalgamation of two initiatives, Rethinking Construction and Construction Best Practice. These two initiatives followed the publications of the Latham (1994) and Egan (1998) reports. In 2001, the National Audit Office produced a report, *Modernising Construction*.

10.9 PERFORMANCE INDICATORS

It is easy, on the basis of subjective judgement and anecdotal evidence alone, to suggest that improvements in processes and practices are being achieved. There is often a reluctance to measure or attempt to quantify such changes. It is also all too easy to distort the data, unless clear and precise guidelines are employed. In some cases in the past, improvements have occurred and their effect has then been attributed to a particular cause. Upon further investigation it has been found that the so-called cause and effect are not linked. For example, the 30% reduction in cost identified by the Latham report (1994) may appear to have been achieved largely through the suppressed costs of both labour and materials during the recession in the mid-1990s. It may be difficult when looking back to the start of the 21st century to identify whether improved methods of working were actually achieved through specific programmes, or whether quite unrelated improvements in technologies was the real reason. The poor time performance of the UK building industry in the 1980s, when compared with that of other countries, was only partly remedied through productivity agreements. Other improvements in time performance were often restricted because of the different regulations and patterns of organisation in the UK.

In this context it is important that the construction industry sets itself clear and measurable objectives. These might be achieved through the use of performance indicators or quantified targets. Measures of improvement will be required in terms of cost, time and quality, relevant to the aims and objectives of the individual client. The targets must be real and composite. They must not be achieved through cutting corners in other areas, such as safety and wages. In order to make such gains last, and thereby add value, continuous improvement must be implemented.

The report, *Rethinking Construction* (Egan 1998) identified a number of measures designed for achieving sustained improvement, as shown in Table 10.3.

Table 10.3 Performance indicators

Indicator	Improvement per year	Definition
Capital cost	Reduce by 10%	All costs excluding land and finance
Construction time	Reduce by 10%	Time from client approval to practical completion
Predictability	Increase by 20%	Number of projects completed within time and budget
Defects	Reduce by 20%	Reduction in the number of defects at handover
Accidents	Reduce by 20%	Reduction in the number of reportable accidents
Productivity	Increase by 10%	Increase in value added per head
Turnover	Increase by 10%	Turnover of construction firms
Profits	Increase by 10%	Profits of construction firms

Source: Egan (1998)

10.10 OTHER INDUSTRY COMPARISONS

It is always relevant when examining a subject like procurement to see how it is done elsewhere. This comparison may be made against similar or competing firms, perhaps in the form of a benchmarking study. Alternatively comparisons can be made with firms or organisations overseas, in countries that mirror UK practices and in countries where different traditions exist. It is also important to consider other industry comparisons, as illustrated in Latham (1994), which compared the performance of the construction industry with that of the motor car industry. Table 10.4 is an adaptation of that comparison. Other comparisons have been made with the aerospace industry (Flanagan, 1999). The outcome of such comparisons may be the development of a guide to good practices found elsewhere but which might have been overlooked. Current comparisons do not place the construction industry in a good light, but act as motivators to help change the culture of the construction industry.

The motor car manufacture and aerospace sectors include the following attributes that are generally absent from the construction industry:

- The recognition of a manufacturing culture
- The integration of design with production
- The importance of the supply chain network
- A focus on innovation and the recognition that this will only be secured through adequate research and development
- An acceptance of standardisation in design, components and assembly across the product range

Table 10.4   Construction industry performance compared to the car industry

	Motor car	Modern buildings		
		Domestic	Commercial	Industrial
Value for money	4	5	3	4
Pleasing to look at	4	4	3	3
Mainly free from defects	5	3	1	2
Timely delivery	4	4	4	4
Fit for purpose	5	4	2	3
Guarantee	5	4	1	1
Reasonable running costs	4	4	2	3
Durability	4	3	2	2
Customer delight	5	3	2	2

Source: Latham (1994)

10.11 PROCUREMENT AND THE COST STUDIES OF BUILDINGS

The different ways in which buildings are procured will have a direct bearing upon their costs, both initially and over their life cycle. It is well understood in economics generally that the absence of competition tends to drive up costs. This is one reason why the UK Monopolies and Mergers Commission is keen to prevent any organisation gaining monopoly status. There is anecdotal evidence that negotiated tender prices increase initial building costs by at least 5%. PFI projects that involve contractors borrowing finance from banks over long periods of time will increase the costs of construction compared with government borrowing the same finance because government is a more secure client. However, in the absence of PFI the Exchequer would be unlikely to be able to fund such projects at the rate that has been achieved by PFI. Cost studies need to ensure that they compare like for like. Design and build, for example, includes design fees and when comparing this type of procurement with more traditional methods, the fees of the designer need to be accounted for. Of particular significance to the employer or client are the overall costs of construction, including items such as VAT, and the long-term cost of owning and operating an asset.

Procurement methods should be designed to offer fairness to all the parties involved in the design, construction and management of a project. Care needs to be exercised in recommending procurement methods that encourage added value and smart ways of reducing construction costs. Methods that are innovative and untested will usually result in a premium being added to cover uncertainty. This should not, however, discourage the search for novel solutions. Employers who have huge building programmes will be prepared to invest in new methods in order to gain longer-term, future benefits. Consideration also needs to be given to the greed of the human race. Even when people have far more than is sufficient they often want even more. Procurement methods must therefore not be lacking in checks and balances.

For each of the different methods of procurement, a budget is set that is based on an estimate. The estimate will include uncertain costs and risks known as contingencies. In addition cost control procedures will be used to ensure that the estimated costs are not exceeded, or that where they are, all relevant parties are clearly informed of the consequences. The cost control procedures in the design stage (pre-construction) and in the construction stage of a project are radically different. These will depend to some extent on the procurement methods that are adopted and how cost has been defined at the outset.

SELF ASSESSMENT QUESTIONS

1. Do the wide variety of construction procurement systems make industry practices too complicated?

2. What criteria would you adopt when recommending a particular procurement route and what would be the implications on the costs of buildings?

3. What are the advantages and disadvantages of PFI/PPP methods of financing public works construction?

4. Why have some of the developments in procurement been introduced in the industry as recently as the past twenty years?

BIBLIOGRAPHY

Ashworth, A., *Contractual Procedures in the Construction Industry*. Pearson 2006.

Ashworth, A. and Hogg, K.I., *Added Value in Design and Construction*. Pearson Education 2000.

Bates, M., *Review of PFI: Summary and Conclusions*. HM Treasury 1997.

Construction Industry Board, *Code of Procedure for Selection of Main Contractors*. Thomas Telford 1997.

Davis Langdon, *Buildings That Work for Your Business*. Director Publications Ltd 2001.

Davis Langdon, *Contracts in Use Survey*. Davis Langdon/RICS 2002.

DETR, *Best Value*. Department of the Environment, Transport and the Regions: HMSO 2001.

Egan J., *Rethinking Construction*. Department of the Environment, Transport and the Regions. HMSO 1998.

Egan, J., *Accelerating Change*. Strategic Forum for Construction 2003.

Fairclough, J., *Rethinking Construction and Innovation and Research*. Department of Trade and Industry. HMSO 2002.

Flanagan, R., *Linking Construction Research and Innovation to Research Innovation in Other Sectors*. CRISP 1999.

Gray, C., *Value for Money*. Reading Construction Forum/RICS 1998.

HM Treasury, *PFI: Meeting the Investment Challenge*. HM Treasury 2003.

HM Treasury, *Value for Money in Assessment Guidance*. HM Treasury 2004.

HM Treasury, *Standardisation of PFI Contracts*. HM Treasury 2004.

HM Treasury, *Quantitative Assessment User Guide*. HM Treasury 2004.

HM Treasury, 'Strengthening long-term partnerships', *PFI/PPP Factsheet*. HM Treasury 2006.

Institute of Directors, *Buildings that Work for Your Business*. Director Publications Ltd 2001.

Latham, M., *Constructing the Team*. HMSO 1994.

Morledge, R., Smith, A. and Kashiwagi, D.T., *Building Procurement*. Blackwell Publishing 2006.

National Audit Office, *Modernising Construction*. HMSO 2001.

National Economic Development Office, *Faster Building for Industry*. HMSO 1983.

National Economic Development Office, *Thinking about Building*. HMSO 1985.

National Economic Development Office, *Faster Building for Commerce*. HMSO 1998.

Partnerships UK, *Investigating the Performance of Operational PFI Projects*. Partnerships UK 2008.

Rethinking Construction, A Commitment to People "Our Biggest Asset". Rethinking Construction 2000.

RICS, *UK and US Construction Industries: A Comparison of Design and Contract Procedures*. Royal Institution of Chartered Surveyors 1979.

RICS, *A Study of Quantity Surveying Practice and Client Demand*. Royal Institution of Chartered Surveyors 1984.

RICS, *Improving Value for Money in Construction. Guidance for Chartered Surveyors and Clients*. University of Reading for Royal Institution of Chartered Surveyors 1995.

CHAPTER 11

TAXATION, GRANTS AND INVESTMENT

LEARNING OUTCOMES

After reading this chapter, you should have an understanding about taxation, grants and investment, relevant to the construction industry. You should be able to:

- Understand the principles of taxation generally
- Identify the various types of taxes applicable to buildings
- Appreciate the types of financial assistance available for development
- Identify key investment criteria
- Compare investment in property with other commodities

11.1 TAXATION

In the costs of buildings, taxation can play an important part and can be beneficial if dealt with correctly, at the right time and by interpreting the rules of taxation clearly. Achieving a tax-efficient design is unlikely to be the main aim of a client but it is a further factor that must be considered and can offer advantages for those wishing to build. Taxation has implications for the construction, fitting-out, repairs, running and maintenance costs of buildings. The possibility of designing buildings to be optimally tax-efficient from the owner's or occupier's point of view can yield substantial benefits. However, as in all cases of taxation avoidance it is rarely a primary objective, or straightforward.

Tax evasion is illegal and, if a perpetrator caught, is punishable by the courts. Tax avoidance is a way of reducing the burden of taxation legally through interpreting the complex laws of tax. Many years ago a House of Lords judgment (*Ayrshire Pullman Motor Services and Ritchie* v. *IRC* (1929)) stated, 'No man in this country is under the smallest obligation, moral or other, so to arrange his legal relations to his business or to his property as to enable the Inland Revenue to put the largest possible shovel into his stores'. A similar theme can be found in this quotation from the USA, 'The legal right of a taxpayer to decrease the amount of what otherwise would be his taxes, or altogether avoid them, by means which the law permits, cannot be doubted' (U.S. Supreme Court Justice Sutherland, *Gregory* v. *Helvering*).

Taxation planning, provided that it is within the law, is not only perfectly acceptable and reasonable, but also a necessity, particularly with the high and diverse levels and incidence of taxation.

11.1.1 Introduction

The influence of taxation and its effects on buildings is constantly changing as a result of revisions in taxation principles and the introduction of new measures or rates from time to time by the Chancellor of the Exchequer. These are often incorporated as a part of the annual Budget presentation. The application of taxation through statute law is also governed by case law that is tested in the courts. As with other types of law, taxation laws are interpreted through custom, legislation and case law.

Taxation may be direct or indirect. Both are types now collected by a single body in the UK, HM Revenue and Customs (HMRC). Direct taxation is levied on incomes and earnings and indirect taxation through the purchase price of those products that are subject to an element of tax. Local taxes, such as council tax and business rates, are in addition and are collected by the local authority in whose area a property is situated. Examples of direct taxes include income tax, capital gains tax, inheritance tax and corporation tax. Indirect taxes include value-added tax, stamp duties and council tax. Direct taxes have a direct relationship with income; indirect taxes are unrelated to income. Taxes may be further classified as:

- Proportional – they are a fixed percentage of price
- Regressive – the percentage is smaller as income increases
- Progressive – which takes a higher percentage on larger incomes

The government adjusts taxation to provide income to meet its programmes and policies of meeting social, political and economic needs. It normally does this through an annual budget and in legislation through Finance Acts, but additional budgets can be introduced at other times in a financial year to meet revised demands. The government has an insatiable appetite for taxes in order to meet the increasing demands being placed upon it by its citizens.

All taxes have some bearing on buildings and property. Stamp duty, for example, is a tax on documents and is charged on the transfer of ownership of all land and buildings above a specified value. Council tax and non-domestic rates (business tax) are annual charges relating to the ownership of buildings. With each of these taxes there is little room for reducing the taxation burden. In connection with the latter, clients considering building in a general location may be advised to build in an area that has lower local taxes. The saving this could bring has to be balanced with the need to be in a particular location, the availability of possible sites and their respective charges.

11.1.2 Changes to capital allowances

HMRC prepares a summary of the annual Chancellor's Budget statement. Reference, where appropriate, should therefore be made in respect of any changes to capital allowances that may be included in the latest Budget.

The two taxes in the UK that are of the most importance in relation to buildings are value-added tax and corporation tax.

11.2 VALUE-ADDED TAX

Valued-added tax (VAT) is charged on the supply of goods and services in the UK, and on the import of certain goods and services into the UK. It applies where the supplies are taxable supplies made in the course of business by a taxable person.

Value-added tax (VAT) was introduced to the construction and property industries through the Finance Act 1972. Since this time the extent and rates of tax have been amended several times. The current position is covered in HM Customs and Excise Notice 700/1. VAT Notice 999 provides a complete listing of all the Customs and Excise VAT guides.

Building work is generally either standard-rated work (currently 15.0%) or zero-rated work. Examples of zero-rated works include residential buildings such as children's homes, old people's homes, homes for rehabilitation purposes, hospices, student living accommodation, armed forces living accommodation, religious community dwellings and other accommodation which is used for residential purposes. Certain buildings intended for use by registered charities may also be zero-rated. Buildings which are specifically excluded from zero-rating include hospitals, hotels, inns and similar establishments. The conversion, reconstruction, alteration or enlargement of any existing building is always standard-rated. All services which are merely incidental to the construction of a qualifying building are standard-rated. These include architects', surveyors' and other consultants' fees and much of the temporary work associated with a project. Items which may be typically described as 'furnishings and fittings' (fitted furniture, domestic appliances, carpets, free-standing equipment, etc.) are always standard-rated, even when the project is classified as zero-rated. The VAT guides referred to above provide examples, but also recommend that individuals check their respective liability with their local VAT office. The rating of some items may be arbitrary and some decisions will need to be tested in the courts. Arguing for a reduced rate of tax runs up against two issues. First, there is a loss of direct tax to the Internal Revenue and second, that there is any such reduction might be seen as a precedent, leading to requests for reductions for other goods and services, a result unlikely to be favoured by HMRC.

11.2.1 Taxation rates

The application of taxation rates to buildings is a complex area and is in a state of constant change as HMRC is constantly reviewing its practices and procedures to ensure compliance with the *spirit* of government policy and to plug those loopholes that are commonly referred to as tax avoidance.

HMRC has issued a revised version of its Notice 78, which explains:

- when building work can be zero-rated or reduced-rated at 5%
- when building materials can be zero-rated or reduced-rated at 5%

- when the sale, or long lease, of a building is zero-rated
- when developers are 'blocked' from deducting input tax on goods that are not building materials
- when a builder or developer needs to have a certificate from his customer
- when a customer can issue a certificate to a builder or developer
- what happens if the use of a certificated building is changed
- the special time of supply rules for builders
- when, on using his own labour, a developer must account for a self-supply charge

The main changes in content from the previous edition are as follows:

- The section on zero-rating the construction of new buildings has been amended to explain more clearly what an annexe to a village hall is and to reflect the change in treatment of civil engineering services on land which is to be sold as serviced building plots.
- The section on the reduced rating of the renovation of empty residential premises has been amended to reflect the reduction in the time that premises need to be empty.
- A new section has been introduced that deals with the reduced rate on the installation in domestic accommodation of mobility aids for the elderly.

11.2.2 Construction services

The construction of a new building and work to an existing building are normally standard-rated. There are, however, various exceptions to this. These are shown in brief in Table 11.1.

Table 11.1   Construction services: VAT exceptions

Construction service	Rate of VAT
Construction of new qualifying dwellings and communal residential buildings, and certain new buildings used by charities	0%
Conversion for a housing association of a non-residential building into a qualifying dwelling or communal residential building	0%
Other conversions of premises to a different residential use	5%
Renovation or alteration of empty residential premises	5%
Approved alterations to listed dwellings and communal residential buildings, and certain listed buildings used by charities	0%
Alterations to suit the condition of people with disabilities	0%
Installation of energy-saving materials; and grant-funded heating system measures and qualifying security goods	5%
Development of residential caravan parks	0%
First time gas and electricity connections	0%
Home improvements on domestic property situated in the Isle of Man (at the time of writing, this reduced rate is due to end on 31 December 2010)	5%
Installation of mobility aids for the elderly for use in domestic accommodation	5%

11.3 CORPORATION TAX

Profits, gains and income accruing to companies which are resident in the UK incur liability for corporation tax. Non-resident companies are immune from this tax unless they carry on a trade in the UK through a permanent establishment, branch or office. Companies residing outside the UK may be liable to income tax at the basic rate on other income arising in the UK, e.g. from the letting of property. The liability for corporation tax is governed by the profits, gains or income for an accounting period. This is usually the period for which the financial accounts are made up (accounting period) and will normally comprise successive periods of twelve months. The amount of profits or income for an accounting period must be determined on normal taxation principles. The rate of corporation tax is fixed for a financial year ending 31 March.

11.3.1 Capital and revenue

Capital expenditure is money that is expended in acquiring assets, or for the permanent improvement of, addition to or extension of an existing asset. Such assets must generally have a useful life beyond one year. New buildings, the alteration or extension of an existing building or the refurbishment of a building can thus be described as a capital works project and therefore capital expenditure.

Revenue expenditure is concerned with the maintenance of such an asset while it is in use. It therefore, by definition, includes those costs which cannot be classified as capital expenditure. These may include local council taxes, annual water and sewerage charges, energy, cleaning, insurances and minor repairs.

Capital expenditure will result in increased amounts for fixed assets on a balance sheet, whereas the revenue expenditure is chargeable to the trading or profit and loss account.

The following principles can be applied in respect of corporation tax.

11.3.2 Capital allowances

While capital expenditure is not allowable in calculating income profits, taxable profits may be reduced in the form of allowances. The law on capital allowances is contained in the Capital Allowances Act 2001. The Act includes provisions for its amendment by subsequent Finance Acts. Allowances may be given for plant and machinery, industrial buildings, agricultural buildings, hotels, etc.

Almost all commercial property owners and users who incur capital expenditure on property are entitled to claim capital allowances or financial reliefs under the current Finance Act. As much as 40% of construction costs can be allowable, releasing much needed capital into the business itself. It is estimated that some 50% of allowable items go unclaimed each year. Property taxation remains a complex area of law. In order to identify and value capital allowances, and to negotiate successfully with the Inland Revenue, companies require specialist help, usually through a combination of accountants and quantity surveyors.

Capital allowances can be applied for the purchase of existing properties, refurbishment by landlords, occupiers or tenants, for new-build schemes and for alterations and extension work undertaken by tenants, leaseholders or freeholders.

The actual cash benefit depends on the marginal tax rate of a business. For example, every £100,000 of plant and machinery claimed by a business will mean a tax saving of £30,000 over time, assuming a 30% corporation tax rate. Similarly a higher-rate income tax payer would save £40,000.

The allowances are calculated on the basis of the following.

Initial allowance   This is an initial sum that is allowed against expenditure on the purchase of an item in any financial year.

Writing-down allowance   The writing-down allowance is a sum allowed by the Inland Revenue that can be offset against income on an annual basis for a specified term of years. This may represent, for example, the theoretical depreciation of an asset.

Balancing allowances and charges   A balancing allowance or charge is provided where, upon the item's disposal, the actual amount is calculated and adjusted to take into account the above allowances that have already been given. If the sale of the asset falls short of the unrelieved expenditure then a balancing allowance is made.

These allowances do not need to be taken in full in order to make the best possible advantage of the reliefs and allowances that are available.

11.3.3 Buildings

Some types of new buildings are eligible to have their full costs offset against corporation tax. The capital amounts and percentage allowances are subject to regular review by HMRC in the UK. The capital allowances may also vary depending on where the building is to be constructed. For example, if a building is to be constructed in an area that has been designated as one where construction is encouraged by government it may be possible for taxation adjustments to be made.

While the respective Finance Acts define the criteria by which a building qualifies for the allowance, subsequent legal case law will be used to interpret the Act. In the event of a disagreement between HMRC and a client, the courts are asked to make a judgement on the circumstances relating to a particular project. This then becomes case law to be used and applied by others. It is important therefore to be aware of what cases are subject to summary judgement (i.e. where a court may decide a claim without a trial when it considers the claim has little prospect of succeeding).

Essentially, all businesses liable to income tax or corporation tax on profits arising can make a claim. The Capital Allowances Act does, however, require further criteria to be fulfilled. For example, to make a claim for plant and machinery, a business must have used this in the course of its trade.

Before a claim is made, the purchase price of the asset has to be split into the three components of land cost, building cost and fixed plant and machinery. Specialist valuations are required for each of the three components and these values must be apportioned using HMRC guidelines, to equal the price actually paid for the entire asset.

Industrial buildings

Industrial buildings are treated differently to other types of buildings. They may be broadly defined as buildings used for the processing or manufacture of goods. They also include buildings that are used for the storage of materials before manufacture and for goods after production. They must have a direct link with production. Wholesale warehouses are therefore excluded from this provision. The building must be used for a qualifying trade, as defined in the current Capital Allowances Act. Any offices that form a part of the factory can be included if they do not exceed 25% of the total cost.

The full costs of construction including professional fees are allowed. Land costs are excluded but the costs of any site preparation (such as dealing with contaminated land) may be included. The full costs of the purchase of a building from a building company (such as a speculative building development) are allowed. Costs expended on an existing building or plant and equipment costs associated with the building are also included. The allowable costs are 20% of the initial costs and a writing-down allowance of 4% until the costs have been fully written down. In the event of a sale, a balancing adjustment is introduced and there will be a possible claim for relief by the purchaser.

Where an area has been designated an Enterprise Zone by the Secretary of State, expenditure incurred on buildings qualifies for an initial allowance of 100%. Where fixed plant and machinery are an integral part of the building, these can also be treated as part of the building for the purpose of claiming Enterprise Zone allowances.

Hotels

Relief is given on the entire capital expenditure, providing the property is a qualifying hotel, as set out in the statute. The claim is similar to that of industrial buildings tax relief.

11.3.5 Plant and machinery

The treatment of capital expenditure on plant and machinery is a very complex area. While the definition of machinery is generally understood, the definition of plant is not and cases concerning plant have come before the courts on many occasions. Plant may include whatever apparatus is used by a businessman for carrying on a business. This will exclude stock-in-trade which is bought or made for sale, but will include 'all goods and chattels, fixed or moveable, alive or dead, which are kept for permanent employment in a business'. The legal opinion was stated in the case of *Yarmouth* v. *France* (1887). The main problem lies in distinguishing the apparatus with which a business is carried out from the setting in which it is carried on. Items forming part of the setting do not attract relief unless they do so as part of the building itself and not as plant. Lifts and central heating systems are treated as plant, but plumbing and electricity systems are not. Specific lighting to create an atmosphere in a hotel and special lighting in fast food restaurants have been held to be plant.

Table 11.2 lists a schedule of items that may be deemed to be 'plant and machinery' for taxation purposes. In the case of *Jarrold* v. *John Good and Sons* (1963) it was stated that allowable items need not be subject to wear and tear, and that they could play a passive role in the operations of the trade. The maintenance of plant is always allowable therefore as a trade expense. Builders' work in this connection, specifically required for the installation of plant items, can be claimed as a capital cost item under the heading of plant and machinery.

Expenditure on computer hardware is a capital expenditure on plant and machinery. Allowances are usually claimed under the short-life-asset rules. Where software is purchased at the same time, HMRC has suggested that this should also be treated as part of this same capital cost. Licences to operate software are treated as a normal expense against profits.

Plant and machinery costs, for taxation purposes, are not treated in the same way as the buildings that house them. First-year allowances can in some cases be 100% of their capital cost. The full amount of taxation relief can therefore be gained immediately on these items. However, it is more usual to allow a first-year allowance and then subsequent writing-down allowances of typically 25% in following years until the item is written down completely.

Different types of properties qualify businesses for different levels of allowances. The following are typical percentages, reflecting the proportion of the total construction cost accounted for by plant and machinery:

- Air-conditioned offices   25–45%
- Heated offices   17–25%
- Hotels   30–50% (plus hotel allowances)
- Industrial premises   5–12% (plus industrial building allowances)
- Fit-out contracts   40–50%

11.3.6 Research and development

Where businesses are involved in technological development, they are able to claim tax relief for property-related research and development costs.

Table 11.2   Schedule of items that may be deemed to be 'plant and machinery' for tax purposes

Items of plant	Comments
1. Blinds	
2. Demountable partitions	
3. Carpet	Excluding other floor finishes (tiles, vinyl)
4. Suspended ceiling	Where acting as part of the air conditioning system (where plenum is used in lieu of ducting)
5. Loose furniture	
6. Screens	
7. Office equipment	
8. Sanitary appliances	Wash hand basins, baths, urinals
9. Hot water system	Includes pipework, tanks, builders' work
10. Heating system	Includes pipework, builders' work
11. Control panels	That is, temperature controls – but not switchboards
12. Air conditioning and ventilation systems	Includes ductwork, builders' work
13. Emergency generators and transformers	
14. Task lighting	Where attached to furniture (ambient lighting, too, if in the form of movable fittings, but only if connected to a circuit not dedicated solely to lighting)
15. Wiring from outlet to loose equipment	No other general electrical wiring included
16. Sprinkler systems	Includes pipework, tanks and builders' work
17. Fire alarms/fire fighting equipment	Includes builders' work
18. Security system	Includes builders' work
19. Lifts and escalators	Includes electrical work and builders' work
20. Preliminaries	Proportion of preliminary costs (foreman, scaffolding, temporary work, insurances)
21. Professional fees	Proportion of total cost

Source: *Facilities Journal*

11.4 DEPRECIATION OF ASSETS

The writing-down allowances referred to above are given in respect of the depreciation of an asset. In practice there may be little relationship between the allowance, the accounting amount for depreciation as shown in a company's books and the actual depreciation of the asset concerned. Assets are frequently written down much more quickly for tax purposes than occurs in reality. An underlying philosophy behind this principle is to encourage companies to invest and in so doing help to improve the nation's overall economic performance. In accordance with this viewpoint, preferential rates are applied to items of plant and machinery since these become obsolete more quickly and are more directly related to economic production than other assets.

The depreciation of buildings includes the necessary infrastructure, together with any demolition that might be required prior to construction and the relevant professional fees that are required. It specifically excludes the cost of land, although it can include the costs of any ground stabilisation or the treatment of contaminated land that may be a necessary part of the building construction.

When new plant or equipment is purchased, its value will begin to decrease from the time of purchase. It may be totally without value within, for example, five years for some items of equipment, possibly a much shorter period for equipment such as computers that change very rapidly, or perhaps a longer period of time for some heavy items of plant installed in a factory. Buildings, however, have tended to appreciate in value over time, and are one of the few items of capital expenditure to do so. Some of this increase is attributable to the land on which the building is placed, rather than the actual building itself. However, the majority of buildings do not remain for ever. Some have a remarkably short life, such as some of the multi-storey blocks of flats that were constructed in the middle of the 1960s and lasted barely 25 years. Most buildings constructed today would be expected to have a life approaching 100 years.

Depreciation is the term given to the reduction in value over time. It is necessary to assess this for the company's balance sheet. A building contractor normally recovers a part of this loss on plant and machinery by including an appropriate amount in the rates charged for doing the work.

There are several different ways of calculating depreciation, in order to distribute the appropriate costs over the expected life of the project. Where depreciation occurs over a long period of time it may be necessary to allow for the time value of money through use of one of the discounting methods. The following are the usual methods of allowing for depreciation.

11.4.1 The straight line method

This is sometimes described as the fixed instalment method. The original value of the asset, less any residual value, is divided by the number of years of its estimated life.

Example

A manufacturer has purchased a new item of capital equipment for £30 000. It has an expected scrap value of £5 000 and an expected life of 5 years. What is its annual depreciation?

$$\frac{30\,000 - 5\,000}{5} = £5\,000$$

While the method is simple to calculate, it has the disadvantage that it does not represent the actual depreciation of an asset. This will be higher during the first few years of ownership and decreases as it reaches the end of its useful life.

If these figures were used in a company's accounts, then it would be necessary to amend the final year's figures by making a balancing adjustment to agree with the actual amounts involved. If the equipment was to be replaced at the end of its useful life then it might be necessary to allocate these amounts to a sinking fund for the equipment's replacement, in which case, due to inflation, the fund would probably be insufficient.

11.4.2 The reducing balance method

Since most plant and equipment incur higher depreciation amounts during the earlier years of their lives, it is desirable that the depreciation calculated models this fact. The reducing balance method reduces the value of an item of equipment by a fixed percentage each year. While the percentage is fixed, the actual amount will vary, decreasing each year. The Inland Revenue for taxation purposes proposes a figure of 25%. For items that have a very short life expectancy this figure may need to be increased. The following example is based upon 25% depreciation.

End of year	Depreciation for year ($£$)	Book value ($£$)
0	0	10 000
1	2 500	7 500
2	1 875	5 625
3	1 406	4 219
4	1 054	3 155 etc.

11.4.3 The depreciation fund method

A fixed proportion of the initial cost is transferred each year from the revenue account to the depreciation reserve. If this reserve is allowed to accumulate with compound interest, it should at the end of the asset's life produce an initial cost less value. An alternative to this method is referred to as the sinking fund method, where the annual sum is then reinvested. The advantage of this method is that it provides the actual cash to replace the asset. A further alternative approach is the insurance policy method, whereby a policy is taken out with an insurance company for the cost of the asset, due when the asset is to be replaced.

11.4.4 The valuation method

The different methods of depreciation are shown graphically in Figure 11.1. Depreciation may also be determined by the process of actual revaluation of the asset at fixed periods of time, normally annually at the end of the financial year. Such valuations are also a useful check on the other methods that might be used to calculate depreciation. In contracting, depreciation might also be used to

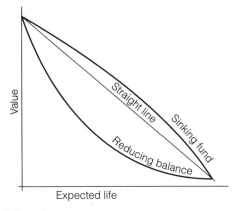

Fig. 11.1 Methods of depreciation

revalue items of plant at the end of a contract, where such costs-in-use can then be attributed to the project.

11.5 TAX-EFFICIENT DESIGN

Businesses that commission a tax-planning design audit of their projects will be sure of maximising their entitlement to any tax relief. The objective of such an audit is to compile the necessary documentary evidence to claim the appropriate tax relief. The sorts of items that may be considered are shown in Table 11.2. Any known precedents in tax case law will be used to support such applications by the relevant authorities.

For an item to qualify for a capital allowance it must pass two tests. The first of these is known as the business use test. It is not necessarily the nature of the item itself that determines eligibility, but the way in which it features in the user's trade. The second test is referred to as the premises test. If an item has become part of the building it cannot be classed as plant. The way in which items are fixed therefore becomes the issue.

It is important, therefore, to consider the following in order to become tax-efficient:

- Always consider capital allowances early
- Obtain a capital allowances forecast during the feasibility stages and factor this into the financial appraisal
- Obtain a forecast of the capital allowances before exchanging on purchased property to determine the increased effects of yields
- On larger schemes consider a tax-planning audit
- Never disregard historical expenditure

In respect of the last of these items, where historical expenditure has been overlooked the allowances are not necessarily lost, but may be deferred to a later period provided that the asset is still owned.

It has been suggested that only about half of businesses actually claim their full entitlement to capital allowances in respect of their construction projects. The possible reasons for this are:

- Poor cost scheduling that does not allow a detailed analysis of the relevant qualifying categories for tax purposes
- No appreciation of how such allowances are valued on second-hand property
- Mistakenly considering the exercise to be a tax matter for an accountant when it really relates to issues of property cost apportionment
- Failure to access the appropriate specialist advice either through ignorance or poor recommendations
- Leaving the exercise too late so that such allowances cannot be properly assessed

In order to make sure that a business receives its correct entitlement, it needs to consult with the appropriate professional advisers. The subject is complex in its nature, demanding a detailed knowledge of the technical issues of buildings, property costs, capital allowances and taxation matters.

11.5.1 The implications of taxation on construction projects

Many clients of the construction industry make investment decisions regarding capital works projects on the best advice available to them at the time. More detailed information will allow them to make these decisions with a greater degree of confidence. The kind of advice they require will be influenced by the effects of taxation and grant allowances. Effective tax planning can, for example, turn an apparently loss-making project into a profitable one. In the case of conflicting or alternative proposals this factor can change the ranking order of the projects under consideration.

The calculation of a company's liability for tax allows approved revenue expenditure to be deducted prior to assessment. Capital expenditure such as expenditure on a new building is not treated in this way for tax purposes. This expenditure is eligible for capital allowances only in certain cases, and where it is approved then such amounts can be offset against pre-tax profits.

Taxation regulations change frequently in their applications, as do tax rates and eligibility. This may be as a result of changes in the Finance Act, other government legislation or a case in the courts. The expenditure of money on plant and equipment is always eligible as a capital allowance even where the building structure is not. Expertise is necessary, therefore, to identify and separate correctly the items or costs of such items within the total capital expenditure. HMRC has offered some guidance on this matter, but its opinions are always open to question on matters of interpretation of the taxation provisions.

Clients cannot expect full guidance from accountants, since their technical knowledge of the construction process and product is extremely limited. In this respect the quantity surveyor is able, through the documentation produced, to identify item by item the various components of a building project for taxation purposes. Every project will have its own peculiarities and will require individual analysis in order to determine the separate classification of each item. Generally, where an item of building work is required solely because of the plant and equipment, the cost of the building work can be included in the plant and equipment category. Quantity surveyors have in the past been rather lax in dealing with this aspect of construction financial management. Clients are now much more aware of the problem, and of the advantages of effective tax planning in this respect. This has been emphasised in connection with whole-life cost planning, and the more elaborate and complex taxation provisions that this involves.

The possibility of designing buildings to be optimally tax-efficient from the owner's or occupier's point of view could yield significant financial benefits. The potential tax avoidance, however, is rarely a primary objective, but is a spin-off from other considerations.

The principal tax affected by premises costs is corporation tax, which is the tax payable on a company's profits in a given accounting period. The rates of tax are subject to annual changes. Generally:

- The highest proportion of capital costs should be allocated to plant for tax purposes
- Appropriate professional fees should be allocated to these items
- Certain types of building are eligible for tax relief, especially if they are located in Enterprise Zones, i.e. areas which are unattractive to investors without an incentive
- Certain items of building engineering services, etc. are classified by HMRC as plant
- Repairs to existing buildings are allowed as a full expense for tax purposes
- In refurbishment projects it is not always easy to distinguish between capital and revenue expenditure. Repair costs include those sums spent on maintaining the original performance of an asset. This is regular expenditure. Capital expenditure is generally a sum of money which has a benefit that extends beyond a single accounting period
- Capital allowances are granted in place of depreciation and can be deducted from the net profit before tax

11.6 FINANCIAL ASSISTANCE FOR DEVELOPMENT

While the various planning regulations are able to prevent undesirable development from taking place they cannot encourage socially desirable development to be undertaken. As part of the planning process the authorities are able to suggest to a developer that certain specified works are carried out as part of the approval of development. These might include, for example, leaving part of a proposed housing development as a public open space. In order to encourage desirable developments to take place in unattractive locations, some form of financial assistance is often necessary. This assistance may be obtainable from several different sources, but especially from government agencies through grants, loans, taxation relief, etc.

The intention of such financial assistance is to support projects in areas where otherwise they might not take place, or in circumstances where there may be little obvious economic benefit to a developer. One of the biggest difficulties for developers is that government policy on these matters is constantly under review and change, in order to meet new demands and to help generate improved economic conditions. Such assistance may be offered from a local or regional authority or a central government department that has responsibilities for the environment, trade, industry, education or employment. In the case of Action for Cities, which was launched in 1988, the promotion included all of these departments working jointly. A major aspect of this initiative is also the effective co-operation between the public sector, local and central government and private business and voluntary organisations. Government invests in urban regeneration in Britain with an expectation of additional sums being made available by the private sector. Capital expenditure which is incurred on new buildings, adaptations to existing buildings, new machinery, plant or equipment, in the right circumstances, is eligible for grant aid.

Financial assistance associated with construction projects arises for several reasons, which include:

- Urban renewal programmes
- Regeneration of industrial areas
- Investing in jobs to benefit areas of high unemployment
- Land reclamation schemes
- Property improvement, such as housing improvements
- Slum clearances and derelict land clearance

The aim of providing such assistance is to encourage private companies or public managing agents to develop areas either as a means of improving the standards and amenities, or through investing in projects that will help in wealth creation and at the same time reduce levels of unemployment in a region. Financial assistance is therefore targeted in those areas where it might be difficult to encourage companies to invest otherwise. While financial assistance may be available for a variety of different purposes, it is in the designated areas that the amounts of such grants are the highest. Higher levels of grant are available for approved schemes in Urban Programme and Assisted Areas. Urban policy initiatives also include Enterprise Zones and Urban Priority Areas and their respective successors.

Investment grants are made by the government, to manufacturing and extractive industries only, in respect of new buildings or adaptations and plant and machinery. The grants are treated as non-taxable capital receipts. Loans are treated in a similar way and may be free of interest. Loans are sometimes offered to companies which, for a variety of reasons, are unable to secure finance in the normal conventional ways through commercial banks. The financial assistance offered may be a combination of:

- Taxation allowances on capital expenditure for buildings and plant. The company in receipt of such an allowance must in the first instance make a profit to secure the benefit

- Low rents or business rates which are offered by a local authority as an inducement to locating a business in their area. This will be offered to a company for a limited number of years
- Grants for capital items to assist the firm in new developments. These grants may be as high as 50% of the total capital costs of building
- Extending, converting and improving industrial and commercial property
- Amenity grants which can represent 100% of the costs associated with providing access roads, car parking and other amenities
- Bridging finance to close the gap between developing a building and its market value
- Interest relief grants to help offset some of the costs of borrowing finance
- Building loans. As well as acting as a guarantor for bank finance for building, up to 90% of the market value of land and buildings may be met through loans charged at preferential rates of interest
- Enterprise Zone benefits, which include 100% tax allowances for money invested in commercial and industrial buildings, and exemption from local property taxes
- Enterprise Zone status also provides for simplified planning procedures for developments
- Subsidies paid to companies which employ additional employees in specified occupations. These, like low rents, are an inducement for a limited period of time only

11.6.1 Regional initiatives

Regional industrial policy operates within a general economic framework designed to encourage enterprise and economic growth in all areas of Britain. However, in some areas specific additional help is required under a regional initiative. Help is thus focused on the assisted areas, which are designated intermediate and development areas. These are based in the existing industrial conurbations around Glasgow, the northeast, Yorkshire and Humberside, Lancashire, the Midlands and South Wales. Also included under the assisted areas scheme are the north of Scotland and Cornwall. Northern Ireland has its own full range of incentives. The Scottish and Welsh Development Agencies also promote industrial development in their respective countries.

Regional initiatives are aimed at those areas of high unemployment caused by the demise of traditional industries or the loss of major employers. In order to obtain assistance the project envisaged must create, or in exceptional circumstances safeguard, jobs within a designated area. The project must also have a good chance of long-term viability. In addition, the greater part of the project costs must be financed by the applicant or from private sector sources. The applicant must also be able to show that without this assistance the project would not take place at all on the proposed basis. A further criterion is that improved economic efficiency and greater security of employment should result. Grants are based on the fixed capital costs of new buildings or adaptations of existing buildings, plant, equipment, machinery, vehicles, etc.

Regional selective assistance

This is designed to meet the capital costs of investment projects which create employment. Eligible projects include new and second-hand plant and machinery, buildings, the purchase of land, site preparation and vehicles used solely on site. Grants can meet up to 30% of fixed asset costs and are available to manufacturing operations and to service businesses which trade in local markets.

Regional enterprise grants

These are designed to encourage growth among smaller businesses, by providing grants for investment and innovation. An investment grant provides 15% of the costs of plant, new and second-hand machinery, buildings, the purchase of land, site preparation and vehicles used solely on site to a maximum grant of £15,000 for businesses employing up to 25 people. An innovation grant provides 50% of the cash, to a maximum grant of £25,000, for new product development or improvements to products and processes in firms of up to 50 people. The majority of manufacturing and service industries are eligible.

The Enterprise Initiative

Through the Enterprise Initiative, the government has provided grants towards the costs of using outside expertise in marketing, product design, quality, production, research and development and business planning.

Training and research grants

These are available to help meet the costs of training. Financial support is also awarded to encourage the formation of research partnerships with the academic community.

11.6.2 Financial assistance examples

The following are just a few examples of situations where financial assistance for building works might be available.

Redundant buildings

Regional development agencies, such as the South East England Development Agency (SEEDA), can provide financial support for the refurbishment of redundant buildings to bring them back into productive use or to enhance their current business use. In order to qualify for grants, the premises must be either completely unused, or unusable in their current state for the purposes for which they are now needed. Grants are not available for refurbishment of current business premises, or for conversion of existing buildings to residential properties. Grants cannot be given

if work on the project has already commenced. Qualifying business activities
will include:

- Manufacturing, servicing and craft businesses
- Office-based activities
- Retail businesses
- Tourism and leisure businesses
- Tourism accommodation will be considered

The Architectural Heritage Fund

The Architectural Heritage Fund (AHF) is a registered charity founded in 1976 to
promote the conservation of historic buildings in the UK. It does this by providing
advice, information and financial assistance in the form of grants and low-interest
working capital loans for projects undertaken by building preservation trusts (BPTs)
and other charities throughout the country.

Only organisations with charitable status are eligible for financial assistance.
Any charity with a qualifying project is entitled to apply for an options appraisal
grant, or a loan, but the other grants are reserved for BPTs. These are charities
established specifically to preserve historic buildings. Financial assistance is available
only for buildings that are:

- Listed
- Scheduled
- In a conservation area
- Of acknowledged historic merit

Projects must involve a change either in the ownership of a property or in its use.
The AHF is unable to assist private owners or buildings in continuing long-term
ownership or use. The activities of the AHF include:

- Advising on the restoration and reuse of historic buildings which are listed or
  in a conservation area
- Advising on the setting up of building preservation trusts, charities specialising
  in historic building regeneration
- Giving grants to charities intending to restore historic buildings
- Making low-interest loans to charities to help them finance projects on site
- Providing information on sources of funds for restoring and repairing historic
  buildings
- Producing helpful publications

Building Preservation Trusts

A building preservation trust (BPT) is a charity whose main aims include the
preservation and regeneration of historic buildings. There are almost 300 BPTs
in the UK. The majority are rooted in their local communities. Some cover
individual towns, cities or whole counties. Others specialise in particular types

of building and a few cover the whole of the UK. Some were formed to save just one building and others, known as revolving fund trusts, a succession of buildings. The AHF maintains a register of revolving fund trusts.

The BPT structure has a number of advantages:

- It is a recognised form of charity, so it is easier to set up than an ad hoc charity
- It specialises in historic building regeneration, and the AHF gives preference to BPT applications if its resources are under pressure
- Buildings then qualify for additional grants from the AHF
- BPTs qualify for membership of the Association of Preservation Trusts
- BPTs will have all the powers needed to rescue historic buildings
- BPTs are established as limited companies, reducing risk to their trustees

11.6.3 The European Union

The European Union (EU) seeks to increase the degree of economic cohesion and to ensure a more balanced distribution of economic activities within the EU. The principal responsibility for helping depressed areas remains with the national authorities, but the EU may complement schemes through aid from a number of sources. The European Regional Development Fund (ERDF) was established in 1975. Its purpose is to contribute to the correction of the main imbalances within the EU by participating in the development and structural adjustment of regions whose development is lagging behind and in the conversion of declining industrial regions. The EU extended eligibility to certain non-assisted areas, with Britain receiving funding for special programmes to improve the environment and to encourage employment initiatives in certain textiles, steel, coal, shipbuilding and fisheries closure areas. The European Coal and Steel Community (ECSC) provides loans and grants to encourage a rational distribution of production and a high level of productivity in the coal and steel industries while safeguarding employment and avoiding unfair competition.

11.7 THE EFFECTS OF TAXATION AND GRANTS ON CASH FLOW

Cash flow calculations must take into account the following factors which have a financial effect on the acquisition and ownership of fixed assets:

- Are grants available from the local authority or central government which will reduce the purchase cost of the asset?
- Are allowances available in the early life of the asset such as accelerated capital allowances? Where these are available they will improve the cash flows in the early years when they are the most valuable.
- What is the method to be used for taking the year-by-year depreciation or writing-down allowances? These reduce the taxable profits in the year in which they are taken and will therefore reduce the cash outlay.

- When assets are disposed of, what will be the effect on tax? The present UK system envisages either a balancing charge or an allowance, depending on how the written-down value compares with the realised sum.
- What is the timing of the cash flow effects considered above? In the UK tax is payable one year after the end of an accounting period. The tax saving of capital spending therefore will not become effective until one year later, when the tax payment is made. Grants and subsidies are usually paid by government and other agencies at some date after the outlay has taken place.
- In the UK it is preferable to have capital expenditure that is appropriate for tax relief towards the end of the accounting period, so that a minimum delay occurs between expenditure and tax relief.

11.8 INVESTMENT

11.8.1 Property generally

Investment in property has traditionally, and during the past few decades, been seen as a good hedge against inflation, and therefore a good investment. It may not represent the best investment that is available. However, as with all investments, performance in the past is not necessarily a guide to its possible performance in the future. There are disadvantages associated with investing in property and the overall risks will be a reflection of national and international markets, government policies, local and regional economies, geography and fashion and demand. Generally, the more risk that is associated with an investment, the higher the yield and return that might be expected. The risk, of course, might become a reality and outweigh any possible returns.

Good property investments have security and return and compare favourably with other types of investment. Good-quality property in desirable locations is in high demand, with little likelihood of possible risks occurring. For example, when leases are not renewed and the property becomes vacant, then it is possible to find new owners or occupants quickly for this type of property. However, property is not easy to dispose of at short notice. In some cases it may take several months to find a potential purchaser who is willing and able to proceed before the owner can realise the capital asset.

Institutional investors are able to identify those properties that return good yields and in many cases will specify the standards of property that they are willing to purchase. While owning the building they will have little interest in its operation. This is facilitated by the fully repairing lease, whereby a tenant agrees to repair and maintain the building throughout the life of the lease. The institutional investors' aims and objectives are shown in Table 11.3.

11.8.2 Investment in property

Investment in property can be described as 'direct', in which case the property is purchased by the owner for occupation and use. Investment in property of this type

Table 11.3   Institutional investors' aims and objectives

High yield
High retained value of the property
Low management costs
Secure tenants
Blue chip tenant companies
Happy tenants
Long leases
Frequent upward-only rent reviews

---

requires careful consideration, since a wrong decision in respect of design or location, for example, cannot easily be rectified at some later date. Direct property investment is really a spin-off from ownership and use. Any investment that accrues is therefore of a secondary interest. Alternatively, a property investment may be described as 'indirect', in which case the interest in the property is solely in respect of the possible financial gain associated with such an interest.

The property sector can be separately analysed into a number of different subgroups according to function, type, size, location, etc. A direct investor would include, for example, an industrial manufacturer. Such a firm would need to define as closely as possible its needs in terms of its property ownership. However, it considers its interest in property largely as a means of production rather than as investment. Other groups, such as pension and insurance fund managers, will be largely indifferent to many of the characteristics associated with a property's use, being concerned only with the investment potential of the property. There are also other circumstances such as sale and leaseback, where an owner–occupier may wish to raise capital on a project but still retain use of it. In these circumstances, an investor will purchase the property and then immediately lease it back to the original owner on agreed terms. This type of arrangement may be preferable to raising the required capital in some other way, often at much higher interest rates. The disadvantages are the loss of ownership and the periodic increases in the rent in the lease terms. An alternative to sale and leaseback is to mortgage the property, the mortgagee having an indirect interest in the property. The purchase of property bonds or shares in property companies is another way of acquiring an indirect interest in property.

The amount of yield or rate of interest received from an investment will vary according to some of the factors described below. Inflation and taxation will also have some influence on interest rates. Changes in taxation structures, especially capital gains tax and the availability of grants or loans, will influence possible investors in property. Over a period of time, like all other forms of investment, prime property will show yields comparable to those of other commodities in the market. At different times, the returns on prime property investments will considerably outstrip those of investments of other types. Prime property enjoys those characteristics that are frequently most sought after and are in high demand, even in times of a general recession. The valuation of prime high street property, for example, is typically twenty times the annual rental income.

11.9 PROPERTY INVESTMENT AND EQUITIES

Insurance companies and pension funds have traditionally invested in government securities, since these have been able to provide the investor with a known rate of return on which they could depend. These companies need such a guaranteed return to enable them to meet all possible claims. In times of inflation, the institutions found that returns above the rate of inflation could also be achieved by investing in equities. In stock market and property jargon, equity is used to mean the interest in a company or a property that bears the full risk and receives the full rewards with no upper limit on potential earnings. It is effectively the entrepreneur's or owner's money, as opposed to that which is borrowed from elsewhere. Thus equities on the stock market are simply ordinary shares. Preference shares are not strictly speaking equities, since they have preferential rights and fixed dividends.

The major institutional investors are always seeking better ways of investment, and industry analysts are able to advise on those areas, industries or companies that provide the best possible return. The institutional investors still invest large amounts where risk is minimal and returns are almost guaranteed. They also take advice in attempting to predict those opportunities where return will be higher and the investment will still largely be secure. Property offers a good location for funds, at times considerably better than other types of investment. The advantages of direct property investment include the following:

- The investor has more control over the property purchased
- Fewer acquisitions need to be made where large sums are involved
- Property values are generally less volatile than stocks and shares, whose prices change daily
- Property companies are often highly geared, making their return less secure than other shares or direct property investment
- When the stock markets are depressed it is still possible to sell property at a reasonable price
- The tenants of property often pay their rent in advance, usually quarterly, whereas the dividends on equities are in most cases paid half-yearly in arrears
- Rent continues to be paid by tenants, even if the company concerned is making a loss
- Rental income is a more secure form of investment since it is paid ahead of bank interest
- Even if a tenant enters into liquidation the investor still retains the property asset
- Following a liquidation, a company's equities will have minimal value, if any
- Most modern property leases have included within their lease agreements the provision for rent reviews, enabling the investor to charge current rates irrespective of the tenant's profitability

Direct investment in property also has a number of disadvantages such as:

■ The time and costs involved in purchase or selling. The volatility in the prices of, for example, stocks and shares is partly offset by their rapid trading capability
■ Property values do not always match those of inflation
■ Of critical importance with property is deciding which is the best property to purchase
■ Changes in technology, design or working and living practices will all affect the property's value. In some cases where these factors have not been properly considered, they may have the effect of making the project obsolete before its decay
■ Changes in communication networks will have a positive or negative effect
■ Town planning decisions may also have an effect upon both the valuation and the demand for a property

11.10 PROPERTY INVESTMENT

Investment institutions such as insurance and pension funds will seek to reduce their risks by spreading their investments across a wide range of companies and industries. Fund managers are constantly transferring funds, hopefully into those areas or sectors that offer the best possible returns. The level of investment reflects the level of performance. The information in Tables 11.4 and 11.5 is based upon a survey undertaken by The World Markets Company PLC. This company measures over 2,000 pension funds, of which 1,336 provided data that contributed to that shown in the tables.

Table 11.4   Annual return across a number of investments 1977–2008 (property is highlighted)

	Annual percentage returns									
	1977	1992	1994	1996	1998	2000	2002	2004	2006	2008
UK equities	14.2	20.8	−5.6	16.8	12.0	−4.1	−22.5	12.9	16.6	−29.7
Overseas equities	11.3	19.7	−3.7	2.6	17.2	−6.7	−24.2	10.3	8.1	−22.5
UK bonds	12.3	19.1	−8.4	7.6	20.6	9.4	9.6	7.4	0.6	1.9
Overseas bonds	9.3	29.1	−5.3	−5.8	11.6	14.6	9.3	4.8	−4.8	17.0
Index linked	8.0	17.8	−8.1	6.5	20.3	4.3	8.7	8.5	2.6	5.0
Cash/other	9.6	13.2	5.4	6.1	7.3	11.5	2.6	6.9	8.3	2.5
Property	17.1	−1.1	12.3	8.6	12.8	10.1	9.1	18.9	19.3	−22.7
Totals	13.0	18.6	−3.9	10.7	14.0	−1.3	−13.9	11.2	10.5	−17.2
Property ranking	1	7	1	2	4	3	3	1	1	6
Retail price index	4.7	2.6	2.9	2.5	2.8	2.9	2.9	3.5	4.4	0.9
Average earnings	8.9	5.0	4.4	4.9	4.5	4.5	4.1	4.7	4.0	1.9

Note: Percentages above the total asset performance in bold
Source: World Markets Company PLC (2008)

Table 11.5 Percentage asset allocation of investments (property is highlighted)

	1977	1980	1985	1990	1995	2000	2005	2008
UK equities	49	45	47	54	53	49	33	21
Overseas equities	5	8	20	18	27	24	32	27
UK bonds	23	22	18	9	6	9	12	20
Overseas bonds	0	0	0	3	4	4	3	4
Index linked	0	0	2	3	4	5	9	12
Cash/other	3	3	3	4	3	4	4	9
Property	20	22	10	9	3	5	7	7
	100	100	100	100	100	100	100	100

Source: World Markets Company PLC (2008)

Table 11.5 illustrates a typical asset mix allocation based upon funds between 1977 and 2008. Whilst the predominant investment of these funds was in equities, the proportion fluctuated marginally between 1997 and 2008. This emphasises the importance of investing in equities for the best expected return over a period of time. Institutional investment in property showed a considerable decline after 1977, although by the turn of the century returns had improved, ahead of other forms of investment.

The analysis in Tables 11.4 and 11.5 obviously takes an historical perspective, and investors are inevitably concerned about the future performance of their investments. All funds state that past results are no guarantee of future performance. Even with the careful analysis of data there is always an element of chance involved and there are losers as well as those who gain considerable amounts through investment. For the past 20 years it has represented less than 20% of all investments.

Whilst the performance of property investments compared with other forms of investment is poor, there are other factors that must also be borne in mind. The distinctive features of property have already been summarised. These include in particular the individual nature of each item of property, the costs involved in the transfer of ownership and the fact that it takes time to complete property transactions. Equities are liquid, allowing for their easy sale and purchase.

Tables 11.4 and 11.5 compare the annual rates of return on different kinds of investment between 1977 and 2008. During the late 1980s there was a boom in property generally, but this was followed by the rapid collapse of the property market that occurred in the early 1990s. The downturn began in 1991 and did not recover, especially in the south of England, until the end of the decade (see Ashworth 2008). Compared with other forms of investment such as equities, for example, property investment consistently under-performed, other than in a few individual years (see Table 11.4).

At the turn of the 21st century, property was enjoying a comeback, albeit probably brief. Equities were performing poorly against the background of a predicted world recession, fuelled by uncertainties in world events. Economic forecasters and the Royal Institution of Chartered Surveyors were all suggesting that a slowdown in UK property prices would occur during the early part of 2002. In some areas of the country, property prices were expected to fall marginally.

However, worse was yet to come and this is not reflected in the above data. In 2008 the slide in property values started to occur. Whilst this had been carefully predicted by some authors (ABN–AMRO 2007), many chose to ignore the signs. This report suggested falls in the valuation of property. It suggested that UK housing, for example, was almost 50% overvalued, compared with 25% in the USA. This report recognised that this seemed difficult to rationalise, considering at the time trends in employment, income and interest rates. Structural factors, such as a lack of supply or migration, did not seem to explain the premium on UK housing, as rents have remained subdued relative to prices. Toxic debts had still to come.

Overseas property has only ever represented a very small proportion of total investments and is now less than 0.5% of total assets. Overseas property investments have performed poorly, being one of the few types of investment to yield returns below the retail prices index (RPI).

Using the RPI as a measure of comparison, property has generally performed well throughout the period under examination. Over the ten-year period between 1991 and 2000, property returns exceeded the percentages of the RPI and average earnings in eight of the ten years. However, measured over a longer period, investment in UK property has overall performed at about only half the rate of United Kingdom equities.

Investment in property is also now much more of an international venture, with Japan and the USA having considerable investments in European property, and the UK investing abroad in the USA and Australia. Recent surveys have indicated that London, amongst European cities, remains the most popular and sought-after location for commercial premises, ahead of Paris, Frankfurt and Brussels. In recent years it has been able to consolidate its premier position. The analysis is based upon a survey of businesses using eleven indicators, such as access to markets, costs and availability of staff, communications, and availability of accommodation. The cities of southern Europe, such as Barcelona, Milan and Madrid, have all become more important in recent years, whereas some of the provincial cities of the UK have declined, relatively, in their importance.

International investors also have to consider currency issues in addition to the other common problems that face investors. For example, during the early 1980s the UK investor could have gained over 18% per annum through currency alone. Over the subsequent period, most of these gains were reversed, and for the three years up to 1987 the loss on the US dollar reached almost 15% per annum. Until 1988, the Japanese yen showed consistent strength against sterling, never falling below 8.5% per annum. In the early 1980s the German mark displayed consistent gains against sterling, but since 1992 it has shown a sharp rise in variability. The refusal of the UK government to join the common European currency (the euro) has sent messages to overseas investors. Clearly, in terms of manufacturing this has presented difficulties, especially bearing in mind the value of the strong UK pound.

11.11 CARBON TAX

Carbon tax, or environmental taxation, based on CO_2 emissions from commercial buildings, industry and the home owner is yet to be introduced. A growing number of pressure groups and environmentalists wish to reduce greenhouse gases and in theory prevent a real change in the climate of the world. In due course a carbon tax may be introduced based on the carbon footprint of homes or businesses (see also p. 21).

In the EU, including the UK, governments are committed to reducing CO_2 emissions throughout industry as part of their environmental policy. Ever-increasing energy demands usually translate into more investment in fossil-fuelled plants in some parts of the world. The introduction of *tax credits* for sectors of industry was designed to reward businesses that were making efforts to reduce their carbon footprint. Reducing emissions can be achieved in a number of ways. The primary key is consumption by the user. The calculation of CO_2 emissions by homes and businesses is already under way. In short, if consumption is reduced, then the footprint is naturally reduced, and money is saved on utility bills.

Commerce and industry are large consumers of energy. Yet in many cases it takes only a small investment by a householder or business in energy efficiency measures to reduce energy usage, and thus reduce carbon emissions. For businesses, government grants are available for energy-efficient greener industrial plants. These grants can help the asset owner to meet the costs of capital expenditure on equipment. Similarly energy efficiency grants are available to owners of residential property; indeed landlords actively seek grants from local authorities and energy companies for the installation of additional insulation in homes.

11.11.1 Could a carbon tax be implemented?

The implementation of a carbon tax would be relatively straightforward for any government to bring into effect. The government regularly receives information about commercial buildings and residential property. Carbon emissions based on a carbon footprint are already being calculated and submitted to a central database. Energy taxation will almost certainly become a reality, when sufficient numbers of properties have been assessed for their environmental impact, in terms of CO_2 emissions. A sliding-scale tax band could then be simply applied on the basis of how 'green' a particular building is. If such a tax were introduced there would be no exemptions from paying some form of tax to the government, however small the CO_2 footprint. To avoid paying the tax is to become more energy-efficient.

Asset ratings for commercial buildings will, in the future, become more important as a property with a poor rating may well prove less attractive to potential clients. Improved environmental ratings for lower CO_2 emissions may one day be a taxable proposition. Climate change may already be under way but businesses may well see being *environmentally friendly* as a way of attracting investors.

Investors may need to invest further financial resources to enhance unsustainable assets in the marketplace. In this environment sustainable assets would be more liquid than their unsustainable counterparts. It is probable that a two-tier market will develop, where energy efficiency will be considered as prime space.

Eventually when buildings have been surveyed to determine their efficiency ratings and the data collated in a central database this information could be made available to property appraisers.

SELF ASSESSMENT QUESTIONS

1. What are the major forms of taxation on buildings and what is their implication for new building projects?

2. 'Planning regulations are designed to prevent undesirable building developments from taking place, whereas government grants are provided to encourage desirable building developments.' Discuss.

3. What are the advantages and disadvantages of investing in property? On the basis of published data would you recommend this today to a potential investor?

BIBLIOGRAPHY

ABN-AMRO, *Euro Vision: Home truths*. ABN-AMRO Bank 2007.

Ashworth, A., *Precontract Studies: Development Economics, Tendering and Estimating*, 3rd edn. Blackwell Science Ltd. 2008.

Brett, M., *Property and Money*. Estates Gazette 1997.

Coke, R., *Property Finance*. Butterworth Tolley 2001.

Crooks, R., *Tolley's VAT on Construction and Property*. Tolley Publishing Co. 2005.

DTZ Debenham Thorpe, *Money into Property*. Debenham Thorpe Limited 2006.

McIntosh, A.P.J. and Sykes, S.G., *A Guide to Institutional Property Investment*. Macmillan 1985.

Ministry of Public Building and Works, *Implications of Taxation Provisions for Building Maintenance*. HMSO 1989.

Nash, T. (ed.), *Buildings that Work for your Business: Building Premises to Enhance Performance*. Davis Langdon and Everest 2001.

Philipp, A., *Pension Funds and their Advisers*. AP Information Services 2007.

RICS, *The Surveyor's Construction Handbook*. Royal Institution of Chartered Surveyors (updated annually).

The World Markets Company, *UK Pension Fund Service Annual Review*. The World Markets Company 2008.

Williams, B., 'Tax efficient design', *Facilities*, May 1983.

CHAPTER 12

COST RESEARCH AND INNOVATION

LEARNING OUTCOMES

After reading this chapter, you should have an understanding about research and its implications for innovation in respect of cost studies of buildings and its relevance to the construction industry. You should be able to:

- Understand the importance of research in terms of making progress and improvement
- Identify the different philosophies regarding building cost research
- Recognise the developments in quantity surveying practice
- Understand the research process
- Identify organisations applicable to this research
- Consider the implications for the future

12.1 INTRODUCTION

Cost research in the construction industry involves the investigation of any matters which affect the costs of construction, either initially or throughout the building's life. The research may be done for the benefit of the client, contractor or developer or to suit the needs of the professionals and, in a broader context, industry and society. Some of the research that has been undertaken in the past has been directly related to improving the quality and scope of the professional service offered to the industry. The broader objective of any research is to provide a better understanding of the subject, based on empirical and reliable evidence rather than rule-of-thumb or experience alone. It is hoped that the better understanding will then allow an improved service to be provided with a greater level of confidence.

Most cost research has been carried out by academics, and the growth of research in this subject generally has coincided with the development and expansion of undergraduate courses in surveying and the need to underpin students' learning. Some of the larger surveying practices have also established their own research departments and programmes, with an emphasis on improving their own knowledge and databases of information, but with a clear focus on better serving the needs of clients in a commercial environment.

The importance of research has been identified in Chapter 3. The following refers to research globally, to the UK in general and to surveying specifically:

■ Technical change is accelerating and progressive businesses tend to adopt new techniques and applications quickly.
■ Research and development are inseparable from the well-being and prosperity of a country and of the businesses within the country.
■ Expenditure on construction research and development increased during the past decade, but in Britain it still amounts to only 0.65% of the construction output.
■ Research and innovation are inseparable.
■ In respect of expenditure on research and development, the British construction industry lags far behind both competitors overseas and other British industries.
■ The value of research and development to the construction industry cannot be overestimated. Research and development are necessary to maintain international competitiveness and success, particularly as the craft-based traditions of construction diminish and the technological base expands.
■ This background of constant change and challenges demands an effective research and development base to introduce change effectively and efficiently.
■ While research in the construction industry is lagging behind that in other industries and other countries, it is far ahead of that in surveying.

12.2 COST RESEARCH PHILOSOPHY

Although the profession of quantity surveying has existed for some considerable time (records go back to the eighteenth century and there is evidence of it much further back), its progress and development in the construction industry have largely been determined by the needs of practice alone. The profession developed largely on the basis of the usefulness of its practices to clients and contractors. Many of its activities have been determined through 'common sense' and the need to achieve solutions to problems simply, within a limited timescale and at an economical price. Too little emphasis has been directed to evaluating these practices, and any improvement has largely been a good idea. For example, there is a limited knowledge base relating the costs of construction and their determinants. Quantity surveying has therefore developed remarkably well as a strong practical profession, but without, until the past twenty years, a defined academic base. The lack of this base has tended to inhibit future research and development, and created suspicion among some practitioners. Without this development and understanding of the knowledge base, its study at undergraduate level may become questionable. This may sound like a self-preservation philosophy, but it is more serious than that. The lack of a research base will inhibit the development of such courses, with a consequence that good-quality students will not apply for such courses and the discipline will enter a downward spiral, lacking in growth, recognition and influence.

A further difficulty facing researchers is where to start. For example, for the measurement of construction works it is generally assumed that the most appropriate or sensible area is the finished quantities of work. Previous research has established that by measuring the major 100 or so items, cost can be predicted as easily as by measuring, say, 2,000 items (see Chapter 4). The Standard Method of Measurement development has therefore sought to mirror this approach in refining the commonly accepted method of measuring building works. Sometimes professional bodies get in the way. Why then have two different methods, if the principles can generally be agreed regardless of the type of project? Practice, however, suggests that many different sorts of methods of measurement are needed and used to suit different circumstances. But what about foreign countries, it may be asked – how do they quantify construction works? The question, however, is of a more fundamental nature. It may not be a question of whether appropriate research can improve the methods of measurement, but whether we should be measuring in this way at all. For example, during the 1970s it was suggested that construction costs in the future might be more accurately and quickly forecast by putting the information into a mathematical model. This was a radical solution and one that seemed far removed from quantity surveying. As the research progressed it became possible that this might achieve the desired aims. But it was ahead of its time, was not understood by the majority of practitioners and was doomed to failure. It did not achieve its desired aims. It did, however, add a new dimension to the way that building costs could be determined, and provided a bit of lateral thinking.

Cost research is akin to a person prospecting for oil in a field. When no oil is found, then the suggestion made is to dig deeper boreholes. However, the real solution may be to move into a new field. It is often assumed that the surveying profession is moving along the right track, but what may be needed is a paradigm shift, into a new way of thinking. Rather than putting on a few more storeys of development, it would be prudent for the profession to insist that its foundations of knowledge were properly in place and tested for their soundness and appropriateness.

12.3 COLLABORATION WITH PRACTICE

A further problem facing the researcher relates to the objectives of the research and the views of practising surveyors. Understanding almost for the sake of it is frowned upon by some in practice. They would like to see researchers helping them become more efficient, developing their opportunities and generally working at the leading edge of the profession. There is an unfortunate gap between practitioners and academics, although in reality all are seeking to achieve the same ends. The researcher who ignores the advice of the experienced practitioner is arrogant. The practitioner who discounts the contribution of research is foolish. A President of the RICS in the 1980s said that, 'Research papers circulated in the industry from academics were often produced without any link with practice, and often made only little contribution to the construction industry. Greater links between practice and education are not only desirable but essential, so that the latter may receive the support and liaison they deserve and the research will then be of a more positive nature.' The majority of serious researchers today will use industry as a base for their information and data and, if they are wise, will work in collaboration with a practitioner wherever this is possible.

The general position of cost research can therefore be summarised as follows:

- Little attempt has been made to verify current practice.
- Understanding is limited to experience and intuition.
- Current practice may be unsoundly based, and thus there are disadvantages of continuing along this route as a priority.
- Future investigations should build on a factual basis, and not on the possibility of false assumptions.
- The profession should be convinced of the necessity of getting the fundamental principles authenticated, since these are often not properly understood.
- Opinions alone may be biased and therefore unreliable.

During the past 50 years, cost research has been undertaken on a small but increasing scale. The range of projects covered has been considerable, indicating that research has been spread over a wide area, reflecting the interests of those involved in this activity. In most subject areas it is accepted that large amounts of research will need to be undertaken to move a discipline forward by just a small step. The nature of research is such that the researcher is never sure what may be found out. It is akin to a ship in uncharted waters.

12.3.1 Quantity surveying practices

Several of the larger surveying practices have established research and innovation sections as an integral part of their work. This has been done in support of the diversification of practices. It is also used to enrich the knowledge base of existing activities in order to deliver highly relevant services. Research activities, wherever these are carried out, enhance an image whilst at the same time they provide a competitive edge in order to deliver client-effective services. In some cases the larger clients of the industry commission research in an attempt to resolve specific problems in the industry. Some practices are able to recoup fees from these activities. Reviewing current practices or finding ways for a better understanding of construction costs can sometimes yield future cost savings that make the initial costs of the research miniscule. Some practices have developed formal links with universities for this purpose, have become members of research advisory teams, allowed researchers access to their data and information or sponsored projects.

Research and innovation are important for the following reasons:

- Developing changes in the way in which practice continues to evolve
- Maintaining a leading edge over competing practices
- Improving the quality of the service provided to clients
- Finding more effective, efficient and economic ways of achieving the same objectives

- Improving the efficiency of work practices
- Extending the services which can be provided
- Developing a greater awareness and practice of new technologies
- Providing a fee-earning capability from these activities
- Enhancing the image of a practice through publicity

Many of the large quantity surveying practices recognise that there is a general shortage of in-depth research in the built environment that can be used to meet the needs of practices and their clients. Their activities will cover a wide spectrum and include the following that are of importance to the cost studies of buildings:

- Cost research
- Establishing and maintaining cost indices
- Cash flow analysis
- Whole-life costing
- Sustainability
- Procurement methods
- Project and building types
- Tax
- Value management
- Cost planning
- Risk analysis
- Due diligence
- Tall, large and complex structures
- International comparisons
- Whole-building cost appraisal

Access to the websites of the larger quantity surveying practices will yield a wide range of different activities in research that is associated with construction costs. They provide reports and brochures on the research projects which they have completed. Many of these can be accessed freely, but for the more detailed analysis a charge is usually made.

Many of these practices will also prepare regular market overview reports to inform clients and the construction industry. Some of these reports in the middle of 2009 were suggesting:

- The recession would be long and deep
- Tender prices may show three years of successive falls
- Tender prices may fall by 10% in 2009 and 6% in 2010
- Tender prices may fall by as much as 25% in the hardest-hit sectors
- Contractors are cutting profits
- Subcontract trade package prices are falling fast due to reduced opportunities and the need to secure future workloads
- A greater number of contractor failures is expected in an industry that already tops the league for bankruptcies and liquidations
- The industry will become more claims-conscious to avert losses
- Clients who are in long-term supply chains will secure greater benefits

- If, as expected, the market begins to recover in 2011, clients should position themselves to take advantage of low prices now and gear themselves up for an uptake of activities
- The euro exchange rate is expected to stabilise at €1.20 by the beginning of 2010

All of the larger practices now have an international perspective as they have offices in many of the larger cities around the world. These practices are therefore able to provide advice at the local level whilst at the same time making comparisons with other countries. A number of practices provide reports that focus on particular countries. These include:

- The countries' economic performance and future projections
- Construction activity
- Indicative building costs
- Economic analysis
- Public and private investment in construction

These practices also provide incisive reports on aspects of construction activity. Some of these include:

- Building Schools for the Future (BSF)
- Cost and time certainty
- Supply chain efficiencies
- Maintenance management
- How to weather the storm
- Project recovery
- Strategic asset management

An interesting factor common to all of these large firms is that quantity surveying has become a subset of their work. Some now appear to be unbranded whilst others use the catch-all title of construction and management consultants.

12.4 DEVELOPMENTS IN QUANTITY SURVEYING

There is no need to recount the development of the quantity surveying profession from contractors' measurers to its now strategic role in the analysis, forecasting and accounting of costs. Other books and papers have developed this history. Construction costs have been almost exclusively within the province of the quantity surveyor. Few designers have sought to acquire the necessary knowledge or understanding, or wish to take on this responsibility. Their role is with design, in making buildings functional and attractive. The quantity surveyor is there to ensure that this is done at a price within the constraints of the funding available and to offer a view on value for money. Back in the 1960s quantity surveyors were being asked why it cost 50% more to build a primary school in one area than another. The then Department of Education began a study of cost planning, realising that its money was being poorly spent to meet different standards and criteria in different parts of the country. The result was that the large amount of school building in the 1960s meant that more could be done with less: a philosophy that has taken some time to develop in other areas of work. Today the quantity surveyor is being asked to examine plans for new buildings for their overall cost-effectiveness, without reducing the project's overall performance or aesthetic appeal. A shift has also occurred from considering initial building costs alone to whole-life concepts. It has been suggested that if construction costs were to be better spent and controlled, the project as a whole would need to be much better managed with clearer objectives. Table 12.1 lists some of the milestones in surveying practice over the past 40 years. The majority of the achievements in practice have originated from professional practice rather than from education or research. This has been largely because both undergraduate teaching and research are still in their relative infancy, and practice has evolved to meet perceived and pragmatic needs and to provide efficiency, effectiveness and economy.

12.5 THE RESEARCH PROCESS

The research process (Figure 12.1) usually commences with a desire by an individual to 'do' research or with industry or universities establishing programmes of study for which they then recruit researchers. The initial starting point for someone new to research is setting down the objectives as a research proposal. In many cases these are too broad, with expectations far outweighing possible achievements. The better proposals are those which are related to a particular problem, where the researcher already has some knowledge and experience. While the objectives should be clearly defined at the outset, it is not uncommon to find these being refined as the project gets under way. It may be necessary to reappraise the problem after an initial investigation and thorough literature search. It is then necessary to determine which methods should be used to best achieve the revised objectives. In the first instance the literature survey will reveal the extent of the available knowledge and which methods have already been used in similar situations. The failure of a previous project should not automatically tell us that an inappropriate method was used. In some cases the researcher may have given up through a lack of determination or through frustration, or information which was then not available may now be available through new information technology. For some projects the literature survey may be extensive, with one research paper leading on to ten new references. The supervisor of the research should be able to point the researcher in the right direction. Invariably some sort of data will be required, and strict codes of practice regarding collection and confidentiality will need to be observed during their acquisition. A brief description of the possible research methods is given later in this chapter. The data will need to be tested for sufficiency and statistical reliability, and against new data if the research findings are to be tested.

An alternative way of considering the research process is shown in Figure 12.2. This considers the different phases of the research process as a continuous event or

Table 12.1 Chronology of some developments in research and practice

Characteristics	Themes
Pre-1960s	
Bills of quantities	Practice-based profession
Final accounts	
Approximate estimating	Post-war building boom
Building bulletin 4, 'Cost study'	
1960s	
Cut and shuffle	Practice-based, but efforts made towards
Cost planning	understanding building costs and the development
BCIS	of cost centres for analysis
Cost analysis	
Elemental bills	
Operational bills	Use of computers for bill processing
Standard phraseology	
1970s	
Data co-ordination	Use of computers for processes other than bills of
Computer bills	quantities
Cost studies of buildings/components	
BMI	
Costs-in-use	Development of undergraduate courses in quantity
Cost modelling	surveying
Bidding strategy	
Computer applications	
Contractors' estimating	
Formula methods of price adjustment	
1980s	
Life-cycle costing	Emphasis shift towards whole-life costing
Coordinated project information	
Cost data explosion	
Project management	Role of QS in project management
Procurement systems	Development of postgraduate courses
EC and world comparisons	
Cost engineering methods	Diversification of professional activities
Accuracy in forecasting	
Industry analysis	
Value analysis/engineering management	Value-added concept
Risk analysis	
1990s	
Quality systems	Decade of quality
Expert systems	
Facilities management	Impetus of the single European market
Commercial revolution	Multidisciplinary working and developments
2000s	
Sustainable development	Rethinking construction
Benchmarking	Lean methods and waste elimination
Whole-life costing	

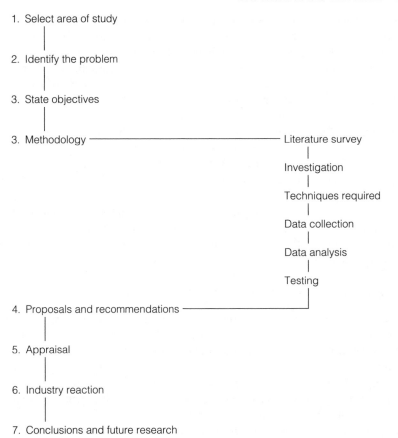

1. Select area of study

2. Identify the problem

3. State objectives

3. Methodology ———————————————— Literature survey

Investigation

Techniques required

Data collection

Data analysis

Testing

4. Proposals and recommendations ————————

5. Appraisal

6. Industry reaction

7. Conclusions and future research

Fig. 12.1   The research process

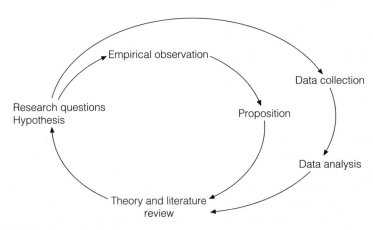

Empirical observation

Research questions
Hypothesis

Proposition

Data collection

Data analysis

Theory and literature
review

Fig. 12.2   The research wheel

research wheel, where processes are repeated over time to establish the validity of a hypothesis. The most common entry point is usually due to empirical observation, where a researcher has identified a problem and wonders whether a solution can be found. The proposition may be based on a hunch that is typically guided by the values, assumptions and goals of the researcher. The proposition must then be set within the broader context of theory and previous research. The researcher must move from the larger context of theory to generate a more specific research question or hypothesis. The first loop is completed as the researcher seeks to discover or collect data that will serve to answer the research question.

The culmination of the research will be a publication which explains its methodology and achievements and what further work should now be undertaken to solve other problems discovered during the research. The research may show that an anticipated solution failed to solve the problem for one reason or another, but recommend a course of action that the researcher thinks could achieve the desired results in the future. The failure to achieve the objectives does not imply that the research was not undertaken correctly, since the negative results may aid the direction of a future research proposal.

Research may be classified under the following headings:

- *Basic research*: Experimental or theoretical work undertaken primarily to acquire new knowledge of underlying foundations of phenomena and observable facts. Undertaken without any particular application in mind.
- *Strategic research*: Applied research which is in a subject area which has not yet advanced to the stage where eventual applications can be clearly specified.
- *Applied research*: The acquisition of new knowledge which is primarily directed towards specific practical aims or objectives.
- *Scholarship*: Work which is intended to expand the boundaries of knowledge within and across disciplines by in-depth analysis, synthesis and interpretation of ideas and information and by making use of rigorous and documented methodology.
- *Creative work*: The invention and generation of ideas, images and artefacts including design. Usually applied to the pursuit of knowledge in the arts.
- *Consultancy*: The development of existing knowledge for the resolution of specific problems presented by clients often within an industrial or commercial context.

12.6 RESEARCH METHODS

Although the methodology of research is important, it can sometimes be overemphasised. It should be regarded as nothing more than the tools of the trade. It is important to be aware of the range of research methods available and to appreciate their relevance to the work being studied in order to meet the objectives of the research. It should also be noted that many researchers have never received any formal instruction in methodology, and yet there is no evidence to suggest that their work is inferior to that of the supposedly better informed people. It may be necessary to combine some of the following methods during the carrying out of research projects.

12.6.1 Surveys

The primary function of surveys is to collect information which can be analysed to produce conclusions. The main purpose of surveys is to describe what is happening and to obtain all the facts which are considered to be relevant. The survey may also explain, by identification or through analysis, the facts which have arisen from it. The different types of survey may be classified as follows.

Observation surveys

The simple process of observing and recording events or situations is probably the oldest form of research. A great deal can be learned by careful observation of the world around us. The nature of the construction industry, however, does make these techniques very difficult to establish.

Questionnaire surveys

These are often used at some stage during a cost research project. The most popular types of questionnaire are those which have what are known as closed questions. The respondent is asked a question and is required to answer by choosing between a number of alternatives. This type of questionnaire is easy to complete and analyse. Questionnaires are more likely to be used when opinions and views are sought than where facts are available. The design of the questionnaire is very important, and expert advice may need to be sought during its preparation.

Interview surveys

These must be properly structured by using an interview schedule, and as such are very similar to questionnaires. They are less formal than the latter, but the researcher can gain other information from the way the questions are answered. The interview also provides more opportunity to obtain qualified answers by probing and prompting.

Sampling

This is an essential part of any survey, since it is unlikely that an entire population can be surveyed. The secret is to select a sample which will be representative or have the same characteristics as the overall population. There are strict rules to be observed when using this statistical technique, and these should be followed at all times.

12.6.2 Experimental research

Experimental research is the traditional type of research used by scientists. The researcher sets up an experiment in order to test a hypothesis or theory.

The experiment is usually designed so that the researcher has as much control as possible over the conditions under which the actual tests take place. For this reason most scientific experiments take place in laboratories. Also, in order to draw sound conclusions from the results the experiment needs to be repeated several times.

12.6.3 Historical research

Historical research is essentially an attempt to describe and learn from the past. It can therefore be purely descriptive, recording the sequence of events and seeking to present the fullest possible picture of the development of something.

12.6.4 Case studies

Case studies are usually used when the researcher is attempting to understand complex organisational problems. They allow the researcher to focus on something which is sufficiently manageable to be understood in all its complexity.

12.6.5 World of facts

The objective of any research should be to provide us with more factual information about a subject, as shown in Figure 12.3. This should then enable us to move away from hypotheses to laws, as shown in Figure 12.4.

12.6.6 Operational research

Operational research was developed during the Second World War to help military planners cope with large-scale logistical problems. Since then the techniques have been applied to a wide variety of organisations. It is concerned with problems of uncertainty. Essentially it is the application of scientific method to management

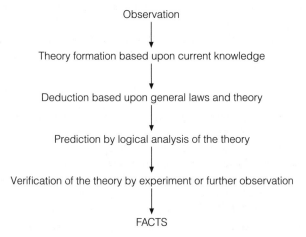

Observation

↓

Theory formation based upon current knowledge

↓

Deduction based upon general laws and theory

↓

Prediction by logical analysis of the theory

↓

Verification of the theory by experiment or further observation

↓

FACTS

Fig. 12.3   World of facts

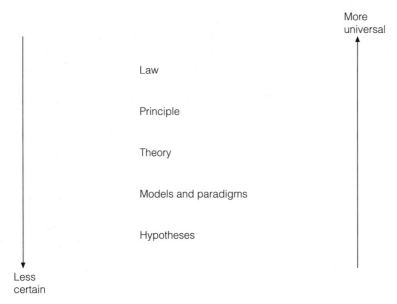

More
universal

Law

Principle

Theory

Models and paradigms

Hypotheses

Less
certain

Fig. 12.4   Research propositions

decision-making. The operational researcher attempts to understand the different forces and relationships which cause organisations to behave in the way they do. When this has been understood the next stage is to construct a model which can be used to explain this behaviour. This model generally takes the form of a mathematical equation (see Chapter 16).

12.6.7 Field testing

Field testing consists of trying out an idea or procedure in real life. The researcher has to convince the construction client to take the risk of trying out the new proposals. A field test is sometimes used at the termination of a research project in an attempt to prove its practical worth. The new proposals may in some instances be run in tandem with an existing system.

12.7 RESEARCH ORGANISATIONS IN THE CONSTRUCTION INDUSTRY

There is a wide variety of research organisations in the construction industry representing its different sectors and interests. Many are concerned with the improvement of materials, goods, components and technology. These include, amongst others, TRADA (Timber Research and Development Association), BDA (The Brick Development Association) and BCA (British Cement Association). Others represent government departments that have an interest in the government estate and the wider aspects of the construction industry generally, such as the Office of Government and Commerce (OGC). The different professional bodies in the construction industry also have research interests relevant to their members' needs. All these organisations provide support and funding for research, but the latter is often small. Evidence suggests that, as a whole, the construction industry spends a disproportionately small amount on research and innovation development. The following represent some important research organisations with research interests in cost studies.

12.7.1 Building Research Trust companies

The BRE Trust Companies include Building Research Establishment Ltd (BRE), and BRE Global Ltd. The BRE Group companies are held as a national asset on behalf of the construction industry and its clients, independent of specific commercial interests, which protects BRE's impartiality and objectivity in research and advice.

The BRE Trust companies' mission is to build a better world with a vision to provide an unmistakable imprint on a highly regarded and sustainable built environment. Their authority is derived from knowledge, independence, objectivity and ethics and a commitment to maximise client benefit through a deep understanding of their needs and aspirations. They place emphasis on quality, sustainability and health and safety. The BRE Trust companies work closely with clients and partners in targeted sectors to help find the best possible solutions and maximise innovation, sustainability and assurance. These solutions include the following:

Building healthy hospitals: That are bright, welcoming, comfortable, efficient, safe, secure and sustainable environments designed to foster patient recovery.

Building new and better homes: Meeting the challenge to provide quality, affordable and sustainable homes in the face of serious skills shortages, while avoiding damaging environmental impacts and interminable planning controversies.

Building inspirational schools: Designing schools with the needs of pupils, teachers and the wider community firmly in mind.

Building retail and leisure facilities: Meeting the growing challenges of large commercial buildings.

Building successful offices: Building the offices that deliver better economic, social and environmental performance.

Infrastructure and transport systems: Helping with the environmental challenges of major infrastructure and transport system regeneration projects.

12.7.2 Engineering and Physical Sciences Research Council (EPSRC)

EPSRC is the main UK government agency for funding research and training in engineering and the physical sciences, investing around £740 million a year in a broad range of subjects which includes construction activities. The EPSRC is one of five councils funded through the Department for Innovation Universities and Skills. Its primary purpose is to sustain standards of education and research through grants and studentships and facilities for academic research. Of particular interest to those in the construction industry are the programmes of research in building, civil engineering, energy in buildings and construction management.

12.7.3 International Council for Building Research Studies and Documentation (CIB)

CIB is the acronym of the abbreviated French (former) name 'Conseil International du Bâtiment' (in English this is International Council for Building). In the course of 1998, the abbreviation was kept but the full name was changed to the International Council for Research and Innovation in Building and Construction. The CIB was established in 1953 as an association whose objectives were to stimulate and facilitate international co-operation and information exchange between governmental research institutes in the building and construction sector, with an emphasis on those institutes engaged in technical fields of research. It has since developed into a worldwide network of over 5,000 experts from about 500 member organisations with a research, university, industry or government background, who collectively are active in all aspects of research and innovation for building and construction.

Its aims are *to encourage, facilitate and develop international co-operation in building, housing and planning research, studies and documentation*. It covers not only the technical but also the economic and social aspects of building and the related environment. The CIB seeks to attain these objectives in three ways: first, through a network of highly specialised working commissions and steering groups which operate in a wide variety of subjects; second, through organising congresses, symposia and colloquia on themes of general and particular interest; third, through its publishing activities, whereby the results of working commissions and proceedings of congresses are made available to practitioners and researchers everywhere.

12.7.4 Construction Industry Research and Information Association (CIRIA)

CIRIA operates across market sectors and disciplines delivering programmes of business improvement services and research activities for its members and those engaged with the delivery and operation of the built environment. It is a member-based research and information organisation dedicated to improvement in the construction industry. Its members include representatives from all parts of the supply chains of the modern built environment, covering building and civil engineering as well as transport and utilities infrastructure. CIRIA's work is recognised as being independent, objective and authoritative.

CIRIA provides a means through which the many different stakeholders in the modern built environment can work together to identify, codify, publish and promote the emerging best practice in the industry. In this way, CIRIA continually seeks to raise the standard of excellence in the broad construction sector beyond the lowest admissible acceptability for designers and constructors set by the framework of legislation, statutory regulations and codes of practice.

CIRIA's primary aims are to improve quality, efficiency, cost-effectiveness and safety in both the provision and the operation of the modern built environment. In pursuit of these aims, it seeks to be the leading provider of performance improvement products and services to organisations involved in procuring, delivering, owning and maintaining the fabric of the modern built environment.

12.7.5 Association of Researchers in Construction Management (ARCOM)

ARCOM brings together all those interested in construction management research. Its aim is to further the advancement of knowledge in all aspects of management in construction by supporting education, dissemination of information and research. It provides a quarterly newsletter to keep members in touch with the latest news from other members. It has an annual conference that brings together researchers from all over the world. Various research workshops provide a means to develop specific subject themes as well as general research methodology. Its publications form a very useful resource for all researchers in this field. These activities help to encourage excellence in the research conducted by its members. It is proud of its tradition of focusing upon researchers rather than institutions. Members include research students, lecturers, professors, consultants and other practitioners. Although it is primarily a UK-based association, it welcomes members from all over the world. Membership of ARCOM is a good way of keeping in touch with the latest developments in construction management research. It provides opportunities for contact between researchers in similar fields and between researchers and practitioners. Members pay an annual membership fee, and membership is open to any interested individual. ARCOM has a useful website providing a range of helpful information.

12.7.6 Australian Universities' Building Education Association (AUBEA)

The principal aim of AUBEA is to promote and improve teaching and research in building through communication and collaboration. It aims to co-ordinate efforts to market the discipline of building to prospective students, research partners and clients. AUBEA is a membership-based organisation, with its members being drawn from the community of universities which teach and conduct research in building, property and construction in Australasia. AUBEA holds an annual conference and maintains a strong connection to industry and in particular the professional institutes, such as the AIB, the AIQS and the RICS.

12.7.7 Associated Schools of Construction (ASC)

ASC is a professional association, based in the USA, for the development and advancement of construction education, where the sharing of ideas and knowledge inspires, guides and promotes excellence in curricula, teaching, research and service. The ASC is the international association representing the interests of both academic and industry professionals interested in institutional construction education. The ASC website is an information resource that collects, analyses, evaluates, packages and disseminates information to facilitate accurate communication and decision-making within academic, industry and government settings.

The membership of the ASC, like the field of construction management, is richly interdisciplinary, drawing membership from such disciplines as architecture, engineering, management, technology, to name only a few. The Association offers a variety of programmes and services designed to help its members serve their customers more effectively and succeed in an increasingly challenging environment of information management and technology. The ASC is dedicated to the professional growth and success of its membership, and is committed to fostering excellence in construction communication, scholarship, research, education, and practice. It held its 45th annual conference in 2009.

12.7.8 The European Council of Construction Economists (ECCE)

Conseil Européen des Economistes de la Construction (The Council of European Construction Economists or CEEC) was formed nearly 30 years ago by the coming together of representatives of various national institutions representing the field of construction economics (quantity surveying to the English-speaking nations) in their own countries with the aim of promoting the profession at a European level.

In the diverse European construction industry, many national industries have already absorbed the positive influence of the profession of construction economist in the control of the economics of construction. That discipline has grown into an independent profession, advising building and infrastructure clients or contractors on how to achieve best value for money in an industry which can be notorious for swallowing up large sums of clients' budgets or causing contractors serious losses due to inadequate estimating. In many of these countries there are recognised degree courses in construction economics, for example in the UK, France, Ireland, the Netherlands, Switzerland and Denmark. Where it is not recognised as a separate discipline, it has become in many cases a specialisation for architects, engineers or project managers, who have used the skills of the construction economist to give added value to their own discipline.

12.7.9 Royal Institution of Chartered Surveyors (RICS)

RICS is the leading organisation of its kind in the world for professionals in property, land, construction and environmental assets. It takes its responsibilities of providing impartial advice to governments and policy-makers seriously.

Through its independent research and reports it aims to inform the thinking of those who influence the development of the places where people live and work. It recognises the huge importance of property, land and the built environment to our economic, social and environmental welfare. The RICS publishes research results through:

- **RICS research papers** – a means by which RICS publishes reports on research and activities relevant to the profession. Papers range from fundamental research through to innovative practical applications of new and interesting ideas.
- **RICS research reports** – the results of research funded or supported by RICS in whole or in part.
- **Conference papers** – through its annual COBRA research conference.

12.7.10 Chartered Institute of Building (CIOB)

CIOB produces research in many key fields within the construction industry. This research can be both member- and public-driven with the aim to share knowledge for the benefit of all participants and interested parties. Some of its recent and relevant publications include:

- *Innovation in Construction – Ideas are the Currency of the Future*
- *Managing the Risk of Delayed Completion in the 21st Century*
- *Skills Shortages in the UK Construction Industry*
- *The Green Perspective: A UK construction industry report on sustainability*

12.7.11 University research

All the university schools or departments of the built environment have developed established research programmes. Many submit to the Research Assessment Exercises (RAE) that are managed by the university funding council, typically on a four-year cycle. Research is carried out, usually by teams of staff that may include research assistants funded by grants from a number of external agencies. Other research training is carried out through specially designed masters programmes or guidance to students studying for PhDs. Much of this work results in publications such as journal articles, papers presented at international research conferences or through other means of dissemination. Present and future activities are helping to formulate a reasonable body of research to underpin the cost studies of buildings.

12.7.12 Department for Business Enterprise and Regulatory Reform (BERR)

Innovation is a key driver of competitive advantage for the sector and of society's need for an efficient and sustainable sector. BERR seeks to help the industry develop and implement a shared vision and strategy for innovation, help promulgate this message through the sector and promote its innovation needs to government and internationally. It seeks to stimulate greater private and public investment to support innovation and assist effective transfer and exploitation of existing knowledge. In 2002, Sir John Fairclough produced *Rethinking Construction Innovation and Research*, his review of government research and development policies and practices.

The findings of the report led to the development of the Construction Research and Innovation Strategy Panel (CRISP), subsequently the National Platform, supported by the Strategic Forum. This body operates as a joint government and industry panel identifying and developing priorities for research funders and helping to set the agenda for the Technology Strategy. BERR is working with the National Platform to develop a future Technology Strategy to frame the research needs for the sector.

12.8 RESEARCH DISSEMINATION

The publication of the findings from research is an important part of the process, whether the research has been done as part of an academic course or to resolve a particular problem identified in industry. The acquisition of knowledge is of importance to the wider audience. This is why universities organise conferences to share research findings amongst the academic community and practitioners alike.

Academic journals are able to publish the results from research in a reasonable amount of depth. The process is rigorous because publication is determined by a refereeing process that provides robustness and reliability, as far as is possible, on the way the research has been conducted and the outcomes of the research project. Trade and professional journals are able to include only brief summaries of research, since for much of their readership an in-depth analysis is not required and space is not available. Some professional institutions, notably the Institution of Civil Engineers, have published specialised academic journals to describe research that has been undertaken. Other professional bodies have in some ways attempted to copy this practice, but with varying degrees of success. All the professional bodies recognise that if a profession is to grow and develop it must encourage research as a way forward. Both the RICS and the CIOB (Chartered Institute of Building) publish occasional research papers rather than a journal, but the processes that they employ are much the same. Government departments also undertake or instigate research that is aimed at their particular sectors and publish the outcomes, often through HMSO (Her Majesty's Stationery Office).

It is only about 20 years ago (1983) that Spon launched *Construction Management and Economics* (*CME*) as the first refereed journal for the construction disciplines. This is edited by staff from the University of Reading with the help of a wide editorial board comprising individuals from around the world. In more recent years (1993) the *Journal of Engineering, Construction and Architectural Management* (*ECAM*) has been successfully developed from a base at Loughborough University. A number of other universities have also been developing journals as they recognise the growth in research output and the need to disseminate results at least amongst the wider academic community. There are also countries overseas that now have similar journals, notably the USA with, for example, a number of journals from the

American Society of Civil Engineers (ASCE). In Australia, the *Australian Journal of Construction Economics and Building* is a recently introduced journal from the Australian Institute of Quantity Surveyors and the Australian Institute of Building and is worthy of consultation. As a relatively young discipline, construction needs outlets for its work as it seeks to build up ap. body of sound knowledge.

In addition, there are now many opportunities for research dissemination through national and international conferences. The most notable of these that are relevant to the subject of this book are those of the CIB (see p. 271), ARCOM (see p. 272) and COBRA (the Construction and Building Research Association). Delegates to these conferences have the opportunity to present their findings (which have been reviewed by a scientific panel) and discuss issues that have been raised. They receive a copy of the conference proceedings in hard copy and also as multimedia CDs. In addition, building networks and sharing ideas informally are most important components of most conferences.

The Co-operative Network of Building Researchers (CNBR) was established in 1992 as an electronic exchange of information for researchers, academics and practitioners. It is managed by staff at the Royal Melbourne Institute of Technology (RMIT) and has over 1,500 members in 150 universities in 40 countries. It aims to provide information and discussion about developments in research and in teaching and learning.

The Centre for Education in Building Environment (CEBE), which is managed jointly between the universities of Cardiff and Salford, has a website and provides publications for the wider academic community.

12.9 CONSTRUCTION RESEARCH AND INNOVATION STRATEGY PANEL (CRISP)

This is an industry-led panel under the aegis of the principal umbrella bodies: the Confederation of Construction Clients (CCC), the Construction Industry Council (CIC), the Construction Industry Employers' Council (CIEC), the Constructors' Liaison Group (CLG), and the Construction Products Association (CPA). CRISP brings together government, clients, industry and the research community to consider research priorities for construction. It has close links with Rethinking Construction (see Chapter 23), including the Movement of Innovation, the Housing Forum and the Construction Best Practice Programme. Under the guidance of the Panel and the Executive, and via task groups, workshops, commissions and other activities, CRISP produces sets of topic-based recommendations.

CRISP has a wide research remit encompassing all aspects of generating and refining the knowledge that the industry and its clients need to improve the performance of UK construction. Its innovation remit is focused on encouraging the development of ideas and tools to enhance the application of new knowledge. CRISP actively seeks to identify key areas of missing knowledge and to promote the development of missing tools. It aims to influence both public and private sector funding agencies, and those who are responsible for the content of research programmes. It serves organisations of all sizes, disciplines and sectors of construction. It focuses on the immediate needs and the longer-term requirements of the construction industry, and it seeks to complement rather than to duplicate the work of others.

CRISP has three main drivers: respect for people, customer needs and sustainable construction. These relate carefully to the individuals who work at different levels in the industry and to the importance of satisfying client needs, and they place particular emphasis upon the longer-term effects and needs of its products. The current main focus of its activities is in the following areas:

- Capturing knowledge
- Changing culture
- Climate change
- Construction futures
- Implementing the Fairclough report

It also has ongoing research interests in the following areas of activity:

- Customer needs
- Sustainable construction
- Design
- Technologies and components
- Process
- Performance
- Information and communication technology (ICT)
- Housing
- Construction research base

This is also within a context of the following related activities:

- The regulatory and financial framework
- Motivation and communication

CRISP has a very informative website (www.ncrisp.org.uk) that clearly explains some of the above policies and practices. Within the above scenarios it has identified the needs of the industry and now looks to research organisations and within the construction industry itself to seek to promote lasting solutions towards continuous improvement.

12.10 LINKING CONSTRUCTION INDUSTRY NEEDS AND CONSTRUCTION RESEARCH OUTPUTS

Research is about the acquisition of knowledge that can challenge existing and outdated practices and offer potential solutions to resolve identified problems. Whilst the publication of research is an important output it is not the only one. For some this is an over-emphasis and over-reliance, but the greatest benefits are trained people. There is the need to get more trained researchers active within the industry because information and technology transfer occurs more effectively through people, as was noted above. In linking research and practice, the important place of academic publications is also questioned. These types of publications are only read by a small circle of individuals and the benefits to practice are therefore relatively minor.

It must also not be easily forgotten that research also services the needs of education. Teaching is an important transfer mechanism, especially at the level of masters programmes. This is especially important for creating a community with a capacity to absorb and apply the results of research outputs. However, academic research is not necessarily designed to be, and often is not, intelligible or indeed usable by the construction industry. Some academic research lacks immediate relevance to industry and practice, and there is a need for better alignment and an understanding of each other's needs. Academic research also often has longer timescales and is averse to quick-fix solutions. It relies upon rigorous knowledge building that can achieve long-term benefits. A distinction needs to be made between knowledge and information. Knowledge requires an interpretive framework to which we bring information.

Strategies must be developed to identify research and innovation needs within the sector and to encourage research councils and funding bodies to focus, although not exclusively, on these needs. This would provide an outward-looking perspective with an external agenda. Practice also requires a self-transformation perspective with an internal agenda. Long-term partnerships between industry and universities are required to avoid a haemorrhaging of knowledge.

12.10.1 Learning from other industries

A considerable amount of research is undertaken by other industry sectors that could have a relevance and application to the construction industry. This derived knowledge and information is not always easily transferred; in some case its relevance is not noticed. In other cases there is the commonly held view that the construction industry is completely different from other industries and therefore much that might be transferred is not relevant. Drawing on the work of other industries could augment the construction industry's performance by maximising knowledge and avoiding a duplication of effort. Of course not all duplication is a waste of time and effort since this might help to authenticate new knowledge and practices. The industries with which the most useful comparisons might be made are the aerospace and automobile industries.

Table 12.2 draws some comparisons between the construction industry and the better-performing industries that include automobiles and aerospace. Whilst the construction industry is achieving continuous improvement, despite its structure, other industries are shown to be more advanced against a number of criteria. Due account must be taken of both the process and the product, and the differences between manufacturing on a client's construction site in all weathers, and the obvious efficiency gains that are easier to achieve and maintain in a single-factory environment. However, the construction industry has at times been slow to recognise that changes in practices can nevertheless be achieved by adopting some

Table 12.2 Comparisons between the construction sector and the best performing industries

Construction industry	Best performing industries
Separation of design and manufacture	Integrated design and manufacturing teams
Limited emphasis on research practice	High recognition of the importance of research and innovation
Focus on initial construction costs	Emphasis on whole-cost considerations
Fragmented supply chains	Small number of preferred suppliers
Teams that change with every project	Team consistency
Adversarial relationships	Trusted supply chains
Bespoke solutions	Prototypes tested before manufacture
Dangerous working environments	Health and safety priorities
Increasing emphasis on environmental issues	Environment considerations at the top of the agenda
Poor image perceptions	Perceived high technology practices

ideas from other industries, some of which have similarities and others of which are markedly different.

In particular Flanagan (1999) has identified the following factors that distinguish the automotive and aerospace sectors:

- A manufacturing culture
- The integration of design with production
- Suppliers being involved in the design team
- A focus on innovation and technology
- Recognition that investment in research and development is crucial in the selling of a product
- An acceptance of standardisation in the design, components and assembly across a product range
- The dissemination of ideas is critical to this process

12.11 THE FUTURE

The role of surveyors has changed considerably during the past 30 years. Even this will probably be overshadowed by the changes which might be expected in the future. The industrial revolution which took place at the beginning of the nineteenth century is being followed by the revolution in commerce that is gaining momentum daily. These changes are a result of the advancement of the paperless office forecast over fifteen years ago. The capability of hardware and software at that time was minuscule compared with the ease, cost, capability and reliability of the systems available today. The following are some of the issues which will face the profession around the turn of the century, and which require research now to provide a sound analysis for the future.

Information technology

The use of computers for preparing bills of quantities in the late 1960s has progressed to such an extent that a designer's drawn information can be converted into a contractor's tender almost at the press of a button. This relies on the architects producing appropriate drawings in the first place, but even in the absence of these the computer will fill in the missing parts with assumptions that can easily be changed to suit the correct design. Where the computer link between drawings and bills has not been made, because of an architect's preference for a particular CAD system, the link between a bill of quantities and a contractor's tender is now available. EDICON (the UK construction industry forum for electronic data interchange) has developed a system which meets this capability. Integrated packages that link pre-contract documentation and post-contract work are now available. Greater use in the future will be made of those computer systems which capture the expertise of the practitioner and refine it for future applications.

Employment

Just as British manufacturing industry cannot compete on costs with the inexpensive capability of the developing world, neither will commerce be able to compete. Because society is moving much faster than at the beginning of the twentieth century, the changes in office practice will occur more quickly than the industrial revolution. The difficulties of this implementation are now restricted to culture change, language and confidentiality over the airwaves. Quantity surveying practitioners realised a few years ago that it was more cost-efficient, for example, to undertake bill production in the north of England than to employ expensive staff in Central London. Access to all kinds of information is now easily obtained through electronic mail and fax machines. Some office processes are likely to be moved to areas of the world where salaries are not as high, but the service provided is just as good.

Diversification

There has been a blurring of professional boundaries over the past few years. The surveyor's role, like many in other professions, has been to diversify into work that previously would have been undertaken by another professional discipline. This is true of all professions both within and outside of the construction industry. The age of the management consultant has arrived, who is able to solve a client's problem or employ the services of someone who will be able to offer the appropriate advice. QS2000 states that 'Significant changes have been occurring in the structure of the profession as a result of wider changes in the industry. . . . Significant but less measurable are changing attitudes to practice and professionalism among, in particular, younger quantity surveyors responding to the more aggressive and commercially minded working environment in the mid-to-late 1980s' (Davis

Langdon and Everest 1991). A recent President of the QS division of the RICS said, 'It is clear that the role of the quantity surveyor is expanding and this is reflected in the growth of the bigger interdisciplinary and international QS practices as well as the growth of niche practices specialising in, for example, taxation advice or dispute resolution' (Powell, C. 1998).

Business orientation

According to QS2000, 'practice is increasingly characterised by a business oriented approach emphasising, for example, rapid turn-round of information and improved quality of communication and presentation'. There is still a dichotomy between business and profession, and this is mirrored in all of the major professions. The driving force in the past was to put the quality of service above profits. Business practice tends to turn this approach on its head.

International factors

It has sometimes been suggested that quantity surveying is a British profession which is not practised beyond our shores. This is of course far from the truth. Its influence in all Commonwealth and ex-Commonwealth countries has been as significant as it has in Britain. Quantity surveying has also been practised extensively in many Middle Eastern countries, where British contractors and consultants have been employed on the development of projects. While the removal of the Iron Curtain separating Eastern Europe has opened up new opportunities, quantity surveyors have been working in Western Europe for many years. The Single European Market which came into operation on 31 December 1992 provides an additional impetus for quantity surveying skills. Organisations such as the European Technical Committee for Construction Economics have existed for some time in Western Europe, as has the American Institute of Construction Economics in the USA.

Quality

The 1990s have been labelled the decade of quality, with many companies seeking to demonstrate that the services they provide are within a quality framework, such as ISO9000. This will become an issue for quantity surveying practices that have not established systems to ensure that the quality of the services provided fits a defined specification. Procurers of professional services may look towards BS kitemarks for assurance on quality performance.

Knowledge, understanding, skills and application

As noted in the Preface to the Third Edition, knowledge is continuing to increase. This is evidenced all around us in an age of knowledge explosion. It is estimated that as much as 50% of all knowledge has been acquired since the end of the Second

World War. Coupled with this, we have developed a better understanding, through research, of processes and procedures associated with cost studies, and the percentage gain is much higher in this subject area. We do not just have the know-how; we have also developed some understanding of the know-why, as a direct result of research activity. At the same time we are acquiring new skills for new applications in an age of rapid change.

CONCLUSIONS

Much of the development in quantity surveying has happened for commercial and pragmatic reasons. This has often occurred in the absence of any research base or market testing. If clients are not interested or will not pay for a service then there is no real point in developing it. The pressures and demands on the scarce use of resources are in evidence worldwide. If quantity surveying is to respond to these in respect of the whole-life costs of construction, it is important that it is supported by research and analysis which is both relevant and rigorous. This should not result only in a response to meet immediate needs; it should be used also to attempt to anticipate future opportunities and threats to the profession. The research effort of the profession in both practice and academia needs to be better harnessed to provide a sound base for all its members immediately and well into the next century. The continuing professional development of all members of the profession must be seen as a priority in these times of rapid change. The aims for research should include:

- Progressing the activities outlined in QS2000
- Developing a strategy for research and a framework for its implementation
- Raising an awareness within the profession of the benefits of a strong research base
- Seeking to persuade government and industry to actively support research
- Stimulating a debate within the profession about the direction of future research
- Encouraging industry, education and research links

SELF ASSESSMENT QUESTIONS

1. Explain why research is of fundamental importance to the subject of the cost studies of buildings.

2. In the context of the studies of buildings, which is the more important, the formulation of basic principles or the application of new techniques and practices?

3. Given the impetus for change in the construction industry, which areas of research should be targeted as the most beneficial for the future?

BIBLIOGRAPHY

Ashworth, A., 'Fifth generation quantity surveyors', *Chartered Quantity Surveyor*, August 1983.

Ashworth, A. and Hogg, K.I., *Added Value in Design and Construction.* Pearson 2000.

Ashworth, A. and Hogg, K.I., *Practice and Procedure for the Quantity Surveyor.* Blackwell Publishing 2007.

Brandon, P.S., 'A framework for cost exploration and strategic cost planning in design', *Chartered Surveyor*, B & QS Quarterly, Summer 1978.

Brandon, P.S., *Computers: Friend or Foe?* RICS 1984.

Brandon, P.S. (ed.), *Quantity Surveying Techniques – New Directions.* Spon 1982.

Brandon, P.S. (ed.), *Building Cost Modelling and Computers.* Spon 1987.

Davis Langdon and Everest, *QS2000: The Future Role of the Chartered Quantity Surveyor.* RICS 1991.

Fairclough, J., *Rethinking Construction Innovation and Research.* Department of Trade and Industry and Department of Transport, Local Government and the Regions. HMSO 2002.

Fellows, R. and Liu, A., *Research Methods for Construction.* Blackwell Publishing 2003.

Flanagan, R., *Linking construction research and innovation to research innovation in other sectors.* CRISP Consultancy Commission 1999.

McCabe, S., *Benchmarking in Construction.* Wiley-Blackwell 2001.

Nisbet, J., 'Identifying the knowledge base', *Chartered Quantity Surveyor*, June 1991.

Popper, Sir K., *The Logic of Scientific Discovery.* Hutchinson 1972.

Powell, C., *The Challenge of Change. The QS Think Tank 1998.* Royal Institution of Chartered Surveyors 1998.

Runeson, G. and Skitmore, M., *Writing Research Reports.* Deakin University Press 1999.

Royal Institution of Chartered Surveyors, *A Study of Quantity Surveying Practice and Demand.* RICS 1984.

Royal Institution of Chartered Surveyors, *The Research and Development Strengths of the Chartered Surveying Profession: The Academic Base.* RICS 1991.

Royal Institution of Chartered Surveyors, *The Core Skills and Knowledge Base of the Quantity Surveyor.* RICS Research Paper 19, 1992.

COST PRACTICE

CHAPTER 13

DEVELOPMENT APPRAISAL

LEARNING OUTCOMES

After reading this chapter, you should have an understanding about development appraisal in the construction industry. You should be able to:

- Understand the general determinants of value
- Understand the different methods of valuation
- Use valuation tables
- Prepare a developer's budget using different methods
- Apply different methods of investment appraisal
- Appreciate the principles of cost–benefit analysis

13.1 INTRODUCTION

Construction projects arise for several reasons. They may be undertaken in the public sector to meet political, social or community needs. They may be undertaken in the private sector for use, or as projects that can be sold upon completion for profit, be rented or leased to some other organisation. Different techniques are available by which to evaluate the original needs of development. In some cases, these rely upon investment appraisal techniques that assess the expected profitability of undertaking such work. Other techniques may also be used that attempt to form a relationship between the benefits that might be achieved by the development and the costs involved with the project. In other cases it is possible to calculate the costs of not undertaking such work and to compare these against the costs of development. Public accountability will also require that funds have been spent wisely on the appropriate developments. In every situation the necessity of understanding the full financial implications is very important, since whether the development is private or public sector funded, there are only limited funds available for investment purposes.

If the project is to be effective then adequate systems of investment appraisal must be adopted at inception, while the concept is still little more than a possible solution to meet either a need or a desire. It has often been suggested that at least 70% of the initial capital costs of construction are already committed to the design once the project leaves the inception stage and enters the next stage in the development process. Development appraisal is the title given to examining the financial implications of a project at this stage. It is therefore important to have some understanding of valuation methods, valuation tables, the developer's budget and other ancillary matters.

13.2 DEVELOPMENT VALUE

The development value of a plot of land is the difference between the costs of development, which will include the costs of the land, construction etc., and the market value of the finished work. The latter is much more subjective and difficult to predict since it is more influenced by the state of the economy at the time of completion, which will be some time away. It will also be in competition with other similar projects that will be available at the time of completion.

There are a wide range of considerations influencing the development of a construction site. These include:

- Type of development envisaged
- Location
- Shape, size, topography, aspect and access
- Ground conditions and site preparation difficulties
- Availability of utility services
- Planning controls
- Legal considerations
- Government assistance that might be available
- The costs of developing the site and its eventual worth

13.3 GENERAL DETERMINANTS OF VALUE

The supply of property is relatively inelastic owing to a number of factors, particularly the physical nature of land and the length of time required for development purposes. The demand for property is also relatively inelastic. It arises from four possible motives:

1. Occupation
2. Investment
3. Speculation
4. Development

Some of the major factors that affect demand, and hence value, are as follows:

- *Economy*: The general state of a country's economy. In a time of economic well-being there is a desire to invest in property.

■ *Structural changes in the economy*: For example, a movement from manufacturing to a service sector economy increases the need for office accommodation.

■ *Costs of ownership*: Significant changes in the costs of ownership, such as rents and taxes. A significant increase in business rates may cause businesses to cease to trade.

■ *Location*: The better located offices are able to charge the higher rents.

■ *Condition*: This factor will have an effect at the margins of value. Where the condition is deteriorating then this will be a more significant factor.

■ *Government*: Providing grant aid and other incentives will affect the worth of the property.

■ *Infrastructure*: The provision of new or easier means of transport will have a positive effect on value where the provision offers benefit to the owner.

■ *Population*: Demographic trends will influence the size of families and hence housing. Greater longevity may require more by way of sheltered housing schemes.

■ *Funding*: Changing the costs of borrowing or the amounts that lenders are prepared to lend will have significant effects upon value.

13.4 INVESTMENT APPRAISAL

In its simplest form, an investment decision can be defined as one that involves a firm in making a financial outlay with the aim of receiving, in return, future cash inflows. Different variations on this definition are possible, such as financial outlay today resulting in a saving of expenditure at some time in the future.

Investment appraisal is an aid to decision-making. Its objective is to achieve the maximum return that can be obtained from investment expenditure. It comprises a range of techniques for sorting, organising and presenting information and alternatives to assist decision-makers to achieve the best value for money from the use of resources. The techniques of investment appraisal are relevant for a whole range of capital investment decisions such as buildings and equipment. Investment appraisal may be used to decide:

■ Whether to invest in new facilities
■ Between alternative methods of achieving a given objective
■ Whether to continue to use or dispose of existing assets
■ Upon the quality and standards of a design
■ Maintenance and service schedules

Investment appraisal therefore has an application throughout the whole-life cost process. This ranges from the setting of strategic priorities to the final details of design and operational practice and the eventual disposal of the site.

It is worth remembering that the different methods used in investment appraisal may produce different solutions, depending upon the objectives that have been established. It should also be noted that these techniques are only a guide for decision-makers; they will not make the decision. However, they will help to make a more informed decision, where judgement then relies upon some form of analysis in addition to other forms of skills and expertise. Such techniques will never replace managerial judgement. The preparation of estimates or forecasts of some future activity will also always include some element of uncertainty and this can only be assessed on the basis of previous performance and expected trends.

The basic steps of investment appraisal normally follow the following sequence.

Define the objectives   It is necessary at the outset clearly to define the expected outcome, and to evaluate how the results are to be measured. The objectives need to be a balance between the general and the specific. If they are too general then they may lack credibility, whereas if they are too narrow, viable options may be overlooked.

Identify options   Several different solutions may be available, all of which achieve the prescribed objectives.

Measure costs and benefits   The correct choice can only properly be made in the light of the full facts. These need to be measured as accurately as possible.

Discount costs and benefits   In order to evaluate present and future costs properly, these must both be transferred to a common time base by discounting.

Consider uncertainties   Some aspects of the proposal will be unknown and it is essential to assess how these may affect the final outcome of the appraisal.

Assess other factors   Other factors which are outside the scope of the analysis may have a bearing upon the final decision. For example, political uncertainty will need to be considered and how this might affect the decision.

Relevant techniques are available to assist and help identify the most profitable of a number of options. These can be used alongside professional judgement, to provide some objectivity in the analysis. The following techniques have been subdivided into two categories of conventional and discounting methods.

13.5 METHODS OF VALUATION

Valuations of land and property are usually undertaken by the valuation surveyor for a variety of different purposes. The purpose of the valuation will affect the assessment of its value, and this may differ because of the assumptions made and also because they are only estimates of value anyway. Valuations are required for statutory purposes in order to assess capital transfer tax or when a public body seeks to acquire land or property by means of compulsory purchase. A valuation may also be required when a purchaser such as an insurance company or pension fund wants to invest their capital. It may also be required during the sale and purchase of property, in connection with a mortgage loan or for determining an

Table 13.1   Methods of property valuation

Comparative method
Contractor's method
Residual method
Profits or accounts method
Investment method
Reinstatement method
Hedonic price modelling

auction reserve. A number of alternative methods (Table 13.1) can be used to estimate either the capital value or the rental value of an interest in land or property. It should be noted that values can vary considerably depending upon the location nationally or even within a small area.

13.5.1 The comparative method

The comparative method is the most popular method used for valuation purposes. Its main uses are in connection with residential property where direct comparisons can be made against other types of property on the open market. The method is only reliable, however, where there are sufficient records of many recent transactions and the properties are in the same geographical area. Other factors that will influence the valuation are the similarity of properties in respect of design, size and condition, and the legal interest. A stable market and economic factors such as lending rates will also affect the reliability of the valuation.

13.5.2 The contractor's method

The basis of the contractor's method is to suggest that the value of a property is equivalent to the cost of erecting the buildings together with the cost of the site. It is an unsound assumption, however, since value is determined not necessarily by the component costs involved but by the amount which prospective purchasers are prepared to pay. Its main use is in connection with valuations for insurance purposes and for buildings such as schools, churches, hospitals etc. for which there may be little in the way of comparative valuations.

It is necessary when using this method that allowances are made for depreciation, since a building that is 60 years old is unlikely to have the same value as a modern building of a similar type and quality. Some of these buildings may be ornate and have been costly to construct, but this will not necessarily be reflected in the value.

13.5.3 The residual method

The residual method is used in those circumstances where the value of a property can be increased after carrying out development work. For example, an old house may have the potential and ability for conversion into flats, when its best potential can be realised. The building is valued on the basis of its future worth after conversion, and the costs of this work together with developer's costs are then deducted. The resulting sum is the value of the property in its original state and is known as its residual value.

13.5.4 The profits or accounts method

Almost all types of property are capable of producing an income under certain conditions, and a relationship will exist between this and the capital value of the property. The profits or accounts method is more appropriate to commercial premises such as hotels, shops and leisure projects than domestic premises. The usual approach is to estimate the gross earnings, deduct expenses, and the balance remaining then represents the amount available for payment of rents. This can then be converted into a capital sum.

13.5.5 The investment method

The investment method can be used in those circumstances where the property produces an income. The income expected must be comparable with that which could be earned by investing the capital elsewhere. In considering alternative investment possibilities factors such as security values, ease of realisation, costs of purchase and selling, and any tax liability will influence competing proposals. The principal investors are pension funds, insurance companies, property companies, historic owners, local authorities and government agencies.

13.5.6 The reinstatement method

The reinstatement method requires the estimation of the cost of rebuilding a particular property and then adding to it the value of the land on which the property stands. It is a useful method for fire insurance purposes, in order to calculate the premium to be paid. It may sometimes appear that the insurance premium should only be based upon rebuilding costs, since the site will remain, even in the event of a fire. It will be necessary, however, to allow for demolition and site clearance costs where the building is to be rebuilt. These costs will also have to take into account possible site damage and temporary works that might be necessary before demolition can commence.

Each of the above methods, other than the profits method, is useful for estimating capital values, whereas the residual method and investment method are not really suitable for the determination of rental value. The demand for a particular type of landed property will be influenced by changes in the size of population, methods of communication, standards of living and society in general.

13.5.7 Hedonic price modelling

Hedonic price modelling is a computer-based system for valuing property on the basis of the different variables involved. It uses the technique of multiple regression analysis to find a formula or mathematical model that best describes the data characteristics that have been collected. The technique is normally used in those circumstances where the relationship between the variables is not unique. This is in the sense where the value of one variable always corresponds to that of another. In order to calculate the value of a property, it is first necessary to identify the variables that might be important. In the case of residential property, variables such as location, type, size, number of bedrooms, garages, central heating etc. are important. Large amounts of data concerning previous transactions are then required in order to discover mathematical relationships. It is unlikely that a perfect relationship will be found, just as it is unlikely that a number of valuers would all predict the same value for a property. The model will be able to predict confidence limits to the results, and where a good model has been constructed then these should allow the value to be stated within tolerable limits. The location of property is the most significant variable that affects value. Knowing a property's postcode will therefore allow a value to be predicted within the above model formation, as long as the data in the model are representative of the value that is being predicted.

13.6 VALUATION TABLES

In order to allow comparisons to be made between money spent or received at different times we need to be able to convert these sums into a common timescale. Valuation tables are used as the means of making this conversion.

In most societies the payment of interest for the use of capital or money is an established part of economic life. The money which is lent is called the principal. The sum of the principal and the interest for any length of time is called the amount. The money paid for the use of the principal is calculated on a percentage rate basis and this is generally calculated using either simple interest or a compound interest formula. The basis of valuation tables is compound interest (see also Figures 13.1 and 13.2).

13.6.1 Simple interest

Simple interest arises when only the original capital invested earns interest. For example, if A borrowed £100 from B with the agreement that 5% interest was payable each year, then at the end of the first year if B paid A £5 the obligation would be met. £5 would be paid each and every year on this basis.

13.6.2 Compound interest

Interest may accrue on interest as well as on the capital. For example, C decides to borrow from D £100 at a 5% rate of interest for two years. At the end of the first

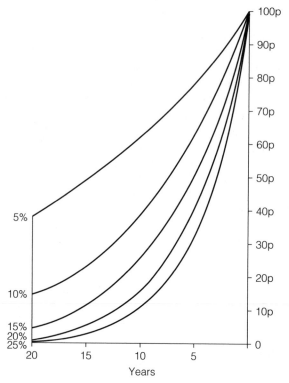

Fig. 13.1   The present value (PV) of £1
Source: Barclays Bank (1974)

year C owes D £105. The interest in the second year is calculated as 5% of the £105, which equals £5.25. At the end of the second year C would repay D £110.25. Using simple interest as a basis this would only have been worth £110.

Calculations involving compound interest involve:

- Compounding, i.e. the way a present sum of money will grow
- Discounting, which is the reverse of compounding. This considers how much a future sum of money might be worth today, given a rate of interest.

13.6.3 Parry's valuation tables

In valuation practice and other studies associated with land and buildings where some aspect of financial analysis is involved, compound interest calculations of a tedious and time-consuming nature are often required. Valuation tables can be used to reduce the time-consuming aspect of such calculations. There are a number of different books of valuation tables now available. The most well known are *Parry's Valuation and Conversion Tables* (The Estates Gazette). These valuation tables were first prepared in 1913 by the late Richard Parry. Although other valuation tables are in common use, Parry's has become ubiquitous in property valuation where calculations requiring interest rates are to be taken into account. The comprehensive

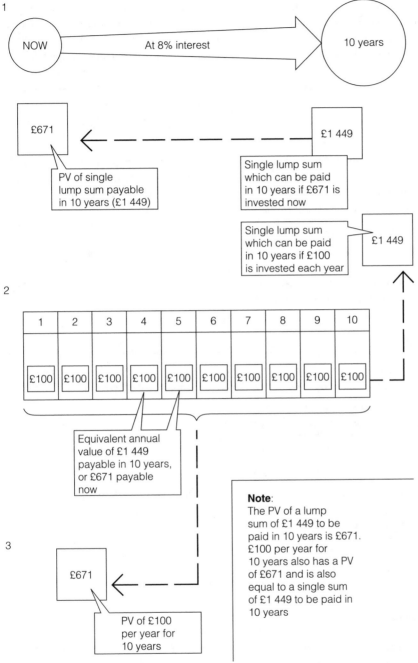

Fig. 13.2   The varying value of money at 8% per annum
Source: *The Decision to Build*, HMSO (1974)

set of tables provide for the requirements of current practice. The later editions of these tables have been prepared by computer and in order to minimise possible error a computer-linked typesetter has been used. Over the years different editions of the tables have seen both the introduction of new material and the removal of tables now thought to be obsolete. In addition to the actual computational values, an explanation of the purpose or use of the tables is provided. The first chapter of the book deals with the construction and use of the tables generally.

1. Amount of £1 table

This is the table which forms the basis for construction of many of the other tables. The multiplying factors given in the valuation tables represent the amount to which £1 invested now will accumulate at compound interest over a given period of time. It is represented by the formula:

amount of $£1 = (1 + i)^n$

where i is the interest rate and n is the number of years. The principal is multiplied by the appropriate figure from the tables for the required interest rate and term of years. The amount of £1 table has multipliers greater than unity. Tables of this type are commonly referred to as accumulating tables.

Example

A builder purchased a plot of land five years ago for £15 000. Assuming that land has increased in value by an average of 6% per annum, what would be its value today?

£15 000 × 1.34 = £20 100
(amount of £1 table, 5 years at 6%)

2. Present value (PV) of £1 table

In this table, the investor is seeking to find what sum must be put into the bank today in order for it to amount to £1 at the end of a given period of time, using compound interest. It is a discounting table and is the reciprocal of table 1 above, and is thus represented by the formula:

$$\text{present value (PV) of } £1 = \frac{1}{\text{amount of } £1}$$

Example

A boiler will need replacing in twenty years' time and this is estimated to cost £8 000. If the average annual rate of interest is 4%, what amount should be invested today in order to be able to make this replacement?

£8 000 × 0.456 = £3 648
(PV of £1 table, 4% for 20 years)

3. Amount of £1 per annum table

This is similar to table 1 above in that it is an accumulating table. The difference is that whereas table 1 represents a once and for all single sum of money, table 3 provides for an equivalent amount annually for the required term of years:

$$\text{amount of £1 per annum} = \frac{(1+i)^{n-1}}{i}$$

Since the formula for calculating the amount of £1 $= (1+i)^n$, then replacing this by A, the amount of £1 per annum can be simplified to:

$$\frac{A-1}{i}$$

Example

What sum will be obtained if an investor puts £200 in his bank account every year for ten years at a rate of 6%?

£200 × 13.18 = £2 636
(amount of £1 per annum table, 6% for 10 years)

4. Annual sinking fund table

This table is the reciprocal of the previous table (amount of £1 per annum). It is used to calculate the annual amount to be saved each year at a given rate of interest in order to meet a known expense at an expected date in the future:

$$\text{annual sinking fund (ASF)} = \frac{1}{\text{amount of £1 per annum}}$$

Example

Extensive modernisations to a client's offices are expected to be carried out in eight years' time. What sum needs to be invested annually at a rate of interest of 5% to cover the future costs of £75 000?

£75 000 × 0.106 = £7 875
(ASF, 5% for 8 years)

5. Present value of £1 per annum table (year's purchase (YP) single rate)

This table is used to calculate the present value of future payments which are made at regular annual periods. The formula is derived from the 'PV of £1 table' by adding together the multipliers for each year. It can therefore be represented by the formula:

$$\text{PV of } \pounds 1 \text{ per annum} = \frac{1 - \text{PV of } \pounds 1}{i}$$

As the number of years approaches perpetuity, the value of the 'PV of £1' becomes so small that it is insignificant. The formula is sometimes referred to as the 'year's purchase' table and can in these circumstances be abbreviated to:

$$\text{YP (single rate)} = \frac{1}{i}$$

Example

A client wishes to know how much must be invested today at 5% rate of interest to cover the average annual payments of £2 000 for energy consumption during the next 25 years.

£2 000 × 14.09 = £28 180
(PV of £1 per annum, 5% for 25 years)

6. Present value of £1 per annum table (dual rate)

The PV of £1 per annum table (single rate) provides for the same rate on both of the following:

■ The interest on the sum invested
■ The ASF to recover the capital value over the term of years

In practice the rate of interest on the loan and the ASF may be different. In these circumstances it is therefore necessary to use a dual rate (DR) table, which allows these rates to be different:

$$\text{YP}_{\text{DR}} = \frac{1}{i + \text{ASF}}$$

Example

Assume that the cost of capital is 12% and an ASF rate of 5% is required to cover future replacements of a boiler plant during the next 30 years. The PV of these replacements is £33 936. What is the annual charge to cover this amount?

In order to convert a PV to an annual equivalent we divide this by the YP factor calculated above, as follows:

$$YP_{DR} = \frac{1}{0.12 + 0.015}$$

(12% inflation) (5% ASF for 30 years)

$$= \frac{1}{0.135} = 7.4074$$

$$\frac{£33\ 936}{7.4074} = £4\ 581 \text{ is the annual charge}$$

7. Annuity £1 will purchase

The term 'annuity' is generally used to mean a series of payments that are to be made during a given period of time at fixed intervals. If these payments are only to last for a fixed period of time then it is termed an annuity certain. The rent from property, either for a fixed term of years or in perpetuity, is an annuity certain. When the period of the annuity is perpetual, the annuity is more properly described as in perpetuity. The annuity £1 will purchase is given by the following formula:

$$A = i + SF$$

Example

A leaseholder paid £3 000 for an interest last month. The lease has a 30 years' unexpired term. It is decided to let the property. What is the minimum rent to be accepted if a return of 10% on the outlay is expected with a sinking fund of 2.5%?

Capital cost	3 000
Annuity £1 will purchase 30 years @ 10% and 2.5%	0.1228
Equivalent annuity	£368

8. Mortgage repayment tables

A mortgage is the annual equivalent of a capital sum lent by a mortgagee, often a building society, normally for house purchase. The mortgager or purchaser agrees to repay the capital borrowed together with interest charged at the society's rate. The parties agree beforehand upon the number of years for which the mortgage will run. The amount of repayment therefore depends upon the size of the loan, the term of years and the interest rate applicable. In Parry's valuation tables the annual equivalent is the sum of twelve monthly payments needed to repay £100 on a monthly basis. This table represents the values in the annuity table (single rate) multiplied by 100 and divided by 12.

$$\text{mortgage instalment} = \frac{(i + SF) \times 100}{12}$$

Table 13.2   Summary of valuation formulae

1. Amount of £1	$A = (1 + i)^n$
2. Present value of £1	$PV = \dfrac{1}{A}$
3. Amount of £1 per annum	$Am = \dfrac{A - 1}{i}$
4. Annual sinking fund	$ASF = \dfrac{1}{Am}$
5. Present value of £1 per annum (single rate)	$PVA = \dfrac{1 - PV}{i}$ or $YP = \dfrac{1}{i}$
6. Present value of £1 per annum (dual rate)	$PVA = \dfrac{1}{i + ASF}$
7. Annuity £1 will purchase	$A = i + SF$
8. Mortgage instalment	$MI = \dfrac{(i + SF) \times 100}{12}$

Example

What is the monthly repayment for a mortgage of £25 000 over 25 years at 12%?

£25 000/100   = 250 × 1.0625   = £265.63
tables based on   mortgage instalment   monthly
units of £100   table   payment

Annual payment = £3 187.56

Alternatively this could have been calculated from the annuity table (single rate):

£25 000 × 0.1275 = £3 187.50

The annuity table is sometimes referred to as the annual equivalent table. Table 13.2 summarises the different valuation formulae.

13.7 DEVELOPERS' BUDGETS

Developers seeking sites for development purposes will need to consider many different factors. They may, for example, be looking for a site that is suitable for one of several different development proposals. Alternatively, they may already have a particular scheme in mind and are seeking a site which is most suitable for this need.

Developers will usually be seeking an overall scheme which is likely to be the most profitable and one that is attractive to potential investors.

When a suitable site has been identified, prior to its acquisition it will be necessary to ensure that planning permission, for the type of development envisaged, will be granted. They will therefore need to make enquiries at the offices of the local planning authority. Permission at this stage will only be given in principle, and it might include conditions to be met if approval is to be obtained. Where permission is not forthcoming then a notice of appeal can be made, if there are reasonable and likely grounds for its success. The type of development that will be allowed is generally quite clear, but there is always room for some debate and discussion on arbitrary cases.

In order to determine whether a scheme is feasible it will be necessary to prepare a developer's budget. This will then provide answers to the following questions:

- How much should be paid for the land?
- What will be the maximum building cost?
- What should the selling price or rental value for the property be?

The developer's budget considers the following items.

13.7.1 Gross development value (GDV)

The total rental value is estimated by comparing the proposed scheme with the rents obtained from similar properties. The net rental value is used after deductions for outgoings such as maintenance, repairs, insurances, management etc. have been made. This then provides the net income from the proposed development. The amount before the deduction of any income tax is used in order to compare this with other non-property investments. In the case of a block of flats or a shopping centre development, where there may be many different tenants, then management costs to cover rent collection, surveying etc. would also need to be deducted. These are currently worth about 2.5% of total rents. The valuer will be able to advise upon the appropriate amounts. These values are more prone to error, however, than are building costs. This is due to the many uncertainties in the property market, not least the need to forecast likely prices and demand at some time in the future when the property is constructed and available for occupation. The valuations of two independent valuers could also indicate some wide discrepancies, since in addition to the calculations involved, valuation is a matter of considerable skill and judgement. Opinions, which may be based upon wide experience, are known to conflict, as illustrated by case law on the subject.

13.7.2 Investment yield

While a large proportion of residential properties are owned by the occupants (discounting any mortgage interest), commercial property is more likely to be rented. The theory is that the profits from commercial enterprise are probably better employed in running and expanding that business at which they are experts than being tied up in property. This also allows better opportunities to move premises when the business changes shape through expansion or contraction.

The net income from the budget is then capitalised by multiplying by an investment yield. This figure should compare with investment yields from other types of investment and may fluctuate considerably in an unstable economy. An appropriate yield, sometimes referred to as the year's purchase (YP), can be obtained by dividing 100 by the interest rate:

$$\text{YP in perpetuity} = \frac{100}{\text{rate of interest}}$$

$$\text{YP of 8\%} = \frac{100}{8} = 12.5$$

This is also sometimes known as the 'PV of £1 per annum', e.g. net income £2,000 per annum multiplied by YP at 8% = £25,000. This capitalised figure is known as the development value.

It should also be noted that office block rents are based not upon gross floor areas but upon net usable floor areas. Some allowance must therefore be made for non–usable areas such as circulation space before calculating the development value.

Example

The rental value of an office block is estimated to be £30 per m². The total floor area is 10 000 m² and the non-lettable area represents 20%. What is the development value if the YP is 6%?

$$10\ 000\ \text{m}^2 \times 80\% \times £30 = 240\ 000$$

$$\text{YP at 6\%} = \frac{100}{6} = 16.67$$

gross development value = £4 000 800

As a principle the greater the expected rental growth, the lower the initial return an investor would be prepared to expect. Conversely an investment where income growth is expected to be small, such as in a building society account, will require a high yield to compensate.

13.7.3 Costs of construction

There are several easy-to-apply methods for calculating the approximate cost of a building. However, while the methods rely upon a simple method of quantification, such as the floor area of the proposed building, the skill in selecting a correct current rate by which to calculate cost is much more difficult. This relies upon a knowledge of current prices and being able to interpret these against the designer's brief and outline drawings. See Chapter 14 for further information on the different methods and techniques that can be used for early price estimating.

During the investment appraisal it is common to use the construction costs at the date of tender, i.e. excluding any increases in cost during construction. This is because present-day rents are also used, on the basis that any increases in either will to some extent compensate each other.

13.7.4 Fees

Charges for the professional services provided will need to be added to the costs of construction. The various professional institutions publish fee scales which can be used as a guide in assessing these costs. The fee scales are based upon a combination of a lump sum and a percentage of the construction costs. The fees will vary depending upon the type and size of project, and the description of the service provided. Professional fees are now calculated on the basis of competition, using the fee scales as a guide. The larger the project and the more repetitive its components the smaller will be the overall fee that is charged. For unusual, complex or difficult projects, which might include specialist professions such as archaeologists etc., then the fees will increase accordingly. An addition of 10% will typically cover design, costing and supervision fees. Value-added tax will also be chargeable but may be recoverable by the client, depending upon the type of project and type of client.

Legal fees will be required for the purchase of the site, and the preparation and agreement of leases or the conveyancing documents. Property agents may also be required in connection with letting and management of the property, or for its disposal to potential owners. Their fees may be typically 2%–3% of the selling price, depending upon the service provided and the number of units involved.

13.7.5 Developer's profit

Property development involves the taking of considerable risks. Where these risks fail to materialise then increased profits are made. Where the risks are greater than expected then profits are reduced. The developer is paid a return on the development to cover the skills, time and risk that will have been incurred in the development as well as for expected profits. This is typically estimated to be about 10%–20%, but is dependent on a wide range of factors such as the type and size of development, the length of the development period and possible competition when completed. The risks to be assessed by the developer and the professional advisers involved include rising costs, the speculative nature of the development and inability to either sell or lease on completion. For example, on the ill-fated Canary Wharf development, even the offer of free leases for up to ten years was insufficient to attract some 'blue chip' companies. The objective of this idea was that, if successful, then other firms would have been prepared to relocate to be close to such companies. The greater risk than anticipated of letting the property caused profits to be wiped away, leaving a trail of debts and bankruptcy for the developer.

13.7.6 Finance (site and buildings development)

The developer prior to commencing with the construction work will need to have already purchased a site. This might have been acquired through retained earnings, in which case there will bank interest accruing. Alternatively, the funds may need to be borrowed in which case there will be interest charges to be added. Land is often purchased at least twelve months prior to starting work on site, to allow for planning permission and the security of the site.

Payments to the constructor will be made monthly, the amount to be paid being determined by the quantity surveyor. Payments for professional fees are often made in two parts to cover pre-contract design work and supervision and administration of the project through construction. The time between completion and letting or selling will vary depending upon the local market and the demand for the type of property that has been constructed. Housing developers aim to complete dwellings in line with sales and will deliberately accelerate or delay construction activity to meet the demand for houses.

The interest added to the developer's budget is often on the basis of the full amount for half the time. While this is only an approximate amount it is adequate for including within the calculation. In the case of long contract periods then compound interest rates should be applied. The interest rates selected will be based upon the opportunity costs of capital, this being a few points higher than the base rates from where the finance probably has to be obtained.

13.7.7 Example 1

A speculative developer is considering purchasing a site for the construction of 40 detached houses. The selling price of the houses is £65,000. The cost of the land, inclusive of legal charges, is £150,000. The developer requires a profit of 16% of the gross development value (GDV). What is the allowable amount for building costs?

Developer's budget
  Gross income
    40 houses × £65 000   = £2 600 000 (= GDV)

Developer's costs
  Land
    Cost inclusive of legal fees   =   150 000
    Short-term finance, required for
      say 2.5 years at a compound
      interest rate of 12% (valuation
      table could be used)   =   49 450
  Fees
    Legal, agent's and advertising, 3% of GDV   =   78 000
  Profit
    16% of GDV   =   416 000
  Total   =   £693 450

By deducting this amount from the GDV, the building costs, professional fees and finance for construction can be calculated.

GDV	=	2 600 000
Development costs (above)	=	693 450
Total	= £1 906 550	

Building costs

Let B be the building costs. Assume that finance will be required for 1.5 years at 12%. Professional fees are assumed to be 10%.

Building costs	$= B$
Finance	$= B \times 0.12 \times 1.50$
Fees	$= B \times 0.10$

$$1\ 906\ 550 = B = (B \times 0.12 \times 1.50) + (B \times 0.10)$$
$$= B + 0.18B + 0.10B$$
$$= 1.28B$$

Building costs $\qquad = \dfrac{1\ 906\ 550}{1.28} = £1\ 489\ 492$

Check

Land costs and finance	199 450
Legal/agent's fees	78 000
Developer's profit	416 000
Building costs	1 489 492
Finance	268 108
Fees	148 949
Total	£2 600 000

The allowable amount to cover the costs of building is therefore £1,489,492. This represents £37,237 per house. If the size of each house is known, say 110 m², this can be converted to a rate per square metre of gross internal floor area of £338.50. A consideration of present-day building costs would determine whether this would be adequate for the type and quality of building that is envisaged.

The developer during the early stages of any development will need to provide answers to the following questions:

- Is there a market for the proposed development?
- What will be the likely selling price?
- How much can be afforded to be spent on the scheme?
- Can the necessary finance be raised?
- What will the finance cost?
- Will the scheme be granted planning permission?

Sometimes the developers are able to predetermine their costs and need to know the likely selling price of the development and whether or not this is achievable. The questions can then be approached in the reverse order:

Land cost and finance	= 199 450
Fees 3% of GDV	= ?
Profit 16% of GDV	= ?
Building cost, finance and fees	= 1 906 550
GDV	= ?

Let $GDV = x$.

$x = 199\ 450 + 0.03x + 0.16x + 1\ 906\ 550$

$0.81x = £2\ 106\ 000$

$x = £2\ 600\ 000 = GDV$

$0.03x = £78\ 000 = $ legal/agent's fees

$0.16x = £416\ 000 = $ developer's profit

The cost of each house is then obtained by dividing the GDV by the number of houses.

13.7.8 Contingencies

Some developers may consider that it is prudent to allow for contingencies, to cover for unforeseen costs, i.e. amounts that cannot be properly estimated or as a margin to the costs identified above. It is common, for example, for contingencies to be included in bills of quantities for unforeseen items of expenditure. Sums of money will be required in circumstances such as for capping a mine shaft that is only discovered when the work starts on site. Building works contingencies are typically about 2%–5% depending upon the degree of certainty of the design and the knowledge of the site. On refurbishment projects they are often much higher since there are likely to be more unknown factors in both the design of the project and the condition of the existing building.

13.7.9 Letting and agents

It is normal practice to assume that the completion of the project will not entirely coincide with the letting or selling arrangements and therefore it is sensible to make allowances for such delays in the budget. Where these delays are considerable because of a downturn in the market, then the developer may need to consider offering inducements in order to dispose of the property. If this is considered to be a possibility at the time of inception, it may be desirable to postpone the project. This happened with a large number of designed office blocks at the end of the 1980s. Alternatively developers may consider that improvements will occur in the market for the property when the time comes to dispose of the finished projects. In any case construction costs are likely to be much lower during a recession and it is possibly a good time to build, if one assumes that the business cycle has not stopped. In prime locations, property is always likely to find a buyer. If the developer considers that extra costs will possibly be incurred then these should be included in the budget.

Letting and sale fees usually occur towards the end of the development. It is common practice on larger developments to appoint two or more agents, with a consequent increase in fees. Letting fees are typically based on 10% of a single year's rent where one agent is employed, and 15% in the case of multiple agents.

13.7.10 Example 2

A speculative developer has provided the following details of a proposed speculative office development and has asked you to calculate the allowable building costs:

Gross floor area
    Non-lettable area 22%
    Estimated rent £60/m²
    Capitalisation of rents 7%
    All outgoings are to be recovered by a service charge

Building contract details
    Period 18 months
    Professional fees 15%
    Short-term finance 15%
    Developer's profit 12% of GDV
    Land cost (including fees) £100 000

The solution depends upon an evaluation of future rents in order to establish the amount that can be available now for building purposes. It should be remembered that rents are determined on the basis of the lettable floor area. Therefore the rental received will be as follows.

Rental received

Lettable floor area $= 10\ 000\ m^2 \times (100 - 22) = 7\ 800\ m^2$

The net income to the developer is therefore:

$7\ 800\ m^2 \times £60/m^2 = \underline{£468\ 000}$

Gross development value
This amount must then be capitalised, i.e. converted to a current capital value. It is assumed for the purpose of this question that the rent will be received in perpetuity. It is therefore multiplied by the year's purchase in perpetuity at the given percentage.

YP in perpetuity at $7\% = \dfrac{100}{7} = 14.286 \times £468\ 000 = £6\ 685\ 848$

No adjustments are to be made for any outgoings, such as repairs, insurances etc., as these will be recovered by means of a separate service charge. If these, or any other management charges, were incurred, then the effect would be to reduce the annual rents and hence the capitalisation amount.

Developer's profit

The gross development value is the same as the capital value. The developer's profit is therefore calculated as

$$12\% \times £6\ 685\ 848 = \underline{£802\ 302}$$

Land costs

The cost of the site, which includes professional fees associated with its acquisition,	=	£100 000
Short-term finance will be required until the development is complete, and then presumably let or sold. This is required for at least the contract period, assuming that the site is purchased at the start of the contract. This may be a conservative assumption, since the land is likely to be purchased much earlier and therefore incur additional interest charges.		
Assume 24 months at 15%	=	£ 30 000
		£130 000

Summary

Gross development value		= £6 685 848
Developer's profit	802 302	
Land cost	130 000	= £ 932 302
Amount of allowable building costs		= £5 753 546

Building costs

Let B be building costs, including any allowances for inflation. Finance will be required at 15% for the 18 months' contract duration. Note that the finance will be required as the work progresses. This is equivalent to the full percentage for half the time. Professional fees will not be paid until the project is completed. Therefore

$$B + (B \times 0.15 \times 1.5 \times 0.5)(B \times 0.15) = £5\ 753\ 546$$

$$1.2625B = £5\ 753\ 546$$

$$B = £4\ 557\ 264$$

This is the amount available for building costs. Dividing this by the gross floor area will provide a rate per square metre. This will then suggest a type and quality of construction that may be possible.

13.8 CONVENTIONAL METHODS OF INVESTMENT APPRAISAL

13.8.1 Pay-back method

The pay-back method is the crudest form of investment criterion but nevertheless one of the most widely used. It is defined as the period it takes for an investment to generate sufficient incremental cash to recover its initial capital outlay in full. A cut-off point can be chosen, beyond which the project will be rejected if the investment has not been paid off. The pay-back method appears attractive because it is extremely simple to apply. Since it takes cash receipts into account, it helps to assess a company's future cash flow (particularly advantageous in times of liquidity crisis). However, it fails to measure long-term profitability since it takes no account of cash flows beyond the pay-back period. It is therefore difficult to make comparisons between projects with different life expectancies using this criterion. The technique also falls short in its application within the pay-back period since no account is taken of the timing of the cash flows during that period. The use of the method is sometimes justified by claiming that it is a 'dynamic' criterion, since projects are adopted only if they are paid off quickly, but this argument does not allow for the fact that highly profitable investments do not necessarily pay off in the initial years although large gains may be reaped later.

Example

A client has the option of investing in one of the following three projects:

		Projects		
	Year	A	B	C
Expenditure	0	60 000	100 000	140 000
Income	1	10 000	50 000	50 000
Income	2	20 000	25 000	50 000
Income	3	40 000	25 000	25 000
Income	4	20 000	50 000	45 000
Income	5	20 000	50 000	35 000

The pay-back periods for each of these three projects are

A	$2 + 30/40$ years = 2 years 9 months
B	3 years = 3 years
C	$3 + 15/25$ years = 3 years 7.2 months

It can be seen that the pay-back period is quick and simple to calculate. However, clear objectives need to be formulated in assessing the competing alternatives. In the above example and using the pay-back criterion, project A would be selected since it has the shortest pay-back period. However, there are other criteria that need to be measured, which might have an influence upon the decision to be made. In the example provided, over the five-year period, project C provides the highest cash profit (£65,000), whereas project B offers the largest percentage profit (200%).

13.8.2 Average rate of return method

The average rate of return is the ratio of profit (net of depreciation) to capital. The first decision that must be made is how to define profit and capital. Profit can be taken as either gross of tax or net of tax, but since businesses are mostly interested in their post-tax position, net profit is a more useful yardstick. However, net profit can be either what is made in the first year or the average of what is made over the entire lifetime of the project. Similarly, capital can be taken as either the initial sum invested or a form of average over time of all the capital outlays over the life of the project. This method takes no account of the incidence of cash flows so that projects with the same capital costs, expected length of life and total profitability would be ranked as equally acceptable. The method can be extended, however, by calculating the net average yield. This is done by subtracting the stream of cash outlays from the stream of cash benefits and expressing the answer as a percentage of the initial outlay.

Example

Four projects all have similar capital costs but different income streams as shown in the following table.

	Year	Project D	E	F	G
Capital cost		100 000	100 000	100 000	100 000
Income	1	115 000	50 000	60 000	
	2		50 000	60 000	
	3		25 000	–25 000	
	4		75 000	25 000	200 000
Total income		115 000	200 000	120 000	200 000

In the above example the average rate of return on project D is 115% of the initial capital expended. The life of this project is one year. In project E, the sum of the positive cash flows is £200,000, but over four years. This is worth on average £50,000 per year, giving a rate of return of 50%. In project F, the total income is £120,000 or £30,000 per annum, giving an average rate of return of 30%. Thus this method does not take into account the timing of the cash flows. Therefore projects with the same capital costs, expected life and total profitability would be ranked equally acceptable. In the above example, project G has the same average rate of return as project E, although no account is taken of its income flows which appear at different years.

Table 13.3 Methods of investment appraisal

Conventional methods	Discounting methods
Pay-back	Net present value
Average rate of return	Internal rate of return
Necessity/postponability	

13.8.3 Necessity/postponability

The necessity/postponability criterion is essentially a negative one. The rationale is that the more postponable an investment is, the less attractive it appears, and so the basis of investment decision-making is the urgency of requirements. Thus, if a project was one which could only be carried out now and could not be initiated at a later date, then it would be chosen in preference to a project which could be undertaken in the future.

13.9 DISCOUNTING METHODS

A vital factor ignored by the conventional methods of investment appraisal is that money has a time value. A pound today is worth more than the same pound tomorrow. A sum of money is worth more today than an equal sum of money at some time in the future even ignoring inflation. This is known as the time value of money. This is because it allows for the possibility of investment or consumption taking place in the intervening period. The present value of a future sum is dependent upon two factors: the rate of interest and the term of years. The further in the future the sum is or the higher rate of discount used then the less will be the present value of that sum. There are two major discounting techniques and these are described next.

13.9.1 The net present value (NPV)

In order to determine the NPV of a proposed investment, the forecast net of tax cash flows are simply discounted to the tune of the initial capital outlay (at a rate chosen to reflect the company's cost of capital) and the value of the initial capital outlay is subtracted. The company's cost of capital is generally set at a level which would give the shareholders a rate of return at least equal to what they could obtain outside the company. The discounting technique can be readily adapted to take account of real-life complications such as cash flows arising in the middle of a year, investment grants, capital allowances, inflation and delays in corporation tax payments. With the help of appropriate tables, the volume of calculation and analysis resulting from these complications is not nearly as weighty as might be supposed.

13.9.2 Internal rate of return (IRR)

The IRR is the most common discounting method of investment appraisal.
It can be defined as that rate of interest which, when used to discount the net
of tax cash flows of a proposed investment, reduces the NPV of the project to zero.
The discount rate which will reduce the NPV of the project to zero can be found
by trial and error: if a negative NPV results, the rate chosen is too high; if a positive
NPV is obtained, then the rate is too low. Although it appears to involve a large
number of calculations, in practice it should never be necessary to carry out more
than two trial discounts, the true IRR then being determined by interpolation.
The IRR depicts the annual rate of return on the capital outstanding on the
investment. Thus, in common with the NPV method, the IRR will generally be
higher if the bulk of the cash flows is received earlier rather than later in the life of
the project, reflecting the fact that more capital will have been recovered in the first
years of the project so that the flows remaining represent a higher rate of return.

13.10 OPTIMAL INVESTMENT CRITERION

Although some of the conventional methods of investment appraisal provide a
useful measure of the vulnerability of investment proposals to risk and liquidity
constraints, as gauges of the profitability of projects they must be regarded as
extremely inferior to the discounting methods because of their failure to recognise
that money has a true value. There are occasions when the IRR is meaningless.
If, for example, a particular project involves heavy net capital outlays towards the
end of the project's life, the IRR could be nonsensical. When appraising independent
projects, where the only decision to be made is whether to accept the project or
not, then both the NPV method and the IRR method will give the same answer.
However, when trying to decide which is the most profitable of two mutually
exclusive projects, then the two methods can give very different answers. The risks
associated with a project are largely dependent on the quantity of capital involved
and the length of the project. By showing a rate per unit of capital per unit of time
of the project, the IRR can show the margin over the cost of capital that is being
obtained in return for any risk taken.

Conventional methods of dealing with risk, such as sensitivity analysis,
probability analysis and game theory can of course be used in conjunction with
discounting techniques.

13.11 SENSITIVITY ANALYSIS

During a residual valuation calculation, a large number of assumptions are made
regarding, for example, the costs of construction and the income that might be
generated from the development. It is necessary to test whether the assumptions
made are likely to have any effect upon the overall viability of the proposed scheme.

There is a need to provide the decision-maker with all the relevant information that may influence the outcome of such a decision. A way of testing the analysis is to repeat the calculations by changing the values that have been allocated to some of the variables, such as discount rates to be used, expected construction costs, profit expectancy etc. This might be a tedious process, but the use of a simple computer program or the use of a spreadsheet will allow such calculations to be repeated and tested with ease.

The first calculation is assumed to be the most likely, but changing the values in the equations in this way will demonstrate just how the solution might be affected if things do not turn out in the way that is expected. It is therefore possible to produce worst and best scenarios, in addition to what is believed to be the most likely. The use of sensitivity analysis will help to determine the possible risks associated with the development.

13.12 COST–BENEFIT ANALYSIS

Cost–benefit analysis is a technique used to evaluate the economics of costs incurred with the benefits achieved. It is mainly used in the public sector in connection with investment decisions where some account needs to be taken of those considerations which are not of a purely financial nature. As such it is an investment appraisal technique. It has its origins in a paper presented by a French economist, Duput, in 1884 on the utility of public works. Since then the technique has been further refined and developed in several other countries for a variety of purposes.

Obvious areas of relevance are health and medical provisions, education and defence. One of the early important areas of application was for water resource development in the USA, with the introduction of the Flood Control Act in 1936. This Act stated that the control of flood waters was in the interests of the general welfare. The construction of a number of dams in a river had multiple objectives relating to power supply, provision of water supplies, improved navigation in addition to flood control. In such cases it was important to take all these wide repercussions of the dam development scheme into account in deciding the viability of the projects.

The Department of the Environment undertook a study of office block schemes in 1971. Part of the study encapsulated the initial capital expenditure and also the costs-in-use. A part of the study was also devoted towards other benefits that could not easily be quantified, such as aspects of the buildings' design, their flexibility to meet changing requirements and benefits accruing to both employer and employee. The latter group of items could only realistically be evaluated by cost–benefit analysis.

It has also been used in the road building programme, where some of the benefits listed include the saving of lives through fewer accidents and a reduction in travel time for commerce and industry. It has been used to assess the need and value of an oil pipeline across Alaska. This cost–benefit analysis study took two years to prepare

Table 13.4   Costs and benefits (new reservoir)

Costs
The scheme
The loss of homes and livelihoods to those whose land is flooded
The loss of their productivity to the national economy
Compensation costs
Possible ecological damage

Benefits
Employment during construction to local people
The Midlands town's water supply
Watersports facilities and angling

and resulted in a 4,600 page analytical report. It has also been used in connection with hospital building, urban renewal, the provision of leisure facilities and many other types of project.

It was used extensively in the construction of the Victoria Line underground railway, where one of the benefits listed included the removal of traffic from street level to below ground, therefore benefiting the movement of traffic such as cars, buses and taxis. Those people actually diverted to the Victoria Line would also benefit; otherwise they would not use it. Their gain included time, convenience, possibly comfort and maybe lower fares. In addition, travellers switching to the Victoria Line from other underground lines would help ease the travel conditions of those passengers continuing to use the other lines. These are commonly referred to as direct and indirect benefits.

The direct and indirect benefits can also be illustrated by the example in Table 13.4, concerned with the building of a dam in a Welsh valley to provide water for a Midlands town.

In the 1960s a Government White Paper gave formal recognition to the existence of cost–benefit analysis and assigned it a limited role in the nationalised industries. However, the whole system of cost–benefit analysis was heavily in doubt after the advice given by a statutory commission on the proposed location of London's third international airport. The decision in favour of Stansted was severely criticised by many people, largely on the grounds that insufficient attention had been given to the analysis of alternative sites and that only partial attention had been given to many of the wider repercussions of the project. The government then set up the Roskill Commission (1971) to investigate the proposal with the specific instruction that cost–benefit analysis techniques were to be used to evaluate the alternative sites. The analysis suggested an inland site, where politically a coastal site on the Thames Estuary had been preferred. The disapproval focused on the marked differences between the measured costs of noise nuisance and the costs of time lost by air passengers both in the air and on the ground. The access costs had been calculated by a simulation model and noise nuisance using the principles of compensation.

As a result of this apparent confusion cost–benefit analysis fell into disrepute during the 1970s, although it was still used extensively in the USA and elsewhere. Much of the criticism was misplaced, being based on a misunderstanding of the role that 'money' played in the application of the technique.

Cost–benefit analysis has matured over the past half century and is designed to provide answers to the following questions:

- Who are to be affected by the decisions which are to be made?
- How much are these people likely to be affected by the decisions?
- When are the effects likely to occur?

Costs are defined in their widest possible sense to include all resources such as labour, materials, land and forgone opportunities. For example, when agricultural land is used for building purposes, the agricultural output from such land is lost. The price of the land will reflect a number of factors other than the agricultural output, such as the needs of the local planning authority, the demand for buildings etc. The real value of the land is therefore equal to the value of the opportunities that are forgone.

The approach to the problem arises from the economic proposition that the community at large has relatively unlimited wants, needs and desires compared with the resources available to satisfy them. These requirements are not all of equal importance and there are therefore choices to be made in the use of resources, and there must be some method of establishing their priority. Cost–benefit analysis does not seek solely to justify an alternative on the basis of immediate costs and benefits, but seeks to evaluate these for the lifespan of the project. The argument therefore is for the best use of limited available resources by the public sector. A public body does not need to balance its income with expenditure in the same context as a private company. It is classified as a non-profit-making organisation and relies upon subsidies from one source and another in order to make ends meet. Many of its services are for social need and are often charged for below accounting cost. Politicians of all shades of opinion, however, may argue that this is not the real or total cost and other factors therefore need to be considered.

Benefits are usually assessed in terms of the value of goods or services that a person would be prepared to give up in order to be able to enjoy the facilities which the decision-makers are contemplating should be provided. Benefits may be divided into two parts: the amount a consumer does give up when enjoying a service, i.e. the price paid, and the extra amount over and above that amount which the consumer would be prepared to give up if necessary.

For example, the construction of a new bypass is estimated to save its users three minutes on a journey. The number of journeys undertaken on this stretch of road is estimated to be a million per annum. This then provides us with a total saving of 50 000 hours. If the average occupancy of the vehicles making this journey is two, then a total of 100 000 man-hours can be saved. These are worth something and can be priced, and the benefits of the scheme can be compared with costs and then assessed in relation to the national economy. The biggest single problem in applying the technique is of course to price the benefits. How much is a man–hour worth? Is a highly paid executive worth the same as a lorry driver, someone who is unemployed or even someone who is retired? In practice there is no real answer to such a question – it rather depends upon one's own assessment and personal judgement. The worth of an average person is therefore used, whatever that might be.

In another example, there may be a choice of several different routes for a road to be constructed between two towns. The different routes will incur different attributes, such as the length of road to be constructed, the loss of certain amenities, the need to demolish property, the different costs involved in acquiring the land, the legal difficulties of acquiring the land. In addition it is necessary to consider the users of the new road. On the longer road the travelling time of the users will be increased and this will also need to be taken into account in the way described above. The various costs and benefits can then be assembled, compared and evaluated.

After identifying the problem and the alternative solutions, a cost–benefit analysis evaluation can be made using the following seven basic steps:

1. Determining the objectives
2. Establishing the extent of the effects of the projects
3. Valuing these effects
4. Fixing a timescale
5. Discounting the value of the effects
6. Evaluating the alternatives
7. The final decision

Cost–benefit analysis can be viewed as another tool in the decision-making process. The following points should be noted:

- It does have a real use if used professionally, but it is also open to manipulation for political ends.
- Cost–benefit analysis does not stipulate that a project should only go ahead if the gainers actually compensate the losers.
- Wide divergences of opinion have been expressed about the role and usefulness of this management technique.
- The benefits to be achieved are sometimes little better than a guess, and the less quantifiable they are, the more questionable they become.
- The money values placed on the intangible elements, such as the value of the 'environment' and the value placed by future generations, are highly speculative.
- Many observers have complained that the methods used are often artificial or arbitrary and provide for no means of checking.
- There are at the moment, however, no better alternative techniques to be used under these circumstances.
- If the constraints and limitations are understood then this provides a more objective approach to be used in the selection of projects than relying totally upon opinion alone.

SELF ASSESSMENT QUESTIONS

1. What are the important factors that affect the market value of property?

2. Prepare a developer's budget based on the following information and using other assumptions as necessary:

   Construction of eight new warehouse units
   Expected selling price of each £1m
   Costs of land £2.2m
   Developer's profit 18%

3. 'The difficulty of applying cost–benefit analysis in practice is the wide range of amounts that can be attributed towards costs used in the calculation.' Discuss.

BIBLIOGRAPHY

Anderson, R. and Dibben, M., *A Practical Guide to Lump Sum Investment*. The Investor's Portfolio 1993.

Brett, M., *Property and Money*. Estates Gazette 1998.

Cadman, D., Austin-Crowe, L., Topping, R. and Avis, M., *Property Development*. Spon 1991.

Darlow, C., *Valuation and Development Appraisal*. Estates Gazette 1988.

Davidson, A.W., *Parry's Valuation and Investment Tables*. Estates Gazette 1989.

DTZ Debenham Thorpe, *Money into Property*. Debenham Thorpe Limited 2000.

Enever, N. and Isaac, D., *The Valuation of Property Investments*. Estates Gazette 1995.

Gilbert, B. and Yates, A., *Appraisal of Capital Investment in Property*. Surveyor's Publications 1989.

Jones, Lang and Wootton, *The Glossary of Property Terms*. Estates Gazette 1989.

Millington, A.F., *Property Development*. Estates Gazette 2000.

Millington, A.F., *An Introduction to Property Valuation*. Estates Gazette 2001.

Wiggins, K., *Discounted Cash Flow*. College of Estate Management, Reading 2000.

CHAPTER 14

PRE-TENDER PRICE ESTIMATING

LEARNING OUTCOMES

After reading this chapter, you should have an understanding about the principles and practice of pre-tender price estimating as used in the construction industry. You should be able to:

- Identify the different methods applied to building projects
- Select an appropriate method to use for different types of project
- Identify the general factors that must be considered
- Recognise the difficulty in selecting appropriate rates or prices to use against measured quantities
- Assess levels of accuracy and consistency
- Prepare an estimate with supporting documentation

14.1 INTRODUCTION

One of the first questions asked by a client who wants a building or structure erected is 'How much will it cost?' If the client is wise, the next question will be 'How accurate is this figure?' The purpose of a pre-tender estimate is to provide an indication of the probable costs of construction. This will be an important factor to consider in the client's overall strategy of the decision to build. The estimate will also provide the basis for his budgeting and control of the construction costs. During the project's development and construction phases this estimate may be reviewed and revised many times.

Perhaps the single most important criterion of the estimate is its accuracy. An early price estimate which is too high may discourage the client from proceeding further with the scheme, and so the potential commission is lost. Alternatively, if the estimate is too low, it may result in an abortive design, dissatisfaction on the part of the client or even litigation. It should be accepted, however, that early price estimates are an approximation and will therefore include some amount of uncertainty. Since estimating the costs of construction is a probabilistic activity, preference should be given to offering a range of estimated sums rather than a single amount. Alternatively, if a single sum is desirable then confidence limits should also be given to provide an indication of its reliability.

14.2 ESTIMATE CLASSIFICATION

Table 14.1 shows the chronological development of the project, and the way in which the pre-tender estimates relate to the plan of work and cost planning process. The purpose of producing a pre-tender estimate can be classified into the following categories:

1. Budgeting – this decides whether the project should proceed as envisaged
2. Controlling – this uses the estimate as a control mechanism throughout the design process
3. Comparing – this uses the estimate as a basis for the evaluation of different design solutions

Pre-tender price estimating methods may also be classified as single price-rate, measured analysis or cost models.

Table 14.1   Estimate classifications

Stage	Activity	Plan of work	Estimating types	Cost planning process
1.	Project	Consultation	Preliminary	Initial estimate
2.	identification	Brief	Feasibility	Firm estimate
3.		Investigation	Viability	Preliminary cost plan
4.	Project definition	Constructional details	Authorisation	Final cost plan
5.		Working drawings	Final budget	Cost check
6.	Project execution	Construction	Control	–

14.3 METHODS

The methods normally used for early price estimating are listed in Table 14.2. Although they are sometimes referred to as approximate estimating methods, this needs to be read in the context of the way in which the projects are quantified rather than in terms of accuracy alone. The degree of accuracy will very much depend on the type of information provided to the quantity surveyor and the quality of pricing information and judgement that is used.

Some of the methods have been discarded, while one of the methods described remains in its development stage. Although methods have evolved over a period of time, changes are slow to take effect owing to the conservatism within the industry. Often surveyors will prefer to continue to use an inferior method for

Table 14.2   Methods of pre-tender estimating

Method	Notes
Conference	Based on a consensus viewpoint
Financial methods	Used to determine cost limits or the building costs in a developer's budgets
Unit	Applicable to projects having standard units of accommodation. Often used to fix cost limits for public sector building projects
Superficial	Still widely used, and the most popular method of approximate estimating. Can be applied to virtually all types of buildings
Superficial perimeter	Never used in practice
Cube	Used to be a popular method amongst architects, but now in disuse
Storey-enclosure	Largely unused in practice
Approximate quantities	Still a popular method on difficult and awkward contracts and where time permits
Elemental estimating	Not strictly a method of approximate estimating, but more associated with cost planning; used widely in both the public and private sectors for controlling costs
Resource analysis	Used mainly by contractors for contract estimating and tendering purposes
Cost engineering	Mainly used for petrochemical engineering projects
Cost models	These methods are still in the course of development.

their approximate estimates, rather than attempt to use an unknown method where the results obtained cannot be easily verified. The attractiveness, therefore, of each of these methods includes its ease of application, familiarity and speed, together with a tolerable level of accuracy.

14.3.1 Conference estimate

Conference estimating is a technique that can be used for the preparation of the earliest price estimate which is given to the client. It is based on a collective view of a group of individuals, and may at this stage not be quantified in any particular way. For the best results, it has been shown that the group concerned must have relevant experience of estimating the costs of similar projects. It is used in circumstances where historical cost data may not be appropriate, as in the case of a prototype project. It also offers a qualitative analysis to reinforce or otherwise a measured estimate.

14.3.2 Financial methods

Financial methods fix a cost limit on the building design, based on either units of accommodation or rental values. The estimated cost of a project may be fixed in relation to the number of pupils who are likely to attend a completed school. The architect must then ensure that the design can be constructed within such a cost limit. In the private sector, projects are often evaluated in terms of their selling price or rental value. For example, in connection with a speculative housing development a market research survey would determine the possible selling price of dwellings on a new estate. The builder would then deduct other development costs and profit from the total selling price, and the remainder would represent the amount to be spent on building. Alternatively, building and other development costs (excluding land) and profit could be calculated and deducted from the total selling price in order to determine a maximum price to be paid for the land. This method is used to avoid or reduce the risk of embarking on a profitless venture. The assessment will take place at the outset, and certainly before site purchase.

14.3.3 Unit method

The unit method of approximate estimating consists of choosing a standard unit of accommodation and multiplying this by an approximate cost per unit. The standard units may represent, for example:

- Schools – costs per pupil place
- Hospitals – costs per bed place
- Car parks – costs per car space

The technique is based on the fact that there is usually some close relationship between the cost of a construction project and the number of functional units it accommodates. Functional units are those factors which express the intended use of the building better than any other. This method is extremely useful on occasions where the building's client requires a preliminary estimate based on little more information than the basic units of accommodation.

The method of counting the number of units is extremely simple, but considerable experience is necessary in order to select an appropriate rate. These rates can be obtained by the careful analysis of a number of recently completed projects of a similar type, size and construction. However, adjustments based on professional judgement will always be needed to take into account varying site conditions, specification changes, market conditions, regional changes and inflation. It is one of the simplest and quickest methods to implement, but it must be used with care. It suffers from the major disadvantage of lack of precision, and at best can only be a rather blunt tool for establishing general guidelines. It is advisable, therefore, to express cost within a range of prices that can be useful for budgetary estimating.

Several cost yardsticks operated by the public sector departments in the past have used this method to test whether an estimated cost is reasonable for a proposed project. The unit cost selected in these circumstances can be described as a socially acceptable sum, which precludes the construction of extravagant schemes at the expense of the wider needs of society. It is a useful method of estimating when dealing with national building programmes where some comparability in the unit cost is required.

14.3.4 Superficial area method

The superficial area method is still the most common method in use for early price estimating purposes. The estimate of cost is easy to calculate and thus is expressed in a way that is fairly readily understood by those in the industry and the average construction industry client. The area of each of the floors is measured and then multiplied by a cost per square metre using the rules outlined in Figure 14.1. In order to provide comparability between various schemes, the floor areas are calculated from the internal dimensions of the building. It is largely a post-1945 method, and became appropriate for projects such as schools and housing where storey heights were similar. Storey heights, plan shape and methods of construction are particularly important when deciding on the rate to be used. Another consideration to favour the use of this method is that rates are readily available from many different sources or, alternatively, they can be calculated very easily from existing scheme cost data.

Three considerations should be borne in mind, however. First, the client may express the project only in terms of the usable space required, and it is necessary therefore to add to this area circulation and other non-usable space to make the building function correctly. Second, in a project offering different standards or types of accommodation it will be preferable to price these independently using different rates. A variety of rates may therefore be required, depending on the different functions or construction of the building. Third, items of work which cannot be related to the floor area will need to be priced at separate all-inclusive rates. The huge range in the superficial area rates presents the surveyor with some problems. At best, therefore, they can only represent guide prices and must be adjusted to suit local conditions on the basis of the surveyor's personal experience and skill.

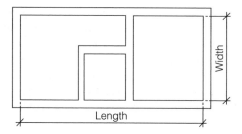

Fig. 14.1 Superficial area method: gross internal floor area calculated by multiplying internal dimensions

14.3.5 Superficial perimeter method

The superficial perimeter method of approximate estimating is a variation on the superficial floor area method. It was devised by John Southwell and published in the RICS paper *Building Cost Forecasting* (1971). Southwell, realising that floor area was the greatest single variable-correlated price, produced a formula that showed an increase in the accuracy of early price prediction. The formula combined floor area with the length of the building's perimeter. This is the second most important variable, and attempts to take into account plan shape when linked with floor area. The wall-to-floor area ratio is known to be an important factor in the economic design of buildings. Tests have indicated that more accurate results can be obtained than when using floor area alone. Because of the reluctance of surveyors to change to this method of approach and of cost data sources to publish appropriate rates, the method has not been used in practice.

14.3.6 Cube rules

The cube method of approximate estimating was used extensively at the beginning of the last century but has since been superseded because of its inherent disadvantages. It was a method extensively used by architects. All architects' offices used to keep a 'cube book' for future estimating purposes. Once the contract was signed the agreed price would be divided by the cubic content and entered into the office price book. The cost of a new job could then be determined by calculating its volume and selecting an appropriate rate from the book. Even with such a primitive method it was necessary to provide some rule for comparable quantification of purposes. The rules of measurement for the cubic content of a building were defined by the RIBA (1954) and are illustrated in Figure 14.2. They are as follows. The external plan area was multiplied by the height halfway from the top of the concrete foundation to halfway up the roof, if pitched, or to 600 mm above the roof if flat. If the roof space was to be occupied then the height of pitched roof buildings was three-quarters-way up the roof. The formula has little to recommend it except uniformity. The allowances for flat and pitched roofs and the measurement to foundation depth are very arbitrary and do not readily correlate with cost. Additional allowances need to be made for projections such as porches, dormers and chimney stacks. Another weakness of this method is that it does not provide any indication to a client of the amount of usable space. It is difficult conceptually to visualise 300 m³, which represents about the size of a typical semi-detached house. It also takes no account of the number of storeys or plan shape and it produces a large cubic quantity that will increase the possibility of further inaccuracy in the estimate.

For example, using the superficial area method the approximate estimate for a church hall might be as follows:

620 m² × £500 = £310 000

Using the cube rules, based on a storey height of 4 m + 0.80 foundation + 0.60 above the flat roof, the same estimate would be calculated as:

3 348 m³ × £93 = £311 364

Fig. 14.2   Cube rules: volume calculated on external dimensions, with height adjustments as indicated

It should be possible to measure the areas and volumes reasonably precisely, but the selection of rates creates considerable difficulty and relies heavily at the moment on the expertise of the surveyor. A £1 error in the rate used in the superficial area method results in a £620 error in the estimate. A £1 error when applying the cube rules creates an error of £3,348. An error of magnitude therefore results.

Where parts of the building vary substantially in constructional method or quality of finish, it is preferable to calculate separate volumes and to apply different rates. The application of cube rates from previous projects does not work quite as well as in the superficial area method. It is also now known that building cost correlates better with superficial floor areas than with volumes.

Ideally, if the cube rules are applied then a very similar building in all respects should be used as the cost database. Otherwise a large number of variables must be considered in order to arrive at anything approaching the correct price. The complexity of modern building is another factor which has contributed to the diminished importance of this method. It is still used to some extent, however, for the valuation of property for fire insurance purposes. Also, the unit of measurement is artificial and relatively meaningless since it relates more to the enclosed void than to the envelope which represents the constructional form.

14.3.7 Storey-enclosure method

In an attempt to overcome the many disadvantages of the other single-price methods of estimating, James (1954) devised a new method using the following rules of calculation:

- Twice the area of the lowest floor
- The area of the roof measured on plan
- Twice the area of the upper floors, plus an addition of 15% for the first floor, 30% for the second floor, 45% for the third floor etc.
- The area of the external walls

The method attempted to take into account:

- Plan shape (by measuring the external wall area)
- Total floor area (by measuring each floor)
- Vertical position of the floors (by using different multipliers for each floor)
- Storey heights (ratio of floor and roof areas for external wall areas)
- Overall building height (ratio of roof area to external wall area)
- Extra costs of providing usable floor areas below ground (by using multipliers)

James claimed that it would perform better in terms of accuracy than the other single-price methods. Lack of use, however, has meant that it has not been possible to verify this claim. The weightings used are highly subjective and are unlikely to apply to every building.

In addition, the quantification does not easily relate to the client's accommodation requirements and as such embodies the same deficiencies as the cube method. By 1954 the limitations of the single-rate approach to estimating were very much apparent, however ingeniously it might be applied. Appropriate rates using this method are almost impossible to obtain, which is a further disadvantage for practitioners. Certainly in those early days credibility was also a factor to be taken into account. It might be more acceptable today to add the areas of walls, floors and roofs and to multiply these by a single all-in rate.

14.3.8 Approximate quantities

Approximate quantities provide a more detailed approximate estimate than any of the methods described above. They represent composite items which are measured by combining or grouping together typical bill-measured items. Whereas the methods described above estimate costs on the basis of measurement and some cost relationship, this method relates cost to the actual work to be carried out. In practice, only the major items that are of cost importance are measured. This method does provide a more detailed and reliable method of approximate estimating, but it involves more time and effort than any of the methods in 14.3.1–14.3.7. No particular rules of measurement exist, and the composite items result from the experience of each individual surveyor. Also, considerably more information is required from the designer if the method is to be applied in practice. The method is therefore suited to a more advanced design stage. It is more reliable, however, when one is attempting to estimate the costs of major refurbishment projects.

Approximate quantities should not be confused with a bill of approximate quantities. The latter would be based on an agreed method of measurement. The former, which is used for approximate estimating purposes, would be much briefer because several of the bill items would be grouped together within a single description. Contractors favour this method when they have to prepare tenders on the basis of a drawing and specification projects.

Using approximate quantities, the roof of a building may be measured and described as follows:

> Three-layer bituminous felt on and including 50 mm prefelted woodwool decking on firrings on 50 × 200 mm softwood joists at 450 mm centres, including vapour barrier and 100 mm fibreglass insulation. −100 m²

The all-in rate for the above could be calculated as follows:

3-layer felt	=	30.00
50 mm woodwool	=	25.00
50 × 75 firrings 2.40 m at £6.00	=	14.40
50 × 200 joists 2.40 m at £7.00	=	16.80
Vapour barrier	=	6.50
Fibreglass	=	11.00
		103.70
Sundries: plates etc. 5%	=	5.19
		£108.89 per m²

The sundries percentage is added to allow for items of work which are necessary but excluded from the analytical cost calculation. A rate of £110.00 would then be used.

Some care must be exercised when using rates from priced bills for approximate estimating purposes. The surveyor should fully examine the entire document to establish how it has been priced, particularly in respect of the preliminaries items and for any discrepancies in rates.

Once tenders have been received, this method allows the approximate estimate to be usefully compared with the lowest tender, and the reasons for any differences can then be quickly assessed. This method also allows the surveyor to become familiar with the prices of construction work, and to develop a 'feel' and expertise in this matter.

14.3.9 Elemental estimating

The first stages of cost planning can be used to determine the approximate cost of a construction project. This method analyses the cost of the project on an elemental basis, attempting to make use of the cost analyses from other similar projects. Cost planning, however, also seeks to do much more. It provides cost advice during the design process, offering the client better value for money. It keeps the designer fully informed of all the cost implications of the design in relation to an approved approximate estimate and likely accepted tender sum. Full cost planning services today would also incorporate the attributes of whole-life costing and value engineering. Two alternative forms of cost planning have been developed, although in practice a combination of both is now generally used. The first form is known as elemental cost planning, where the project must be designed within an overall framework of a cost limit. It is often referred to as 'designing to a cost'. In practice it is more appropriate to public sector projects, which often incorporate some form of cost limit. The alternative form is comparative cost planning, where alternative designs can be examined within an economic context. This method is referred to as 'costing a design'.

14.3.10 Resource analysis

Resource analysis is the method that is traditionally adopted by contractors' estimators to determine their individual rates for measured items in bills of quantities. Each individual measured item is analysed into its constituent parts of labour, materials and plant. Each part is then costed on the basis of outputs, gang sizes, material quantities, plant hours etc. Particular emphasis is placed on such project features as type, size, location, shape and height as important factors affecting the contractor's costs. In theory the contractor will make extensive use of feedback, although some evidence suggests that the whole process is largely determined by value judgements on the basis of previous experience. Alternative analytical methods may calculate resource costs on the basis of operations rather than individual bill items.

Resource estimating is not strictly a pre-tender method of price prediction, because of the amount of time and the type of data required. It will, however, find application in circumstances where, for example, a new material or construction process is envisaged. In these circumstances, where existing cost data are not available the design team may have few alternatives available other than to refer to resource-based estimating.

14.3.11 Cost engineering methods

Three methods used for capital cost estimating in the process–plant industry are as follows.

(i) Functional approach

Bridgewater (1974) states that 'the average cost of a functional unit in a process is the function of the various process parameters'. The estimated cost may therefore be represented in the following way:

$$\text{cost} = F(Q,\ T,\ P,\ M,\ \text{CCI})$$

where Q is the capacity throughout, T is the temperature, P is the pressure, M are the materials of construction and CCI is the construction cost index.

(ii) Factor estimating

This method relies on costing only a portion of the scheme and then multiplying this by a factor to obtain the total cost. Zimmerman (1965) has called these ratio–cost factors. Thus the total cost of a building project may be estimated by multiplying the cost of the shell by, say, 1.6. A range of factors have been derived empirically for different sorts of fixed capital equipment. For example:

Solid processing plant:	factor = 3.9
Solid–fluid processing plant:	factor = 4.1
Fluid processing plant:	factor = 4.8

Several different factors are now widely recognised. Some use a single factor; others apply different factors to the various parts of the project.

(iii) Exponent estimating

The costs of similar items of plant or pieces of equipment of different sizes vary with the size raised to some power. Jelen and Black (1983) expressed this mathematically as follows:

$$\frac{C_2}{C_1} = \left(\frac{Q_2}{Q_1}\right)^x$$

where C_2 is the cost of the desired capacity Q_2 and C_1 is the cost of the known capacity Q_1. A frequent value of x is 0.6, and so the relationship is often referred to as the six-lengths rule. The exponent x can be determined by plotting actual historical costs for the equipment or plant.

These methods can also be used for estimating the costs of building and civil engineering works.

14.3.12 Cost models

Cost modelling is a more modern method that can be used for forecasting the estimated cost of a proposed construction project. Although cost models were first suggested during the early 1970s, there is still only scant evidence of their use in practice. However, considerable research has been undertaken in an attempt to convert the theories into practice (see Chapter 16).

While bills of quantities, cost plans etc. are models of cost, they are not generally considered as such in the context of modern cost modelling.

The use of the computer has allowed more numerical methods such as statistical and operational research techniques to be applied to the forecasting of construction costs. Without computer facilities such applications would not be possible. These models attempt to formulate a better representation of construction costs than do their predecessors, by trying to discover the true determinants of construction costs. There is little evidence at the present time, however, that cost models offer any superiority over the traditional methods in terms of forecasting performance. During the early phase of their development it was assumed that estimating generally was solely a numerical process. This assumption is now believed to be erroneous, and the models, to have any chance of future practical application, must consider the input and expertise of the surveyor or estimator.

During the early days of cost modelling, multiple regression analysis was thought to be the most appropriate technique. Later researchers favour simulation as a more realistic approach to the problem.

Multiple regression analysis is a technique that will find the formula or mathematical model which best describes the data collected in terms of a dependent variable. The dependent variable in a cost model is cost, i.e. the estimate or the prediction of the tender sum. The recommendation of this technique was based on the theory that reliable estimating requires a sound knowledge of previously achieved performance (see Chapter 16).

A simulation model seeks to duplicate the behaviour of the system under investigation by studying the interaction of its components. In this way it copies the process involved and seeks, through a better understanding, to improve the quality of the estimate.

14.4 GENERAL CONSIDERATIONS

The selection of appropriate rates for pre-tender price estimating depends on a wide variety of factors. Some of these can be adjusted objectively, but in many circumstances only experience and 'feel' for the project can help to choose the appropriate rate.

14.4.1 Market and contract conditions

When preparing an estimate, the rates and prices used will normally be obtained from previous projects or historic cost data. The approximate estimate, however, is a forecast of the tender sum at some future date. It is therefore necessary first to update the prices to current pricing levels by using a tender price index. It will also be necessary to take into account the increased costs of labour and materials which have already been announced but have not yet been brought into operation. Allowance must be made for changes in contractual conditions, type of client, labour availability, workloads etc., and the general buoyant or otherwise state of the industry.

14.4.2 Design economics

Where changes in the design occur, such as in shape, height or size, some adjustment will need to be made to the rates used in the approximate estimate. The nature of the building site may also affect the design and the way in which the building is constructed, and thus affect its cost. The type of constructional details selected for the design must be examined in the context of the existing cost information.

14.4.3 Quality considerations

The rates from existing projects are based on a defined standard of quality. If this standard is to be increased or decreased then a change in the proposed estimate rates will be necessary. It may be necessary to make adjustments on a presumed increase in quality standards, by indicating general improvements throughout. Alternatively it is possible to be more precise by choosing, for example, a higher quality of external facing brick, in which case the rates in the estimate can be adjusted more objectively. For example:

Rate per m² gross internal floor area	£751.35
Floor area	1 200 m²
Area of external facings	400 m²

Assume an **increase** in the quality of the bricks of £40 per 1 000.

59 bricks per m²	2.36
+ 7.5% waste	0.18
	2.54
+ 10% overheads and profit	0.25
Increase in cost of facings	£2.79
2.79 × 400 m² (wall area) ÷ 1 200 m² (floor area) =	0.93
Original rate per m² GIFA	751.35
New rate to be used in estimate	£752.28

14.4.4 Engineering services

Engineering services represent an ever-increasing proportion of building projects. Their cost importance emphasises the need to consider them separately from the remainder of the building. On large schemes, specialist quantity surveyors will be employed to offer guidance, particularly at the approximate estimating stage. The provision of air conditioning within a building, for example, may increase the cost of the project considerably.

14.4.5 External works

Because of the considerable differences that often exist between building sites, there is little cost relationship between the external works element and the actual building. It is generally necessary, therefore, to include these costs as a separate item in the estimate. The size of the site and the work to be carried out are important factors to consider.

14.4.6 Exclusions

The proposed estimate of cost should clearly identify what has been included by way of specification, and also what has been excluded. A client may be forgiven for assuming that a £1m estimate includes all their expenditure concerned with the project. They are unlikely to be forgiving when they subsequently find that some items of expenditure have been excluded. Obvious examples of such exclusions include professional fees and expenses; value-added tax; land costs; interest charges; loose furniture and furnishings; and items of special equipment that may be required in workshops and laboratories. It is also necessary when comparing projects to consider items which may be included in one project but not in another. This is particularly important in projects such as shops and offices, where the fitting-out can often form the basis of a separate contract.

14.4.7 Price and design risk

Estimates are prepared on the basis of a combination of three factors: quality, quantity and price. The first two are associated with the design, which is in a general state of change up to the signing of the contract. The design will also have an impact on the construction methods used by the contractor. At the outset of the scheme the design will be represented by little more than sketch plans and elevations and, by necessity, these will be considerably refined during the design process. The costs risk associated with design will thus be much greater at inception than at the tender stage. A larger percentage will therefore need to be added to cover the design risk at inception than at much later stages during the design process. The percentage to be added will be influenced by the type of client, the type of project and the general familiarity of everyone concerned with the design. In some cases, therefore, the design risk percentage will be almost negligible, whereas in other circumstances it may represent 30%–40% of the estimated costs.

The price risk factor is largely related to the market conditions that are prevalent during the design stage. A more volatile market than usual will result in larger percentages being added to the estimate for the price risk. The approximate estimate which may be prepared and then revised at different stages during the design process is really an attempt to forecast the tender sum. In periods of high inflation of 24% per annum, this means that costs are rising by 2% per month. An estimate prepared in January for tender expected in May will therefore need to add 10% just to cover inflation. The particular time of year will also be important, since costs do not rise uniformly each month. Estimates which overlap the construction industry pay awards in July will also need to allow for larger increases than usual. Fixed price contracts will generally require larger percentages to cover the price risk factor, which is fully taken into account on the contractor's tender.

14.5 OTHER FACTORS

14.5.1 Selecting a method

Pre-tender estimates will be required initially as a guide price and then at various stages during the design process as the scheme develops. The method used for forecasting cost at these various stages will depend on a number of factors.

During the early stages of the design period, when little is available in the way of drawings, details and other scheme data, a single price-rate method may be preferred. In the case of projects which may be based on rather strict cost guidelines, such as housing or school buildings, it may be possible to work to a higher level of accuracy. Where an expenditure limit is predetermined, this approach will certainly be desirable during the early stages of the design. There is little point, however, in attempting to prepare approximate quantities for a project which is yet to have its basic design concepts developed. The only merit of this is where some close correlation may exist with a previous project. It may also be desirable under some circumstances to be able to offer some element target costs to which the designer can work and evolve the design.

The choice of a method for preparing the pre-tender price estimates will therefore depend on the following:

- Time available
- Project information
- Cost data
- Preference and familiarity
- Experience of the surveyor

14.5.2 Usage

The method of pre-tender estimating which is still the most widely used today in practice is the superficial floor area method. This is due in part to its familiarity to its users and its relationship to the client and designer's use of space, but it is also because it is relatively easy to obtain appropriate cost information for use in the estimate. The unit and financial methods are used where cost limits are introduced, but the former is restricted to projects which incorporate some functional unit to which cost can be attributed.

Approximate quantities are generally measured for awkward projects, particularly refurbishment work, or in circumstances where the designer requires a breakdown of cost. Evidence also suggests that quantity surveyors prefer to use this approach wherever possible for pre-tender estimating purposes where time and design information are available. They are often used at the later stages of project design and in conjunction with cost planning. The latter technique is now supposedly extensively used on building projects, and is finding some acceptance and application on engineering projects.

The superficial perimeter method and the storey-enclosure method have both had minimal use in practice, as have the cube rules.

Cost models have been used on major highway projects, but extensive development is such as to prohibit widespread usage at the present time. They have additionally not been sufficiently researched and developed as of now, and are likely to find general acceptance only where they can demonstrate some potential advantages. They are often complex and therefore difficult to understand, and this discourages potential users.

14.5.3 Quantifying

The quantification of the construction work in a pre-tender estimate should be relatively straightforward. The methods have been devised on the basis of speed and some assumed cost relationship between quantity and price. Simple rules have been devised to allow for comparability and to ease the facility for using cost data from outside sources. Some care, however, needs to be exercised. There may be some items of construction work which do not easily relate to the method used for preparing the estimate; for example external works are known to have little relationship with the superficial area of the building. In other cases it may be simpler to allow for additional costs than to attempt to adjust a single price-rate for some abnormal part of the project.

Extensive refurbishment projects may present special problems of quantifying and it is therefore important to examine the drawn information properly in order to avoid omitting required items of work. It is also sometimes difficult with this type of scheme to know in advance the full extent of the work which will be required. A larger percentage contingency sum must therefore be allowed than is usual.

14.5.4 Data requirements

In theory, any of the methods can be applied on the basis of a minimal amount of design data or cost information at any stage during the pre-tender process. Each method is therefore appropriate for giving the client or designer cost advice at inception. However, the more information that is available, the better will be the quality of the estimate. The more precise the information, the more accurate the estimate should be. The quantity surveyor should have available relevant and current cost data in the form of price books, cost analyses and priced bills of quantities. Some form of drawn information from the designer, together with brief specification notes, will also be required. Where the latter cannot be provided, the surveyor should make appropriate assumptions and include these as part of the estimate. An approximate estimate providing no indication of quality standards is so open-ended as to be almost meaningless.

14.5.5 Computers

All the methods of pre-tender price estimating, with the exception of cost modelling, have been developed as manual processes. The generation of cost models has arisen only because of the ready access to computers. The computer, however, can be successfully used in conjunction with any of the methods described for the rapid retrieval of cost data, or for calculating an approximate estimate using predetermined parameters. The advantage of using a computer is that an estimate, to any level of detail, can be updated quickly and efficiently for changes which may be necessary in a revised design.

The application of computers to estimating has resulted in two schools of thought:

1. Computerised traditional methods which allow the user to estimate using known techniques. Speed in preparation and updating, and access to a wider database, are claimed as advantages
2. Computerised statistical techniques which require the surveyor or estimator to use new methods of application such as cost modelling

Computer developments in the future affecting pre-tender estimating have been identified in the following two areas:

(i) Computer-aided design (CAD) – these systems allow the designer to model the project in three-dimensional form. It is a very simple matter then to allow its cost to be generated and an estimate produced. The difficulty at the moment is in linking this to a sufficiently reliable cost database which can be accessed automatically. We have already observed that the difficulty of pre-tender estimating is associated not so much with measuring expertise as with pricing expertise. The future may, however, see the surveyor interacting directly with a CAD system in the designer's office.

(ii) Expert systems – these transfer the expertise of the surveyor to the computer, and this information can then be used by any member of the design team. Although this would appear to be a relatively simple matter, in practice it is complicated because of general differences of opinion among the surveyors.

14.5.6 Accuracy

Early price estimates are a forecast of a contractor's tender sum. Clients may also require a forecast of the expected final cost of the project. The client, the size and the nature of the project concerned will have some bearing on its accuracy. It has been suggested that designers in the future may have to redesign at their own expense should their estimate exceed reasonable levels of inaccuracy.

It has sometimes been argued that an acceptable margin of error should equal the percentage difference between the highest and the lowest tenders. Tenders submitted by a number of firms will vary for reasons unconnected with accuracy. It might be argued that any margin of error is unacceptable, but this is also unrealistic. The margin of error in practice is that which the profession at large would consider to be reasonable. The client expects professional advisers to be experts, and will therefore judge them accordingly.

It is becoming increasingly apparent that to predict costs accurately is a problem which is common to all industries. Throughout the construction industry there are extra difficulties due to the complexity and uncertainty of the type of work involved. It should also be noted that an early price estimate is a forecast of a contractor's tender, which is in itself subject to some error.

Any improvement in the accuracy of early price forecasting for construction works is desirable, and is likely to result in a consideration and combination of the following three factors:

1. An improvement in the quality of designer's information, since a vague design can result only in an inaccurate estimate. Evidence is already available to suggest that a properly completed design not only will reduce initial costs, but will also have the effect of shortening the construction period on site and improving the quality of the works.
2. A reappraisal of the methods currently used for estimating, in an attempt to discover more correctly the determinants of construction costs. Attempts have

been made during the past decade to examine new approaches to early price estimating.

3. To identify the qualities in the surveyor or estimator which contribute towards accuracy in estimating and to consider how they might be improved.

The following factors have some influence on the accuracy of estimating the costs of construction work:

- The effects of differing availability of design information
- The amount, type and quality of cost data that are easily accessible
- The type of project – some types of scheme can be estimated with a greater degree of consistency
- Project size, in terms of value – a tentative finding is that accuracy improves by a small amount as projects increase in size
- The number of bidders on competitive projects
- The stability of market conditions
- Personal factors – familiarity with a particular type of project or client has been associated with a substantial improvement in forecasting accuracy
- Proficiency in estimating is said to be a result of skill, experience, judgement, knowledge, intuition, feel, academic background, personality, enthusiasm, hunch and a 'feeling at the back of the head'
- Sheer quantitative experience alone has not been found to be generally correlated with estimating accuracy – an overall sensitivity to price levels, however, is particularly advantageous

It should also be noted that the inconsistency exhibited in estimates is perhaps a more serious problem than accuracy alone. Figure 14.3 shows three typical sets of circumstances. Category C applies to the forecasting of construction costs, so that while some forecasts may be spot on target, others are wide of the mark.

There is general wide agreement over the shape of the accuracy cost curve. Some disagreement does exist, however, over the percentage points quoted. Figure 14.4 shows a typical range of data which can be expected. These classifications are based on the categories outlined in Figure 14.1. The figure is based on the views of those who have offered an expert opinion coupled with the measurement of actual samples of estimates from practice. Often the empirical evidence indicates that the levels of estimating are lower than is presumed. These percentages will also depend on the extent of the inclusion of specialist work quotations previously known or determined by the quantity surveyor.

It needs to be noted that there are real practical difficulties in making any comparison of estimating accuracy.

14.5.7 Measuring accuracy

The difference between an early price estimate and the accepted tender from a contractor represents an inaccuracy. This can be expressed in percentage terms. However, little can be inferred from a single observation. It is therefore necessary to

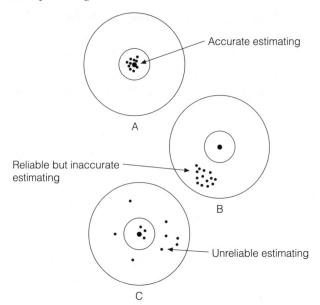

Fig. 14.3   Accuracy and reliability of estimating: category C estimating is prevalent in the construction industry

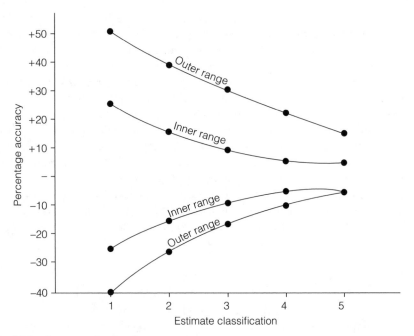

Fig. 14.4   Accuracy of estimating

Table 14.3 Comparison of early price estimates and accepted tender sums (in £)

Project	Early price estimate	Accepted tender sum	Percentage variation x
A	220 000	202 000	+8.91
B	175 000	199 500	−12.28
C	398 000	345 000	+15.36
D	274 000	256 000	+7.03
E	194 000	228 000	−14.91
F	122 000	127 500	−4.31
G	312 000	352 500	−11.49
H	178 000	162 000	+9.88
J	422 000	371 000	+13.75
K	184 000	169 000	+8.88
L	512 000	470 000	+8.94
M	276 000	264 000	+4.55

accumulate a number of observations for analysis purposes. The first means of assessment is to calculate the arithmetic mean. In these circumstances this is a preferable statistic to the median or the mode. A second method of interpreting the data is to consider the spread or range of the estimating accuracy. The range represents the difference between the highest and lowest values. A better method of measuring the variability is to use the standard deviation. A further statistic is the coefficient of variation.

Table 14.3 represents the early price estimates and accepted tender for twelve projects. The percentage variation for each of these projects is also shown. Table 14.4 provides a statistical analysis of these data. These are examples of statistics that might be used. Interpretation depends on many factors, one of which is sample size.

Table 14.4 Statistical measures of accuracy

Mean $\dfrac{\Sigma x}{n}$	$= 10.02\%$
Range	$= 4.31 - 15.36$
Standard deviation S	
$\sqrt{\dfrac{\Sigma(x - \bar{x})^2}{n}}$	$= 3.68$
Coefficient of variation	
$\dfrac{\text{standard deviation}}{\text{mean}} \times 100$	$= 36.73$

14.5.8 Human aspects of estimating

Forecasting the likely tender costs of the successful contractor is a difficult task, relying on a good understanding of the methods involved, the current state of the market, a knowledge of current prices and a kind of intuition that cannot be easily acquired. Pretender price estimating is a mixture of science and art; and both are required for a successful outcome to be achieved. Some individuals will have a good knowledge of likely construction costs and prices for particular projects, such as secondary schools constructed in a particular location or region of the country. Those who are involved with projects in this way are likely to be more reliable than others who might only become involved with such types of projects less frequently.

It has already been noted that the contractor's estimate is in itself only a forecast of the likely cost of the project. If this was a precise forecast then it would no longer be an estimate of cost. Those involved in attempting to forecast the successful contractor's bid will therefore need to make some allowance for this. Since it is an unknown factor it can represent, at best, nothing more than a guess.

There has now been a significant amount of research undertaken in this area of understanding in different parts of the world. Much has been aimed at improving the quality and accuracy of pretender construction estimates. It is unlikely, for the reasons already stated in this chapter, that any improvement in the levels of forecasting will be achieved unless the human interaction is removed and the process becomes an entirely mechanical or numerical process. This is unlikely to happen under present circumstances.

Figure 14.5 illustrates that estimating the costs of construction works is a combination of professional judgement and technical skills. Both are needed if satisfactory results are to be achieved at the highest level.

14.5.9 Legal aspects

A surveyor's estimate must be reasonably accurate or the surveyor may be liable for negligence. The test of reasonableness relates to the amount of skill or expertise which would be usual in such a profession. Quantity surveyors are looked on as experts on building costs, and must therefore perform their duties in the manner expected. Failure to possess the usual skill or expertise or careless use of it will be a cause of negligence. The client when seeking professional advice expects that this will be sound and reliable. If a surveyor delivers an estimate greatly below the sum at which the work can be expected, and thereby induces someone to undertake construction work, there may be liability for negligence. The owner needs to know in a general way the costs involved, and looks to the quantity surveyor before embarking on what could be an expensive exercise in wasted plans and other expense. Surveyors should therefore avoid giving estimates 'off the cuff': where the available information is insufficient, either an opinion on costs should be withheld or it should be severely qualified.

Fig. 14.5   Judgement and technical skills

14.6 PREPARING THE ESTIMATE

The method to be used for the preparation of the pre-tender estimate will to some extent depend on the type of project concerned and the amount of information which can be provided by the client's designer. The more vague this information, the less precise will be the estimate. The estimate for a complex refurbishment project, where some of the detailed work will be unknown until the contractor starts on site, may need to utilise a form of approximate quantities. It could be prepared on the basis of the superficial floor area, but the surveyor would need to take into account the condition of the existing building as well as the client's proposals. The quantity surveyor would need to be particularly skilled in choosing an appropriate rate to be used. If the surveyor is familiar with the type of work envisaged, the designer and previous other projects for the same client, then this approach may prove to be satisfactory.

The preparation of the estimate commences with a scan of the drawings, which are likely to be only in outline form. On the basis of past experience the surveyor may be able to offer a budget figure on a purely subjective assessment of the scheme. However, some types of measurement are likely to be made using one of the methods previously described. Ready access to current cost data such as previous priced bills, cost analyses and price books is required. The pricing is really the skilful part of the operation, since it will not be sufficient to identify a 'typical' price but a forecast of a contractor's likely tender sum. It is usual to include a contingency sum as part of the estimate, and this will decrease with later estimates of the scheme as the design becomes more formalised and the period up to the submission of tenders becomes shorter. It should never be entirely removed, however, until the completion of the final account. Allowances for inflation may be added in two parts, one up to the tender date and a second for the contract period. In periods of variable inflation the forecasting of such sums is particularly hazardous.

Pre-tender estimates should exclude VAT, even on projects for which it will be charged. It can represent a considerable sum of money, and clients should be left in no doubt about it even where they are able to reclaim it from HM Revenue and Customs. The client will also want to know the total budget for the project inclusive

Table 14.5   Construction budget and programme

CLIENT   . . . .

PROJECT   . . . .

Budget serial number . . . .

Price base date . . . .

Costs (for basis see sheet . . . .)

New building work . . . .

                         . . . .

                         . . . .   _____   . . . .

Site works   . . . .

                         . . . .

                         . . . .   _____   . . . .

Alterations   . . . .

                         . . . .

                         . . . .   _____   . . . .

CONSTRUCTION COSTS (at stated price base)   _____   . . . .

Estimated inflation to tender date of   . . . .

CONSTRUCTION COSTS AT TENDER   . . . .

Estimated increased costs payable during the contract period   _____

CONSTRUCTION COSTS AT COMPLETION   . . . .

PROFESSIONAL FEES AND EXPENSES OF ALL   _____
CONSULTANTS

VALUE ADDED TAX: on construction . . . %   . . . .

                              on fees   . . . %   _____

TOTAL BUDGET OF CONSTRUCTION WORK
AND FEES   £ ____

Expected range of total budget £ . . . . . to £ . . . . .

Exclusions from budget

. . . .

. . . .

. . . .

Source: Adapted from *Precontract Cost Control and Cost Planning*, RICS of all costs associated with its construction. Some forecast of all the professional fees involved and their relevant VAT should also be included.

The estimate must be clear on all items which are excluded. Generally the distinction between client's and builder's items will be clear to both parties: where confusion could occur, these excluded items should be listed with the estimate. It needs to be clear that the total budget of construction work generally does not include the purchase of the site and associated fees or the provision of loose furniture, which would normally be provided directly by the client concerned.

Table 14.5 has been adapted from *Precontract Cost Control and Cost Planning* (RICS). The estimate should always be dated and should relate to drawings or schedules and a standard of specification. Three separate 'costs' have been identified:

(i)   The probable construction costs at the tender stage
(ii) The forecast construction costs at the final account stage
(iii) A total budget based on all costs associated with the construction

Example

The following is an example of how an estimate might be prepared using the superficial floor area method for the new hall extension shown in Figure 14.6.

Estimate for Church Hall extension based on drawing in Figure 14.6 (in £)

1. Gross internal floor area of extension
   $18.00 \times 7.00 = 126 \text{ m}^2 \times £1\,000$ = 126 000
   (This illustrates the ease of quantification and the difficulty of pricing, taking into account the seven factors described earlier. Adjustments are often made on the basis of the surveyor's 'intuition'.)
2. Remove existing font and make good (item)   =   1 000
3. Break through to form new door openings, including joinery work and finishings   =   4 000
4. Construct new staircase, including cutting out existing timber floor =   3 000
5. Refurbishment of toilet areas
   $3.80 \times 3.50 \text{ m}^2 \times 2 \times £500$ = 13 300
6. Site works 37 m² @ £150   =   5 550
                                     152 850

Contingencies 3%   4 585
TOTAL   £157 435

Note
It may be necessary to add price and design risk where the architect's intentions are vague. It would also need to be pointed out that the above excludes VAT and professional fees. A brief specification would be attached to the estimate in order to qualify the surveyor's interpretation of the client's intentions and the architect's design. This would be as shown on pp. 342–343.

Fig. 14.6   Meeting hall extension

Specification notes

Substructure	Strip foundations, concrete floor on hardcore
Roof	Timber joists, beams, 3 layer felt on plywood
External walls	Facing bricks £250/1 000, block inner leaf
Windows and external doors	Softwood, polycarbonate sheeting
Internal partitions	Folding screens
Internal doors	Plywood flush
Finishings	Plaster, plasterboard, emulsion, vinyl tiles
Electrical	Strip lighting
Heating	Gas water heating off hall system
External works	Paving, minimum landscaping

Construction budget

Construction costs (at stated base date)	=	157 435
Estimated inflation to probable starting date	=	500
Construction costs at tender	=	157 935
Estimated increases during construction	=	Included
Construction costs at completion	=	157 035
Professional fees and expenses of all consultants	=	12 500
		169 535
VAT on construction 17.5%	=	27 481
VAT on fees 17.5%	=	2 188
Total budget on construction work and fees	=	£199 204

Exclusions from budget: loose furniture and fittings

CONCLUSIONS

The main advantage of using any one of the methods described above in order to obtain an estimate of expected cost is the speed of the process. The computer may improve on this, but only marginally so. The difficulty with all the methods used is in the selection of the appropriate rate to be applied. Considerable expertise on the project concerned is therefore desirable to obtain consistently reliable results. A ready access to current cost information together with a knowledge of future trends is essential in order to achieve acceptable results.

SELF ASSESSMENT QUESTIONS

1. Identify methods that can be used for the early price estimating of building projects today and select a method that you consider to be the most appropriate giving reasons for your choice.

2. There are seven factors to consider when preparing an approximate estimate. Show how these would be applied to the preparation of an approximate estimate for a typical project of your choice.

3. Describe the factors that create inaccuracy in early price estimates and explain ways in which accuracy might be improved.

BIBLIOGRAPHY

Ashworth, A., *Cost Models, their History, Development and Appraisal*. Technical Information Service Paper No. 63, Chartered Institute of Building 1986.

Ashworth, A. and Skitmore, R.M., *Accuracy in Estimating*. Occasional Paper No. 27, Chartered Institute of Building 1982.

Ashworth, A. and Skitmore, R.M., *Accuracy in Cost Estimating*. Ninth International Cost Engineers Congress, Oslo 1987.

Bridgewater, A.V., 'Rapid cost estimation in the chemical process industries', *Third International Cost Engineering Symposium* 1974.

Building Cost Information Service, *Quarterly Review of Building Prices*.

Davis Langdon, Cost Models in *Building* (weekly).

Dinatale, J., 'Assertive concept estimating procedures', *Transactions of the 7th International Cost Engineering Congress* 1982.

Drew, D. and Skitmore, M., 'The effect of contract type and contract size on competitiveness in construction contract bidding', *Construction Management and Economics* 15(5), 1997.

Fortune, C.J. and Lees, M.A., 'The relative performance of new and traditional cost models in strategic advice for clients', RICS Research Papers 1996.

James, W., 'A new approach to single price-rate approximate estimating', *Chartered Surveyor*, 1954.

Jelen, F.C. and Black, J.H. (eds), *Cost and Optimization Engineering*. McGraw-Hill 1983.

Morrison, N. and Stevens, S., *Cost Planning and Computers*. Property Services Agency. HMSO 1981.

Neufville, R. et al., 'Bidding models: effects of bidders' risk aversion', *Journal of the Construction Division*, No. 103, American Society of Civil Engineers 1977.

Nisbet, J., *Estimating and Cost Control*. Batsford 1961.

Precontract Cost Control and Cost Planning. Quantity Surveyors Pamphlet No. 2. Royal Institution of Chartered Surveyors 1982.

Royal Institute of British Architects, *RIBA Rules for Cubing Buildings for Approximate Estimates*, Ref. D/1156/54 1954.

Skitmore, R.M., *The Influence of Professional Expertise in Construction Price Forecasting*. Department of Civil Engineering, University of Salford 1985.

Southwell, J., *Building Cost Forecasting*. Royal Institution of Chartered Surveyors 1971.

Zimmerman, O.T., 'Use of ratio cost factors in estimating', *Cost Engineering*, Vol. 10, 1965.

CHAPTER 15

COST PLANNING

LEARNING OUTCOMES

After reading this chapter, you should have an understanding about the principles and practice of cost planning of buildings. You should be able to:

- Appreciate the historical development and significance of cost planning
- Understand the process and procedures that are used
- Recognise the implications of cost planning in terms of value for money
- Understand quantity and quality considerations
- Understand the BCIS system and descriptions of elements
- Prepare a cost plan

15.1 INTRODUCTION

Clients who are interested in building development will choose to discuss their project with an architect or surveyor or with a building contractor. One of these becomes the lead consultant who is then responsible for appointing the remainder of the design and construction teams. Whilst clients come in many different shapes, sizes and varieties, they can be broadly classified in one of three categories:

1. Clients who have a clear idea of their building requirements in terms of space, design and aesthetics, but are generally uncertain of the cost implications
2. Clients who have a very large amount of money to spend and a project in mind but are unsure of exactly what that amount of money will purchase
3. Clients who know their building requirements and who also have a good idea what this might purchase as well as limited sum of money to spend on the project

Clients who are involved in capital works projects on an occasional basis will usually be in categories 1 and 2, although there is often a mismatch between the project and the funding that is required. Clients who build on a regular basis will have a good idea of their expectations in terms of design and cost. It should be recognised that more than 80% of project costs are determined during the initial stages of the design. Skilful and professional estimating and cost planning add real value by enabling informed decision-making from the outset.

Cost planning in one form or another can be used effectively to evaluate any of the solutions to the above. Before this technique was developed, in the 1950s, final design decisions had to be taken and tenders invited and then abortive measures were often used to equate the design requirements with the funding that was available. The final solution was often not what was required or desirable, resulting in a somewhat dissatisfied and disillusioned client. Designs might be quickly redrawn to reduce them to match the available funding.

The cost planning that is used today might more aptly be renamed value planning, since although there is an emphasis on ensuring that tender sums equate with budget estimates, the other intention is to provide a balanced design and value for money. Where addendum bills of quantities had to be prepared, these often resulted in making savings in the least inconvenient manner, so that the scheme could then proceed as quickly as possible. These savings often produced imbalances in the design which would not have been recommended or realised during the design stage. The only other alternative was to return to the start of the design process and commence the design all over again. This was usually not a feasible suggestion or cost-effective for either the designer or the client. There was also no guarantee anyway that this would provide an acceptable cost solution. Of course the worst which could happen was that the project would be abandoned and this was satisfactory neither to the designer, who lost the fees for the supervision work, nor to the client, who had to pay for design fees for a project that was not to be realised. The general advantages claimed for cost planning therefore include the following:

- The tender sum is more likely to equate with the approved budget estimate.
- There is less possibility of addendum bills of quantities or re-costing being required.
- Cost-effectiveness and a value-for-money design are more likely to be achieved.
- A balanced distribution of expenditure is likely to produce a more rational design.
- Cost considerations are more likely to be taken into account because of the greater involvement of the quantity surveyor during the design process. This will result in an easier preparation of the tender dociments, since the quantity surveyor will be more familiar with the project and have a greater understanding of what the designer is attempting to achieve.
- The amount of pre-tender analysis by the architect and quantity surveyor should enable more decisions to be taken earlier, resulting in a smoother running of the project on site.
- Cost planning provides a sound basis for comparing different projects.

Cost planning, it is claimed, also produces a number of disadvantages which must be weighed against the above. These disadvantages in the main are that the method of design working is reorganised, and it can be argued that this provides a new discipline of thought patterns which in the long term are really advantageous. For example, it has been claimed that a major cause of delay on building projects is the inadequate preparation of the design during the pre-tender stage. Cost planning seeks to redress this. The argument that the quantity surveyor tells the architect or engineer what they can do is entirely untrue, unless the quantity surveyor is the lead consultant. The designers are informed of the cost implications of the design, which they find useful. The designers may be more restricted in methods of working and may need to redraw more details and to make design decisions earlier. The quantity surveyor will be more involved during the design process and be more aware of the costs accruing from different design alternatives.

15.2 HISTORY AND DEVELOPMENT

Before the Second World War quantity surveying was largely an accountancy function. An approximate estimate of the building cost was prepared using one of the methods described in Chapter 14. Quantities were then measured from the working drawings to provide a bill of quantities, and after completion of the works on site the final account for the project was agreed. No attempt was made to relate these costs or to consider any cost implications or to make comparisons with other similar schemes.

In the early 1950s the then Ministry of Education had an extensive school building programme, due to a number of schools having suffered war damage or building neglect, an increasingly young population and the philosophy of the 1944 Education Act. Proposals for new school buildings were sent by church and local authorities for approval, but since cost limits had yet to be devised, enormous variations in building cost and cost per pupil occurred in different parts of the country. The stabilisation of school building costs was therefore made a priority by the Ministry of Education. It was observed that some similarity in the distribution of costs among the various elements occurred. It was suggested that future school building could possibly be cost-planned on the basis of this information. This would also allow the architect to relate design more easily to the costs allowed. The process claimed two advantages over the existing system. First, a greater amount of reconciliation would occur between budget estimates and contractors' tenders. Second, which was less tangible and more subjective to assess, was that of achieving a balanced design and value for money.

The ideas were tested and found to offer the suggested measures of improvement. In the ensuing years various local authorities, notably Hertfordshire County Council, began to develop the process now known as cost planning. In 1956, the RICS set up its Cost Research Panel, which was later instrumental in developing the Building Cost Advisory Service (now BCIS). About the same time the RIBA had its own Cost Research Committee.

Other methods of cost planning were devised, such as the percentage allocation method. This assumed that specific elements of similar buildings are proportional in cost. In projects other than those of a repetitive nature this assumption is ill-founded. Cost planning therefore developed largely along two lines, although in practice the method generally used today is a combination of these alternatives.

The Ministry of Education method became known as elemental cost planning and is described more fully in Department of Education and Science, *Building Bulletin No. 4 – Cost Study* (1972). The RICS method is referred to as comparative cost planning.

15.2.1 Elemental cost planning

This method is sometimes referred to as target cost planning since a cost limit is fixed for the scheme and the architect must then prepare a design not to exceed this cost. The cost plan is therefore the architect's design in financial terms. During the design process, although the architect is able to enjoy a large amount of freedom of expression, a final scheme must be prepared within the designated cost limit. The quantity surveyor is able to assist in choosing the correct economic framework in order to bring the project to a successful conclusion. This method was more appropriate to the public sector, where limits were placed on the amount of money that could be made available for a project.

The cost limit can be calculated using the financial method of approximate estimating, which may be based on the finances available for building using the developer's equation. Alternatively the interpolation method may be used, where the cost is based on other similar schemes, taking into account spatial requirements and quality standards.

15.2.2 Elemental and comparative cost planning

Whereas elemental cost planning is often referred to as *designing to a cost*, comparative cost planning can be described as *costing a design*. Although elemental cost planning can involve examining the cost implications of various design solutions, the selected scheme in practice cannot exceed the cost target. Comparative cost planning, however, allows various solutions to be evaluated, and the architect and quantity surveyor together with the client then decide on the most appropriate scheme taking into account all considerations. Such a scheme may of course be the most expensive, but the decisions are then taken in the full awareness of the cost consequences. This method therefore developed more with the private sector in mind, which did not have the same requirement to apply a cost limit to a proposed scheme.

In practice the cost planning process in use today is a combination of the best parts of the above two methods.

15.3　THE COST PLANNING PROCESS

The cost planning process consists essentially of three phases. The first of these involves the establishment of a realistic first estimate. The second stage plans how this estimate should be spent among the various parts or elements of the project. The final stage is a checking process to ensure that the actual design details for the

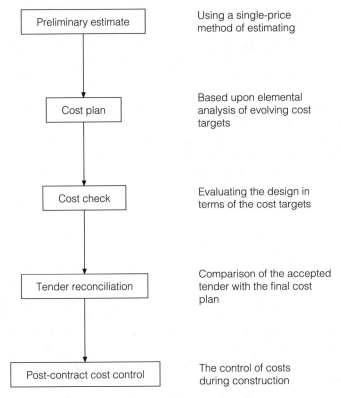

| Preliminary estimate | Using a single-price method of estimating |

| Cost plan | Based upon elemental analysis of evolving cost targets |

| Cost check | Evaluating the design in terms of the cost targets |

| Tender reconciliation | Comparison of the accepted tender with the final cost plan |

| Post-contract cost control | The control of costs during construction |

Fig. 15.1   Cost planning during the design and construction phases

various elements can be constructed within the cost plan. These various stages are now examined in more detail, and shown diagrammatically in Figure 15.1.

15.3.1 Preliminary estimate

Prior to commencing cost planning and the expense of both architect's and quantity surveyor's fees, it is necessary for an indication of cost to be formulated. Since no drawings of any substance will have been prepared, the pricing information provided by way of estimates can be assumed only to be guide prices. The full range of the methods that can be used for this purpose are described in Chapter 14.

15.3.2 Preliminary cost plan

The preliminary cost plan is really the first phase of the cost planning process, with the main purpose of determining a cost target for the scheme. The estimate provided above, which has been accepted by both the client and the architect, may need to be modified in the light of the architect's preliminary investigation of the works and the elementary design and drawings. A sum will eventually be agreed that will form the cost target for the whole scheme. Alternatively, a cost limit, if one is in operation, will become the cost target. This is largely the extent of the quantity surveyor's work at this stage other than to advise on the costs of alternative methods of construction or matters which have a contractual implication. There is no real point at this stage in formulating elemental targets, since design decisions still have to be made and the whole process could be a rather pointless exercise. If, however, the proposed scheme is of a similar type to one already constructed, and a previous analysis is available and going to be followed in principle by the architect, then these cost targets can be utilised, and they must of course be suitably updated for inflation or regional factors. This is the pattern often adopted for schools, local authority housing and hospital schemes where more standardisation in design and layout is envisaged.

The preliminary cost plan, which might more correctly be described as an elemental estimate, requires the measurement of the project very broadly in perhaps a simplified analysis format.

15.3.3 The cost plan

Once the sketch design has been completed and approved by the client, the task of allocating sums to the various elements can take place. The sum of these element cost targets should of course total the cost target for the whole scheme. In the case of elemental cost planning where cost limits apply, it is particularly important that this is so. The methods described later under elemental cost synthesis will be used in order to arrive at the element sums. The following information will generally be required for the preparation of a cost plan:

- Drawings, which should at least include plans and elevations in sketch form
- An indication of the materials to be used and the standard of finishings expected
- Contractual information such as the method of securing tenders, contract period and probable start date
- A comparable cost analysis from a previous project
- Other analyses which can be used to provide 'a second opinion'

Once the cost plan has been prepared it will be in everyone's interest if this is adhered to as far as possible and it will make the process of cost control much more effective. Prior to allocating the costs to the various elements, a deduction is generally made for 'price and design risk'. The amount of this sum will vary depending on the experience of the designer and the quantity surveyor and their familiarity with cost planning procedures. The percentage adjustment can vary considerably and will be influenced by the type and complexity of the design, the nature of the client, an assessment of price trends and the delay expected prior to the receipt of tenders.

15.3.4 The elemental cost plan

Costs may be expressed in three ways for each of the elements: the total cost, the cost per square metre GIFA or the element unit rate. It has been found by

experience that the most convenient way is the second of these choices.
The total cost of the scheme, after allowing for the price and design risk, is divided
by the total floor area (GIFA) to give the cost per square metre. This sum is then
allocated to each element by using a previous cost analysis, approximate quantities or
inspection. It must also be remembered that where historic data are being used these
should be updated. For example, the area of the upper floors may be calculated as
360 m². If the cost of construction is comparable to that in the analysis, then it is
a simple matter to multiply this by the updated element unit rate in order to arrive
at the cost target for this element. Assuming that a current rate from the analysis is
£40 per m², then this gives a cost target for the element of £14,400. Alternatively
we could use the analysis and arrive at this sum by proportion.

Existing analysis

Total cost	*Cost per m²*	*Element*	*Element*
of element	*GIFA*	*unit quantity*	*unit rate*
£27 200	£26.67	680 m²	£40.00

This analysis refers to a three-storey building of 1 020 m². The quantity factor (QF)
can therefore be calculated as:

$$\frac{\text{upper floor area}}{\text{GIFA}} = \frac{680 \text{ m}^2}{1\ 020 \text{ m}^2} = 0.667$$

The proposed project, however, is only a two-storey building with a QF of 0.50.
The cost per m² GIFA is therefore calculated as:

$$\frac{\text{QF proposed}}{\text{QF existing}} \times \frac{\text{cost per m}^2}{\text{GIFA existing}} = \frac{\text{cost per m}^2}{\text{GIFA proposed}}$$

$$\frac{0.50}{0.667} \times 26.67 = \text{£20.22 per m}^2 \text{ GIFA}$$

Multiplying this by the total floor area of 720 m², we arrive at the total cost target
for the element of £14,400. Where the floor construction is dissimilar it will be
necessary to use one of the sources of cost information (Chapter 4) and build up
a new all-in rate. Each of the elements can, with varying degrees of difficulty,
be dealt with in this way. Where specialist work is involved, the quantity surveyor,
unless the work is familiar, should seek specialist advice and obtain quotations
where appropriate.

The second stage in this process is first to ensure that the sum of these element
costs agrees with the total target costs, including the price and design risk. Some
adjustments here and there will be necessary, and these can usually be allowed
for where they are of a minor nature in the latter items. If the costs do not equate
then some rethinking of the scheme design or specification will be required. It is
dangerous, however, and not recommended simply to reduce an elemental cost for
some arbitrary reason. Second, it will be necessary to ensure that the elemental costs
are realistic and achievable in the present-day construction industry. Finally, the

cost plan will be discussed with the architect to consider any possible changes in the quality or quantity of work.

15.3.5 The comparative cost plan

Comparative cost planning endeavours to evaluate different options which satisfy the requirements of the client's brief. The comparative cost plan does not seek to enforce rigid cost limits for the design of particular elements, but rather to maintain flexibility in possible design solutions. It must, however, be accepted that there is in reality a limit to the amount of expenditure on the part of the client. In some cases cost comparisons of whole buildings of widely differing designs may be undertaken. This is particularly relevant for private sector clients who may not have established a pattern of design or construction in the past. Comparisons may also be required for individual elements and this may involve the quantity surveyor in giving rather more than a subjective opinion. However, the procedures involved are similar to those of elemental cost planning.

15.3.6 Cost checking

During the development of the architect's design it is necessary, if the total cost is not to be exceeded, to cost check this as it evolves. Since the architect will design in elements, it is convenient for the cost checks to be carried out on this basis and compared with the cost targets in the agreed cost plan. Cost checking can be a time-consuming process, particularly if major discrepancies occur between the quantity surveyor's cost plan and the architect's final design. Time and effort may be saved where the surveyor has provided the architect with a detailed cost plan and specification descriptions, and where the architect has not diverged too far from these assumptions. If the cost plan is based on a previous similar project for the same architect and client, then the cost checking should be considerably reduced. A one-off design with no similar project to compare with can necessitate great care during the cost-checking process if one is to achieve the desired results.

The cost checking of an element may reveal that the final design is cost-comparable to that originally envisaged in the cost plan. In this case no further action will be necessary on the part of the design staff. It must also be appreciated that the cost checking of an element may need to be carried out several times, as the architect considers different solutions to a design problem. When a cost check reveals a difference from the target cost of the element, the architect must be informed accordingly and may then choose to proceed in one of the following ways:

- Redesign the element so that the target cost for this and the whole scheme is not changed
- Approve the change in the element cost but, in order that the total cost of the scheme remain unchanged, re-examine other elements in order to produce cost-saving measures. This of course must be a realistic proposition

■ Approve the change and accept that the total cost of the scheme will now be increased accordingly. This may be possible, but if this decision is taken the client must be informed accordingly

The cost checking should be carried out as soon as the details are received from the architect to avoid any abortive design work where an element proves to be too costly. The advantage to the client of this process is an awareness of the cost implications of all the design decisions, and a reasonable assurance that his budget estimate will not be exceeded. The amount of time and care spent by the surveyor will depend to a large extent on the importance of the individual element. The greatest attention needs to be given to the cost-sensitive elements, but this does not imply that the minor elements can be ignored. The pricing of the cost checks will be carried out using rates which reflect current market prices. During the later stages of the process, estimates for specialist works can be replaced with firm quotations.

Once the working drawings are available, the quantity surveyor's main role will be to prepare the tender documents. It is a hectic time in the office, with very little time being available for other activities. However, if the whole process of cost checking is to be effective then the tender documentation must be delayed until the final cost check is carried out. It is no use waiting until after the taking-off is complete, since the architect may have revised the details to make the scheme more elaborate and costly. This could result in a few rubbings-out on the part of the architect, but a completely wasted set of measurements on the part of the quantity surveyor.

In some instances it may be possible for cost checking to be carried out on the basis of a visual inspection alone, but care needs to be exercised to ensure that nothing has been overlooked. A nil change return will have to be reported. The usual method of cost checking generally requires the preparation of approximate quantities which can then be priced accordingly. A further and final independent check is often carried out by pricing the tender documents prior to the receipt of tenders.

15.3.7 Tender reconciliation

If the process of cost checking has been carried out thoroughly, then the receipt of tenders should provide few surprises. It is nevertheless useful to undertake some form of reconciliation to highlight any discrepancies between the final cost check and the cost analysis of the accepted tender. This comparison will also provide some input to the cost planning of future projects. If there are discrepancies between the two sums, explanations will not be difficult to find. Errors on the part of both the surveyor and the builder's estimator can occur and there will be occasions where the contractor has deliberately distorted some of the prices for possible future gain. If revisions to the scheme are necessary then the surveyor is likely to be able to do these in a more professional manner. The quantity surveyor will of course include all matters of relevance in his tender report to the architect.

15.3.8 Post-contract cost control

When the contract has been signed, it is sometimes supposed that the cost planning function is now complete. The design, which is now more likely to have stayed within the budget, is evidenced by an acceptable tender sum. The client's main concern, however, is to ensure that the final account should also come within this budget figure. There are various mechanisms that the quantity surveyor can use to make sure that this is achieved, and these are discussed in Chapter 21.

15.4 COST LIMITS

Cost limits in one form or another are used by the majority of clients, both in the public sector and those who are privately funded, for their construction projects. The 'money no object' approach to building projects can be enjoyed only by the very few, and even then some consideration of value for money will be exercised. The great pyramid builder, Cheops, had as his motto, 'I don't care how long it takes or how much it costs'. His resources seemed to be limitless. Few clients today are in this fortunate situation. Although cost limits do in essence aim to limit the capital expenditure on building projects, they also attempt to be cost-effective by balancing costs with space and quality requirements. Public sector accountability will always seek to impose some expenditure limit on the costs of capital construction projects. Such limits have in the past tended to be rather inflexible, and current theory and practice by recent government departments have been to replace them with a block grant system of funding. The organisations concerned with capital building programmes have therefore been given more discretion in the expenditure of money on these projects. Management must, however, set the objectives of such projects clearly in terms of financial allocations, if the project is to be considered successful in these terms. In practice, anyway, a variety of different projects will be competing for the available funds, and some equitable method of expenditure allocation must therefore be used. The public sector, in particular, has to decide between a number of conflicting social proposals in deciding which objectives will be met and how the funding will be made available.

Cost limit systems were first introduced into the UK shortly after the end of the Second World War in order to limit expenditure on public works projects. There is sufficient evidence to suggest that they served the purpose for which they were designed. They provided a framework, albeit sometimes rather complex, for economy in both design and construction in the boom in the building industry during that period. The system provided for the overall control of national expenditure, with appropriate adjustments allowed in the calculation for regional variations in building costs. Adjustments to the cost limit included dealing with abnormal costs such as for unusual site conditions and external works.

Cost limits continue to be applied to educational establishments, health authority projects, public sector housing schemes and other public sector projects, such as magistrates' courts where similar purpose buildings are constructed throughout the country.

15.4.1 Factors to consider

Expenditure limits on capital building costs generally determine the maximum cost which can be allowed in order to satisfy a particular function. The expenditure limit is often linked to both the spatial requirements and a minimum standard in terms of quality of construction. The expenditure limit imposed will largely depend on the functional requirements of the type of building under consideration. Although this may be readily determined on a unit analysis basis, in practice it is more usual to break down the job into certain component parts, each with its own appropriate costs. The factors which will determine the costs of any project can be summarised as follows:

- The number of occupants
- The activities of these occupants
- The space required for these activities
- The division of this space into rooms or zones
- The circulation area requirements
- The required environmental conditions
- The type of structure needed to suit these requirements
- The location of the project

Space is not available here to describe the procedures which are generally in use. Reference should therefore be made to the appropriate documents prepared by the public sector offices dealing with, for example, housing, schools and hospital buildings.

15.4.2 Value-for-money considerations

It has been generally accepted that some form of cost or expenditure limit is necessary for public building projects in order to achieve an economical design solution. It is equally recognised that further refinements in design or layout are likely to have only a marginal effect on initial construction costs.

The following points should be considered in respect of value for money:

- Cost limits are primarily concerned with initial building costs and not with total costs-in-use. It could be argued therefore that they might give very poor value for money in the long term. Certainly there is a greater emphasis today on improving the initial design in order to reduce, minimise or eliminate the future costs of owning and operating a building.
- In the absence of cost limits, housing standards and expenditure could reach high and unacceptable proportions. This could result in a substantial reduction in the housing programme, because a limited amount of cash is available for housing. This would have as a consequence a lengthening of housing waiting lists and large rent increases to users.
- In order to build within the prescribed limits, increased site densities may become necessary. This may result in more three or four-storey blocks being constructed. These have the disadvantages that they are more expensive to construct initially

than equivalent two-storey dwellings, are more costly to manage and maintain and are less desirable to tenants.

- Cost limits must be given some credit as one of the reasons for refinement of layout, economy of specification and elimination of waste in the design of many publicly financed building projects. In order for schemes to obtain approval, design layouts and aesthetics have to be more carefully thought out.
- The use, however, of unrealistic cost limits can encourage poorly designed buildings that have high costs-in-use, stereotyped solutions and cheap elevational treatments.

15.4.3 The effects on running costs of cost limits

A cost limit may vary from a carefully computed sum of money based on a total-cost approach to building, to a figure determined exclusively by the funds the client is willing to make available. In practice, although complicated formulae and analysis may be used, these are devised in recognition of limited funds being available for capital expenditure. The difficulties are compounded by the accounting system used by the client organisation. Capital expenditure is usually obtained from a different fund from that of revenue expenditure. The allocation of money to these separate accounts may be made on a rough, arbitrary basis. The source of funding is therefore dependent on the classification of the expenditure. Many local authority projects are funded, in terms of their capital cost, from central government but are maintained by local government. There may therefore only be a limited incentive on the part of central government to increase its own capital expenditure in order to reduce future long-term savings in maintenance and running costs by local government. Where local government aims to reduce future costs-in-use then it may be necessary to add an appropriate contribution to the capital costs to facilitate this approach.

This raises a further question. Will an increased capital expenditure automatically result in a reduction in future maintenance and running costs? Often these future costs are taken into account only on a subjective basis rather than by some form of analysis. This indicates that those who have set the cost limits have to some extent taken notice of the maintenance standards achieved in practice. Often the problem in using cost limits is that they are too infrequently revised to take into account general rises in building prices. There is the growing realisation that the setting and revision of cost limits is a complex technical problem requiring a substantial professional input from quantity surveyors. In some instances the cost limit may have been unsoundly based owing to a lack of relevant cost data in the first place. The provision of improved quality of data for cost-in-use is an ongoing quest.

The view is sometimes held that to spend more initially on providing a high-quality construction will reduce future costs in some positive way. Although this may be true in some cases, in others it is misleading. For example, the commonly held view that pitched roofs are a better economic proposition than flat roofs may be untrue. They may be more attractive and are certainly less troublesome, but economically they may be more expensive, taken over the whole-life cost of a building. Changes in the opportunity cost of capital will further distort the true picture. Floor finishes which have a higher capital cost may also require higher sums for cleaning and any future repairs. Sophisticated engineering services, which are very expensive, will cost a large amount for their annual running and maintenance charges. Indeed a high-quality, high-capital-cost building – like a Rolls-Royce – will often demand a high expenditure to keep it in good order and repair. Also changes in costs and technology, even for the foreseeable future, are not in any way guaranteed.

The situation is further complicated when taxation and capital allowances need to be considered. It is often argued that, because capital costs are non-tax-deductible but annual running costs are eligible for offsetting against tax, little attempt should be made to reduce the latter if this requires further expenditure on capital costs. In 1969, the then Ministry of Public Building Works produced a paper entitled *A Study of Maintenance Costs of Crown Office Buildings* which showed that the above viewpoint was in error. The introduction of VAT on maintenance work, but not on capital new works, may have affected the issue in some way. Where, however, VAT is reclaimable or can be offset, the situation remains largely the same as in 1969.

The only real way of deciding on the correct balance between capital and recurring costs is by whole-life cost analysis. Sensitivity analysis must also be incorporated to determine the effects of changes in the life of components and interest rates.

Even speculative developers must take into account some recurring costs, since these could have an adverse effect on the possible sale or lease of the property.

Cost limits have also had some impact on architectural design, constructional detailing and specifications. Although they have often tended only to attempt to relate spatial design and cost, they have also impinged on the quality of work. They are therefore blamed for recurring maintenance problems, such as the ingress of water, condensation, deterioration of building materials and other similar defects. Architectural procedures have failed to match the techniques employed in cost control. Modern buildings have therefore been criticised because of their costly maintenance.

There is therefore the need to place a greater emphasis on the total cost evaluation of construction projects and to strike the correct balance in the design solution. Cost limits certainly do have some influence on future costs, but the project and the individual circumstances will also determine whether or not the costs are adverse. Cost limits are also essential where there is some restriction on the available funds.

15.5 ELEMENTAL COST SYNTHESIS

Cost analyses are widely used in the preparation of cost planning. Their costs, however, will need to be adjusted to take into account the various design factors such as size, shape and height. It is also probable that some difference in quality will be envisaged. Construction costs are therefore a combination of price, quality and quantity.

15.5.1 Pricing conditions

One of the difficulties the cost planner faces is identifying the pricing level in the existing cost analyses and the pricing level that should be applied to the proposed project. Some of the difficulty can be explained on a national basis by referring to the market conditions index, but this does not necessarily imply or refer to specific projects in a particular area. The market conditions index will, however, provide a trend that will show whether building prices are increasing or decreasing relative to building costs. An index that is likely to be of greater benefit is the tender price index, which is a measure of how the relative price of building is changing. A description of this, together with the way it is applied in practice, is given in Chapter 7.

The projection of this index up to the date of tender can be achieved only by one of the methods of extrapolation described in Chapter 7. A local knowledge about the availability of work and the order book of suitable contractors, however, will greatly enhance this information. The adjustments of prices from a cost analysis or the application of new rates to approximate quantities are to a large extent a matter of personal judgement, skill and experience. These are qualities which are not easily quantifiable.

15.5.2 Quantity considerations

The quantity of an element is generally referred to in a cost analysis as the element unit quantity. Within a fully detailed cost analysis it is possible to identify separately the quantities and costs of the various types of construction within an element. There are basically three ways in which adjustments for quantity can be made during cost planning.

Approximate quantities

This is the most popular method, and involves measuring the work in the manner described in Chapter 14. If the intention is to use element unit rates then it is essential that the quantifying is done in accordance with the BCIS rules. It may, however, be necessary to use price books, and when building up composite rates it should be remembered to include the 'non-measurable' items within these rates.

Proportion

The existing cost analyses are used as the basis for calculating the elemental costs in the cost plan. This is the preferred method since the cost plan will automatically include the requisite allowances for all the elements in the proposed building. It necessitates the use of quantity factors. An example of how these are utilised is given in the example in connection with the external walls of a building (see pp. 359–361).

Inspection

Some elements are very difficult to quantify at the cost planning stage, and in these circumstances the only course of action is to assess their values subjectively. For example, the preliminaries are usually incorporated on the basis of a percentage initially, which may bear only a slight relationship to their eventual costs. Also, during the early stages of cost planning in particular, it is sometimes difficult to obtain realistic targets for elements such as the external works or building services. Cost analyses are generally very helpful in these situations where it is possible to examine a number of different projects. Personal judgement and experience are important in arriving at appropriate amounts.

15.5.3 Quality considerations

Quality is a very difficult consideration on which to put a value. Although it is commonly believed that quality is explained in the specification clauses, this may be difficult to justify when the same clauses are provided for jobs of widely differing quality requirements and realities. In addition, the standards of quality provided will to some extent be influenced by the designer and the client. Quality, as explained in Chapter 6, is a combination of variables and attributes. In terms of cost planning the former are easier to take into account, since the latter are much more prone to subjective judgements. Some qualitative adjustments, such as a more expensive facing brick or the use of slates on a roof in lieu of concrete tiles, can be taken into account in a similar manner to adjustments for quantity.

Example

Figure 15.2 shows the plan and elevation of a proposed two-storey office building together with design data for the external walls. Also included are the relevant details for this element from an existing cost analysis; this information will be utilised in the cost plan. Two alternative methods can be used to calculate the cost targets.

1. Using approximate quantities
Area of external walling in project A including the slate facings and allowing 20% for windows and external doors:

$$(30 + 12) \times 2 \times 5 \times 80\% = 336 \text{ m}^2$$

This can then be multiplied by the element unit rate of £42.21 to give £14 182.

2. By proportion
The alternative method is to utilise more of the design data, particularly the wall-to-floor area ratio.

PROPOSED PROJECT A – Current tender index 247

Cavity external walls, external facing bricks
£180 per 1 000, and slate facings where indicated

EXISTING ANALYSIS – Tender index 227

	Total cost of element	Cost per m² GIFA	Element unit quantity	Element unit rate
External walls Facings £130 per 1 000	£34 108	£33.44	808 m²	£42.21

Fig. 15.2   Project A

$$\text{existing analysis} = \frac{\text{wall area}}{\text{GIFA}} = \frac{808 \text{ m}^2}{£34\ 108 \div £33.44} = 0.7922$$

$$\text{building A} = \frac{\text{wall area}}{\text{GIFA}} = \frac{336 \text{ m}^2}{30 \times 12 \times 2} = 0.4667$$

$$\frac{\text{new wall/floor ratio}}{\text{existing wall/floor ratio}} \times \begin{array}{c}\text{cost per} \\ \text{m}^2 \text{ GIFA} \\ \text{(existing)}\end{array} \times \begin{array}{c}\text{GIFA} \\ \text{(proposed)}\end{array} = \begin{array}{c}\text{total cost} \\ \text{of element} \\ \text{(proposed)}\end{array}$$

$$\frac{0.4667}{0.7922} \times £33.44 \times 720 \text{ m}^2 = £14\ 184$$

The minor difference between the two is due to rounding errors in the calculation. The lower cost per square metre GIFA of £19.70 and the smaller wall-to-floor ratio indicate the proposed building has been designed to a more economical shape and spatial layout.

The above cost target still requires adjustment for inflation, the slate facings and the higher quality of external facing bricks. As far as the last is concerned we will assume that the higher cost is solely in respect of materials, with no change anticipated in the costs of bricklaying.

Inflation
This can be allowed for by multiplying the element cost by the change in the tender price indices:

$$14\ 182 \times \frac{247}{227} = £15\ 432$$

External facing brickwork
The cost of the facings in the existing analysis was £130 per 1 000 but this has just been updated for inflation. Their relative cost is now therefore:

$$130.00 \times \frac{247}{227} = £141.45$$

The additional cost of the external facings for an improved quality of brick is therefore £180.00 − £141.45 = £38.55 per 1 000. Allowing for 59 bricks per m², 7.5% for waste, 10% for overheads and the profit, the extra material costs per square metre are as follows:

$$\left(\frac{£38.55}{1\ 000} \times 59 + 7.5\% \right) + 10\% = £2.69 \text{ per m}^2$$

Multiplying this by the external wall area of 336 m² adds an extra cost of £904, bringing the external wall element up to £16 336.

Slate facings
The element unit rate, i.e. the cost per square metre of walling, is now as follows:

Original analysis	42.21
Inflation	3.72
Higher quality	2.69
	£48.62

If the current rate for slate facing and an inner-leaf construction to match the remainder of the project is £110 per m², then this can be added to our cost plan as follows:

area of slate facings × difference in rate = extra cost

7.00 × 5.00 m × £110.00 − £48.62 = £2 148

The total cost of this element is now £18 484, and the full details can be presented in BCIS format as shown in Table 15.1. No attempt has been made to adjust the above data in respect of the subjective variables which can only be altered by the intuition or experience of the quantity surveyor concerned. Such factors as a change in the method of obtaining tenders, the competitiveness of the existing cost data which have been used, the keenness or otherwise of contractors likely to be tendering for this job, or simply a feeling that the element cost should be higher or lower come within this remit. These are often introduced into the cost plan by the prefix 'say', e.g. 'say +£x or x%' due to, for example, a sudden increase in work in the area.

Table 15.1   Summary of external walling costs

Element and design criteria	Total cost of element (£)	Cost per m² GIFA (£)	Element unit quantity	Element unit rate (£)	Specification
external walls / GIFA = 336/720 = 0.4667	18 484	25.67	336 m²	55.01	Generally cavity wall construction, facing bricks £180 per 1 000

		Area (m²)	All-in unit rate (£)
External walls (£)			
Cavity wall with brick facings	14 634	301	48.62
Cavity wall with slate facings	3 850	35	110.00

15.6 EXAMPLE OF A COST PLAN

A client is considering the construction of an office building and has approached the architect about possible designs and costs. The client's site, after a first inspection, appears to present no problems for foundations and not to place any restrictions on the possible construction method to be used. The client has indicated a requirement of about 2,000 m^2 of usable floor space, and the site size and planning regulations suggest that this is best achieved in a three-storey building. The client also wants the building to be fitted out and requires a form of demountable partitioning. Loose fittings and office furniture will be provided independently by the client. Since it is a project for owner-occupation, the specification quality, particularly in respect of finishings, is likely to be higher than had it been of a purely speculative nature.

The quantity surveyor, in addition to referring to previous cost analyses, will need to be aware of other factors affecting costs, particularly the market conditions prevalent at the time of expected tenders. It is also necessary to consider the locational aspects, and to refer to indices for updating the data. Perhaps one of the biggest difficulties facing the surveyor is finding an appropriate analysis that will assist the cost planning process. It is worth remembering that a familiar analysis is preferable to using an unknown one from a filing system, even though the latter may appear at first glance to be more comparable with the proposed scheme. It is also considered advisable to use one analysis as a basis, rather than to attempt to use several different analyses throughout the process.

Figure 15.3 and the accompanying cost analysis (Table 15.2) show the details of a speculative-built five-storey office building in East Anglia. It is the most appropriate analysis available. The total GIFA of this project is 2,100 m^2 and its tender sum was accepted at £581,637 (excluding contingencies).

The proposed project is shown in Figure 15.4 (Table 15.3 – see p. 378 – shows a cost plan) and it should be noted that 20% has been added to the net usable space to provide a GIFA of 2,400 m^2. The client's site is in the northwest of England. The method of obtaining tenders and other contractual arrangements is comparable with the original project. The cost analysis will therefore need general adjustments

Plan

Total floor area (GIFA) = 2 100 m^2
External wall area = 13 320 m^2
Wall-to-floor ratio = 0.628

Fig. 15.3 Existing project X

Table 15.2   Details from existing cost analysis (project X)

Element		Total cost (£)	Cost of element per m² GIFA (£)	Element unit quantity	Element unit rate (£)
1	*Substructure* Strip foundations and cavity wall Reinforced concrete ground slab on hardcore	71 060	33.83	420 m²	169.19
2A	*Frame* Not applicable	–	–	–	–
2B	*Upper floors* Precast concrete hollow units	221 957	105.69	1 640 m²	135.34
2C	*Roof* Asphalt on precast concrete units	95 550	45.50	420 m²	227.50
2D	*Stairs* Precast concrete, straight flight	121 600	57.90	8 No.	15 200.00
2E	*External walls* 270 cavity wall external facings £130 per 1 000	200 260	95.36	950 m²	210.80
2F	*Windows and external doors* Aluminium double-glazed in hardwood subframe windows and external doors	129 611	61.72	370 m²	350.30
2G	*Internal walls and partitions* Half-brick and one-brick walls Half-brick 450 m² £55.00 One-brick 740 m² £110.00 WC partitions 125 m² £65.00	114 275	54.42	1 315 m²	86.90
2H	*Internal doors* Glazed firedoors 22 No. £800.00 Plywood-faced flush doors 30 No. £450.00 WC partition doors 15 No. £250.00	34 850	16.60	101 m²	345.04
3A	*Wall finishes* Plaster and emulsion paint	63 277	30.13	2 916 m²	21.70

Table 15.2 *(cont'd)*

Element		Total cost ($£$)	Cost of element per m² GIFA ($£$)	Element unit quantity	Element unit rate ($£$)
3B	*Floor finishes* Thermoplastic tiles on screed	88 374	42.08	2 060 m²	42.90
3C	*Ceiling finishes* Patent suspended ceiling	114 639	54.59	2 060 m²	55.65
4	*Fittings and furnishings* Sundries	9 100	4.33	–	–
5A	*Sanitary appliances* Standard fittings	22 475	10.70	–	–
5D	*Water installations* Hot and cold services	19 470	9.27	–	–
5E	*Heat sources* Gas ducted warm air	270 400	128.76	6 300 m³	10.66
5F	*Space heating* Ducted warm air				
5G	*Ventilating system* Extract fans to toilets	8 600	4.09	–	–
5H	*Electrical installations* Lighting and power	157 037	74.78	–	–
5J	*Lift and conveyor installation* Passenger lift	87 300	41.57	1 No.	87 300.00
5K	*Protective installations*	–	–	–	–
5L	*Communication installations*	–	–	–	–
5N/O	*Builder work and attendance* For services and specialist work	–	–	–	–
6A	*Site works* Tarmacadam paving	67 257	32.02	–	–
6B	*Drainage* Combined system	16 708	7.96	–	–
6C	*External services* Incoming mains Preliminaries	9 800 91 919	4.67 43.77	– –	– –
		$£$2 015 519	$£$959.77	–	–

Total floor area (GIFA) = 2 400 m²
External wall area   = 1 260 m²
Wall-to-floor ratio   = 0.525

Fig. 15.4   Proposed project Y

for inflation and locational factors. Inflation can be dealt with by using a tender price. The index value of 240 is relevant to the existing analysis and 262 for the proposed project. The variation factors are currently 0.92 for East Anglia and 1.03 for the northwest of England. All these data can be obtained from the BCIS. Using the above information, the cost analysis can be adjusted as follows:

$$\frac{\text{updated cost}}{\text{analysis}} = \frac{\text{cost}}{\text{analysis}} \times \frac{\text{current tender price index}}{\text{existing tender price index}} \times \frac{\text{NW regional index}}{\text{EA regional index}}$$

$$= 518\ 637 \times \frac{262}{240} \times \frac{1.03}{0.92}$$

$$\frac{\text{revised cost}}{\text{analysis}} = £633\ 874$$

15.6.1 Proposed cost plan

1 Substructure

It is assumed, since no other information is available, that a similar type of foundation construction will be used to that described in the available cost analysis.

The difference in the number of storeys could result in a less expensive type of foundation, but this is not anticipated at this stage. The cost analysis will first need updating for inflation, and then adjusting for locational differences.

$$\text{Original element cost} \times \frac{\text{Proposed index}}{\text{Existing index}} \times \frac{\text{Northwest}}{\text{East Anglia}} = \text{Revised element cost}$$

$$71\,060 \qquad \times \qquad \frac{262}{240} \qquad \times \qquad \frac{1.03}{0.92} \qquad = \qquad \underline{£86\,849}$$

The ground-floor slab area of the original building is 420 m². It has a perimeter foundation length of 88 m and internal foundations of 52 m. By comparison, the proposed project's quantities are 800 m², 150 m and 82 m respectively. Using current cost information, the proportionate values of these three items are in the following cost ratio: slab = x, external foundation = $3x$, internal foundation = $1.5x$. (This comparative information can easily be obtained from price books but is subject to change over time.) This can be transformed into a formula as follows:

$$420x + 3(88x) + 1.5(52x) = £86\,849$$
$$762x = £86\,849$$
$$x = £113.97$$

The slab cost therefore represents £113.97, the external wall foundation £341.91 and the internal wall foundation £170.96.

Check: 420 × 113.97 = 47 867
       88 × 341.91 = 30 088
       52 × 170.96 = 8 890
                     86 845 (error due to rounding)

These figures can then be used to calculate the proposed substructure elemental cost.

Slab   = 800 m² × 113.97 =   91 176
Perimeter foundation = 150 m × 341.91 =   51 287
Internal foundation = 82 m × 170.96 =   14 019
                                         £156 482

This gives us a rate of £65.20 per m² GIFA (156 482/2 400 m²).

2B Upper floors

The cost of the upper floors element is dependent on several factors (see Chapter 9 on cost analysis). Where the type of construction used in a proposed building project is identical to that used in an existing analysis, the variable changes may be limited to span and loading. Each of these factors, however, is considered to be comparable in connection with the proposed project. Longer spans or increased loadings, of any type, will often result in a change in the method of construction, the incorporation of supporting members (beams) or a deeper floor construction. None of these is envisaged.

Area of each upper floor = $\quad$ 30 × 14 $\quad$ = $\quad$ 420 m^2

Four upper floors $\qquad\qquad\qquad$ = 1 680 m^2

Less staircases area $\quad$ = $\quad$ 4 × 10 m^2 = $\quad$ −40 m^2

$\qquad\qquad\qquad\qquad\qquad\qquad\qquad$ = 1 640 m^2

This gives an element unit rate of: $\dfrac{£221\ 957}{1\ 640\ \text{m}^2} = £135.34$

This is updated as previously: $£135.34 \times \dfrac{262}{240} \times \dfrac{1.03}{0.92} = £165.41/\text{m}^2$

The proposed structure is three storeys in height (two upper floors) each with an area of 800 m^2. Staircases and lifts account for 10 m^2 per floor giving a total element unit quantity of 1,580 m^2. The element total cost in the proposed building is therefore calculated as:

$\quad$ 1 580 m^2 × £165.41 = $\underline{£261\ 348}$

2C Roof

The roof construction on the proposed project bears no resemblance to the roof construction included in the original analysis. The original analysis cannot therefore be used and it is necessary to build up a composite rate from other information such as priced bills, price books, other cost analyses, etc. based on the proposed method of construction. The architect has suggested a superior type of felt roofing on plywood on softwood joists with a substantial amount of thermal insulation.

Roofing analysis

(1) General area: $\qquad\qquad\qquad\qquad\qquad\qquad\qquad\qquad$ (per m^2)

$\quad$ Three-layer high-performance felt $\qquad\qquad\qquad$ = $\quad$ 80.00

$\quad$ 25 mm insulating board underlay $\qquad\qquad\qquad$ = $\quad$ 14.06

$\quad$ 24 mm plywood roof boarding $\qquad\qquad\qquad\quad$ = $\quad$ 23.95

$\quad$ 50 × 75 mm (average) softwood firrings 2.40 linear metres at £10.00 $\ $ = $\quad$ 24.00

$\quad$ 50 × 200 mm softwood joists at 450 centres 2.40 linear metres

$\qquad$ at £25.00 $\qquad\qquad\qquad\qquad\qquad\qquad\qquad\qquad$ = $\quad$ 60.00

$\quad$ Vapour barrier $\qquad\qquad\qquad\qquad\qquad\qquad\qquad$ = $\quad$ 17.50

$\quad$ 150 mm insulating quilt $\qquad\qquad\qquad\qquad\qquad$ = $\quad$ $\underline{9.55}$

$\qquad\qquad\qquad\qquad\qquad\qquad\qquad\qquad\qquad\qquad$ $\underline{£239.06}$

(2) Eaves: $\qquad\qquad\qquad\qquad\qquad\qquad\qquad\qquad\qquad$ (per m)

$\quad$ 300 × 25 mm softwood eaves board and decoration $\qquad$ = $\quad$ 16.00

$\quad$ 50 × 50 mm tilting fillet $\qquad\qquad\qquad\qquad\qquad$ = $\quad$ 8.00

$\quad$ Aluminium roof trim $\qquad\qquad\qquad\qquad\qquad\quad$ = $\quad$ 10.00

$\quad$ Felt upstand $\qquad\qquad\qquad\qquad\qquad\qquad\qquad$ = $\quad$ $\underline{20.00}$

$\qquad\qquad\qquad\qquad\qquad\qquad\qquad\qquad\qquad\qquad$ $\underline{£54.00}$

Roof area 800 m^2 at £239.06 $\qquad\qquad\qquad\qquad\quad$ = $\quad$ 191 248

Roof perimeter 150 m at £54.00 $\qquad\qquad\qquad\qquad$ = $\quad$ 8 100

Sundries, rainwater goods etc., say 2% $\qquad\qquad\qquad$ = $\quad$ $\underline{3\ 984}$

Element cost $\qquad\qquad\qquad\qquad\qquad\qquad\qquad\qquad$ $\underline{£203\ 332}$

It is always prudent to include a small amount for work which will eventually be measured in a bill of quantities or work package and is absent from the all-in rates. The sundries in the above would allow for gutters and rainwater pipes and roof ventilators should these be required.

2D Stairs

There are two staircases per floor on the original project, making eight flights in total. Adjustments may need to be made for both quantity and quality. It is preferable to consider this element using the unit rate per flight of £15,200. The proposed project has four flights (two per floor). The £15,200 is adjusted for inflation and location as previously:

$$£15\ 200 \times \frac{262}{240} \times \frac{1.03}{0.92} = £18\ 577$$

It can also be adjusted for varying heights, although the cost importance is more likely to relate to the number of flights than to their size. Price will change with quantity but not in direct proportion. In this example, however, the price has been adjusted in direct ratio to the height. The width and going are assumed to be similar. The storey heights are slightly different. The existing project is 3.00 m and the proposed project is 2.80 m. An adjustment is made as follows:

$$£18\ 577 \times \frac{2.80}{3.00} = £17\ 339$$

There is no indicated difference in terms of quality, but it should be remembered that most of the cost of this element anyway is connected with the structural work. The cost target for the element is thus

$$£17\ 339 \times 4 = £69\ 356.$$

2E External walls

The existing plan is of a more regular shape, but building height is greater than that proposed. The external walls element cost could be adjusted for these two factors on a largely subjective basis. The output of the bricklayers will theoretically be lower on the proposed project owing to the greater number of changes in direction or corners or walls. Changes in direction of brick walls is a significant factor affecting the bricklayers' productivity. Some savings would be expected on height money, and it is assumed that these might cancel each other out. Individual judgement is necessary here. Different quantity surveyors and indeed building contractors' estimators may come to slightly different conclusions on each of these aspects.

The element unit rate for the external walls first needs amending for inflation and location:

$$£210.80 \times \frac{262}{240} \times \frac{1.03}{0.92} = £257.64 \text{ per m}^2 \text{ of external wall construction.}$$

The area of the external wall area in the proposed building is 1,260 m^2. This includes the windows and external doors area, which is equivalent to 18% of the total walling area. The area of external walls is therefore 1,033 m^2. This area could be calculated using the wall-to-floor area ratio:

external wall area = GIFA × wall/floor ratio × wall percentage

1 033 m^2 = 2 400 × 0.525 × 82%

The external facing bricks on the original project were £130 per 1,000, but these have been adjusted for inflation in the above calculation to:

$$£130 \times \frac{262}{240} = £141.92 \text{ per } 1\ 000$$

The external facing bricks on the proposed project are £185 per 1,000. They are assumed to be of a similar type and therefore the costs of bricklaying will be unchanged. There are 59 facings in a half-brick wall, and with 7.5% wastage and 10% overheads and profit this produces an additional cost of £3.01, to account for the higher quality of brick.

The proposed project also incorporates 50 mm polystyrene slab insulation in the cavity, and this is adds another of £11.00 per m^2.

The rate to be used in the proposed analysis is therefore as follows:

Original analysis	=	210.80
Updating for inflation	=	19.32
Locational adjustment	=	25.20
Higher quality of brick	=	3.01
Wall insulation	=	11.00
	=	£269.33

If this is multiplied by the area of the external walls element (1,003 m^2), a cost target for this element is calculated as £270,138.

2F Windows and external doors

The existing analysis specification for this element is considered to be satisfactory for the proposed project. The anticipated change in the elevational appearance, however, is expected to add about 7.5% to the costs of the windows. The cost plan for this element is therefore calculated as follows on the basis of an elemental area of 18% of the external walling area:

$$1\ 260 \text{ m}^2 \times 18\% \text{ (windows and external doors area)} \times £350.30 \times \frac{262}{240} \times \frac{1.03}{0.92}$$

	=	£97 101
Plus 7.5% added cost for improved design features	=	7 283
		£104 384

2G Internal walls and partitions

The speculative built office block largely provided for the structural internal walls alone. This gave the impression of an open-plan design. Half-brick partitions were constructed for store rooms, around the staircases and lift shafts and to form toilet areas. Prefabricated WC partitions were used in the toilets. The proposed office building is to be constructed on a cellular layout and it is thus necessary to measure some approximate quantities for the cost plan. The information is rather scant, with room layouts, as yet, rather vague. One-brick load-bearing walls have been planned on a form of grid layout and are relatively easily identified. Some half-brick partitioning is used in a similar position to that described previously. The WC partitions are the same as those envisaged earlier. There is also to be the provision of some demountable partitions. The sketch drawings are unclear in this respect, but it might be roughly assumed that 450 m² will be required (150 m² per floor). This information can be summarised and costed as follows, using the data from the original cost analysis:

One-brick wall 740 m² × £110.00	=	81 400
Half-brick wall 330 m² × £55.00	=	18 150
WC partitions 125 m² × £65.00	=	8 125
		£107 675

$$£107\ 675 \times \frac{262}{240} \times \frac{1.03}{0.92} \qquad = £131\ 600$$

Demountable partitions 450 m² × £75 (at current rates)	=	33 750
	=	£165 350

2H Internal doors

The existing analysis provided for three different types of internal door, and as the design was open plan, the number of these was minimal. The proposed project incorporates a wider variety of door types and also a much greater number of doors. These include:

Glazed fire doors to ground floor and entrance vestibule	16 No.
Hardwood faced flush doors to some specialist offices	20 No.
Plywood veneered flush doors to other offices	32 No.
Doors in demountable partitions	21 No.
WC partition doors	15 No.
Softwood fire doors to other floors	12 No.

It is preferable, wherever possible, to use data from the existing analysis as a basis. It is desirable also, because of the vagaries of tendering, that the same analysis is used as far as possible to capture differences in prices. It is anticipated, however, that the quality of door furniture will be much higher on the proposed project. The unit costs from the existing analysis are updated as follows:

$$\text{Glazed fire door} \quad \pounds 800 \times \frac{262}{240} \times \frac{1.03}{0.92} \qquad = \pounds 97\,775$$

$$\text{Plywood flush door} \quad \pounds 450 \times \frac{262}{240} \times \frac{1.03}{0.92} \qquad = \pounds 54\,999$$

$$\text{WC partition door} \quad \pounds 250 \times \frac{262}{240} \times \frac{1.03}{0.92} \qquad = \pounds 30\,554$$

The doors not included in the analysis will be priced from current cost data, or on a pro rata basis using the above cost data as a basis.

Glazed fire door 16 No. @ £97 775	=	15 645
Hardwood flush door 20 No. @ £549.99 + 30%	=	13 199
Plywood flush door 32 No. @ £549.99	=	17 599
Doors in demountable partitions 21 No. @ £400	=	8 400
WC partition doors 15 No. @ £305.54	=	4 583
Softwood fire doors 12 No. @ £750	=	9 000
	=	£68 426

In addition, the hardwood doors, on the proposed project, are likely to incorporate a better quality of ironmongery. The flush doors in the original analysis will be fitted with locks and levers for a prime cost of about £50. The hardwood fire doors will have pull handles, push plates, kicking plates and closers for a prime cost of about £100. The costs of ironmongery can vary immensely and careful guidance will need to be obtained from the architect. An increase of 50% in the quality of this ironmongery is envisaged.

Doors costs	=	68 426
Higher quality:		
Hardwood flush doors 20 No. × £50 × 50%	=	500
Fire doors 16+ 12 No. × £100 × 50%	=	1 400
	=	£70 326

3A Wall finishes

The speculative building allowed for two coats of plaster and emulsion paint finish throughout. The height of this finish up to the level of the suspended ceiling is 2.70 m. The total area is based on the area of external walls plus twice the area of the internal walls, making due allowance for the height factor change. The inside of the lift-shaft does not, of course, have a wall finish.

The proposed project quantity is calculated in a similar manner:

External wall area	= 1 033 m²
One-brick internal wall 740 m² × 2	= 1 480 m²
Half-brick internal wall 330 m² × 2	= 660 m²
	3 173 m²

The storey heights are smaller here but the room heights remain the same:

$$3\ 173\ \text{m}^2 \times \frac{2.70}{2.80} = 3\ 060\ \text{m}^2$$

The lift-shaft wall area is approximately 50 m^2 in total. Therefore the total wall area will be 3,010 m^2 requiring finishing and decoration. The foyer area and boardroom will be panelled in hardwood and some of the offices will be wallpapered.

The element unit rate from the analysis is updated accordingly:

$$£21.70 \times \frac{262}{240} \times \frac{1.03}{0.92} = £26.52$$

The following will therefore need to be measured and priced using approximate quantities and current rates:

Hardwood panelling	180 m^2 × £125.00	=	22 500
Plaster and wallpaper	300 m^2 × £35.00	=	10 500
Plaster and emulsion	2 530 m^2 × £26.52	=	67 096
	3 010 m^2		£100 096

3B Floor finishes

The original analysis, which is of a speculative-built project, showed thermoplastic floor tiles throughout. The proposed project is for owner-occupation and therefore the quality of this element should be in context with schemes of a similar nature. The client has indicated a desire to construct an impressive entrance foyer area and this measures approximately 10 × 8 m. A heavy-duty contract carpeting is suggested on an underlay and a 50 mm cement and sand screed. A similar quality finish is envisaged in some of the more important offices and the boardroom (approximately 200 m^2). The second floor will use the thermoplastic floor tiles, the remaining areas being divided equally between this type of tile and carpet tiles. Hardwood skirting boards will replace the softwood type in the carpeted areas. The data from the analysis are first updated as follows:

$$\text{Element unit rate} = £42.90 \times \frac{262}{240} \times \frac{1.03}{0.92} = £52.43$$

The approximate total floor finish area of 2,340 m^2 is therefore divided as follows:

Thermoplastic tiles and screed	1 384 m^2
Carpet tiles and screed	604 m^2
Contract carpet and screed	352 m^2

In the carpeted areas the approximate length of hardwood skirting board is calculated as 200 m. Since some of this work is not relevant to the original analysis, one of the various sources of cost data must be used to obtain current rates. The total element cost is therefore calculated as follows:

Thermoplastic tiles	1 384 m² × £52.43	=	72 563
Carpet tiles	604 m² × £60.00	=	36 240
Contract carpet	352 m² × £90.00	=	31 680
Hardwood skirting	200 m × £15.00	=	3 000
			£143 483

3C Ceiling finishes

The original analysis allowed for a basic-quality suspended ceiling on a metal suspension system. The proposed project is to include an improved quality of tile generally with a superior finish in the boardroom and selected offices. The differences in cost between the two types of tile are medium quality plus 5% and superior quality plus 10% above the basic tile finish. The cost of the basic tiles can be updated as follows:

$$£55.65 \times \frac{262}{240} \times \frac{1.03}{0.92} = £68.01$$

Medium quality tiles	1 988 m² × £68.01 + 5%	=	141 964
Superior quality tiles	352 m² × £68.01 + 10%	=	26 333
			£168 297

4 Fittings and finishings

Since the original analysis was of only a speculative nature, the fitting-out was to be done by the client. The costs allocated to this analysis therefore covered only items of a minor nature such as pipe casings, meter cupboards, etc. However, even in the proposed project, the majority of the fittings will be of a loose furniture type and provided independently by the client from a preferred office furniture supplier. Also, at the cost planning stage, the actual details of these items will be largely unknown, particularly where the project is of a one-off nature by the client. A high-class reception desk fitment is required in the foyer area and a small sum should be included for fitted shelving. The fittings cost from the original analysis is updated and it seems appropriate to include this on a floor area proportion:

$$£9 100 \times \frac{262}{240} \times \frac{1.03}{0.92} \times \frac{2 400 \text{ m}^2}{2 100 \text{ m}^2} \qquad = \quad 12 710$$

Foyer desk, say	=	5 000
Shelving 80 m × £15.00	=	1 200
	=	£18 910

5A Sanitary appliances

No difference is expected in the quality of the sanitary appliances to be used. In the absence of further details, therefore, the ratio of cost to floor area is assumed to be relevant. The original analysis also included the disposal pipework and the associated builder's work.

$$£22\ 475 \times \frac{262}{240} \times \frac{1.03}{0.92} \times \frac{2\ 400}{2\ 100} = \underline{£31\ 393}$$

5D Water installations

The hot and cold water services supply is calculated in exactly the same way as the above:

$$£19\ 470 \times \frac{262}{240} \times \frac{1.03}{0.92} \times \frac{2\ 400}{2\ 100} = \underline{£27\ 196}$$

5E/F Heating

A similar type of heating installation is envisaged to that in the original analysis. The updated cost of this is as follows:

$$£270\ 400 \times \frac{262}{240} \times \frac{1.03}{0.92} \times \frac{2\ 400}{2\ 100} = \underline{£377\ 692}$$

It is assumed that similar temperature and control levels will be maintained. Adjustments to the cost of the heat source should therefore be made for changes in size (volume of air to be heated) and changes in the insulation provision. The volume of the original building is 6,300 m^3 and of the proposed building is 6,720 m^3. The latter offers improved insulation values in both the external walls and the roof. These two factors will alter the heat source and the space heating requirements and hence the associated costs. The costs of the heating, however, are not linearly related to these variables. It may be preferable, if possible, to obtain new boiler capacities and heat emitter sizes from the building services engineer and to re-price these by using approximate quantities. Using this information and suggestions from the building services engineer, a prime cost target sum of £350,000 will be included for this element in the cost plan.

5G Ventilating system

The cost of this minor element is treated in exactly the same way as the sanitary appliances element:

$$£8\ 600 \times \frac{262}{240} \times \frac{1.03}{0.92} \times \frac{2\ 400}{2\ 100} = \underline{£12\ 012}$$

Although air conditioning is now becoming standard in new high-quality office buildings in the UK, it was not envisaged for this scheme. Air conditioning would have increased the overall cost considerably.

Central systems or multi-splits are the fully tailored systems to suit buildings generally, and come in a variety of types, and from many manufacturers. As an approximate guide, anything from £65–155 per m² based on the area to be treated, rather than gross floor area, should be used.

Every application and building is different. Firms will carry out a survey, offer advice and provide a quotation on the available options and respective costs. For the proposed project of 2,400 m² the cost would therefore be in the region of £156,000–£372,000. These figures could be reduced for those areas where full air conditioning is not required.

5H Electrical installations

The sum included in the original cost analysis was based on a prime cost sum, inserted by the quantity surveyor in the bills of quantities, based on quotations from electrical contracting firms. There is some difficulty in reconciling the speculative and owner-user office blocks for this element. The cost relationship relates primarily to the number of points provided, but in the absence of data on this, the cost per square metre GIFA offers a good indicator of costs. An extra allowance of 5% has been included to cover a superior provision and for the probability of a dedicated system for computing and similar equipment which was excluded from the speculative building.

$$£157\ 037 \times \frac{262}{240} \times \frac{1.03}{0.92} \times \frac{2\ 400}{2\ 100} = £219\ 348$$

Allowance for increase in quality in the proposed project = £219,348 + 5% = £230,315.

5J Lift installation

The existing analysis includes a single passenger lift suitable for twelve persons and the lift serves five floors. It should be remembered that the cost-significant items related to this element are the cars and motor control. It is expected that the lift in the proposed building will serve fewer floors but will be of a higher quality. Any adjustment to the original elemental costs will be highly subjective at this stage.

$$£87\ 300 \times \frac{262}{240} \times \frac{1.03}{0.92} = £106\ 697$$

The building services engineer may be able to offer some advice about possible proposals. Some allowance needs to be made for the higher quality and it is suggested that £125,000 be included as the appropriate amount in the cost plan.

6A Site works

The constructional details for this element are probably one of the later sets of information to be finalised. The area of the site less the plan area of the building will provide the area to be considered. It is known that the proposed project is to be sited on an area of approximately 3,000 m². The major provision is for car parking and an access road, paving around the building and perimeter fencing. The remaining areas are to be landscaped. There is little cost relationship between this element and the GIFA of the building since it depends very much on the size of the site available, the plan size of the building and the external works provision. The most appropriate method of cost planning is to approximately establish the client's intentions and then to measure approximate quantities and price the work accordingly. The cost analyses from existing projects will provide some guidance to allow for a comparison to be made. However, on the basis of what might be roughly expected it is probably better to price the work using approximate areas and current prices, as follows:

Flagged paving	150 m² at £60.00	=	9 000
Car parking	750 m² at £50.00	=	37 500
Driveway	300 m² at £50.00	=	15 000
Landscaping	1 000 m² at £15.00	=	15 000
Fencing	220 m² at £40.00	=	8 800
Signboard at	1 No.		1 500
Miscellaneous/sundries	Item		5 000
			£91 800

6B Drainage

The relationship between the drainage costs and the GIFA is also negligible. The critical question to be resolved is whether a combined or a separate drainage system is required. The average depth of trenches, the number of manholes to be provided and the sewer connection are important cost variables, but little will be known about these at the cost planning stage. Any special requirements such as intercepting manholes or drop-shaft manholes should be noted.

The data from the existing analysis first need to be updated as follows:

$$£16\ 708 \times \frac{262}{240} \times \frac{1.03}{0.92} \times \frac{2\ 400}{2\ 100} = £23\ 338$$

This sum can be approximately separately analysed between the following drainage components:

Manholes 7 No. at £750	=	5 250
Drains 134 m at £100	=	13 400
Sewer connection and manhole	=	3 000
Sundries (Miscellaneous items, Testing, etc.) 7.5%	=	1 688
		£23 338

This now provides us with approximate all-in rates which can be used in connection with the proposed analysis. Alternatively, although not necessarily

Table 15.3   Cost plan for proposed project Y

Cost plan summary   CIFA 2 400 m²
proposed project

Element	(£)	Total cost (£)	Cost of element per m² GIFA	Element unit quantity	Element unit rate (£)
1	Substructure	156 482	65.20	800 m²	195.60
2A	Frame	–	–	–	–
2B	Upper floors	261 348	108.90	1 580 m²	165.41
2C	Roof	203 332	84.72	800 m²	254.17
2D	Stairs	69 356	28.90	4 No.	17 339.00
2E	External walls	270 138	112.56	1 033 m²	261.51
2F	Windows and external doors	104 384	43.49	227 m²	459.84
2G	Internal walls and partitions	165 350	68.90	1 645 m²	100.52
2H	Internal doors	70 326	29.30	80 No.	879.08
3A	Wall finishes	100 096	41.71	3 010 m²	33.25
3B	Floor finishes	143 483	59.78	2 340 m²	61.32
3C	Ceiling finishes	168 297	70.12	2 340 m²	71.92
4	Fittings and furnishings	18 910	7.88	–	–
5A	Sanitary appliances	31 393	13.08	–	–
5B	Services equipment	–	–	–	–
5C	Disposal installations	–	–	–	–
5D	Water installations	27 196	11.33	–	–
5E	Heat sources ⎱	3 500 000	145.83		
5F	Space heating ⎰				
5G	Ventilating system	12 012	5.01	–	–
5H	Electrical installations	230 315	95.96	–	–
5J	Lift and conveyor installation	125 000	52.08		
5K	Protective installations	–	–	–	–
5L	Communication installations	–	–	–	–
5N/O	BWIC	–	–	–	–
6A	Site works	91 800	38.25	–	–
6B	Drainage	29 025	12.09	–	–
6C	External services	11 977	4.99	–	–
	Preliminaries (8%)	211 218	88.01	–	–
		£2 851 438	£1 188.10	–	–

preferably, standard pricing data could be used. The proposed cost plan for the drainage can thus be assembled as follows:

8 No. manholes at £750	=	6 000
180 m drains at £100	=	18 000
Sewer connection and manhole	=	3 000
Sundries (Miscellaneous items, Testing, etc.) 7.5%		2 025
		£29 025

6C External services

Provisional sums were included for service supplies for water, gas and electricity in the original bills of quantities. Unless there is information to the contrary, similar updated sums should be included in the proposed cost plan. It will be worth checking with the supply undertakings whether their charges differ from those in the area of the existing cost analysis:

$$£9\ 800 \times \frac{262}{240} \times \frac{1.03}{0.92} = £11\ 977$$

Preliminaries

The original cost plan included 8% for preliminaries. This is always a difficult element to price, especially since it is used sometimes to decrease or increase a tender sum at the last moment and thus bears no resemblance to the actual site on-costs involved. Head office charges and profit have been included in the above rates, although in some countries these costs are excluded and shown as a separate item to reflect the market conditions prevalent at the time the cost plan is prepared. Unusual projects in terms of site access and egress or proposed short contract periods that might necessitate lots of overtime working would need to have this element adjusted accordingly.

15.7 CONSTRUCTION BUDGET

Client:	ABU Properties PLC	
Project:	Known as the 'Y' scheme	
Budget serial no:	001	
Price base data:	July 2010	
New building work		2 708 011
Site works		143 246
Alterations		—
Construction costs (at stated base date)		2 851 438
Estimated inflation to probable tender date of October 2010		40 000
Construction costs at tender		2 891 438

Estimated increased costs payable to contractors during construction period	60 000
Construction costs at completion	2 951 438
Professional fees of all consultants including expenses	295 144
Costs (less exclusions)	3 246 582

Exclusions from budget:
  Value-added tax
  Loose fittings and furnishings
  Site cost

SELF ASSESSMENT QUESTIONS

1. Describe the two processes of cost planning and the extent to which each achieves its objectives.

2. Show, by use of examples, how pricing, quantity and quality considerations are applied during the process of cost planning.

3. How appropriate are the elements listed in a cost plan to today's modern buildings and is it possible to adopt this model in a universal context?

BIBLIOGRAPHY

Building Research Establishment, *Cost Planning: An Introduction*. BRE 1998.
Department of Education and Science, *Building Bulletin No. 4 – Cost Study*. HMSO 1972.
Flanagan, R. and Tate, B., *Cost Control in Building Design*. Blackwell Publishing 1997.
Fortune, C.J. and Lees, M.A., 'The relative performance of new and traditional cost models in strategic advice for clients', RICS Research Papers 1996.
Neville, R.E., 'Why cost plan?', *QS Weekly* (various, 1979–1980).
Nisbet, J., *Estimating and Cost Control*. Batsford 1961.
Royal Institution of Chartered Surveyors, *An Introduction to Cost Planning*. RICS 1976.
Royal Institution of Chartered Surveyors, *Precontract Cost Control and Cost Planning*. RICS 1982.
University of Reading, *Cost Planning and Computers*. PSA 1981.

CHAPTER 16

COST MODELLING

LEARNING OUTCOMES

After reading this chapter, you should have an understanding about cost modelling in the construction industry. You should be able to:

- Understand the types and purpose of models that are available
- Describe how the different models are classified
- Identify the current trends in cost modelling
- Understand how the different model types are constructed
- Evaluate the advantages and disadvantages of traditional estimating and cost models
- Assess the value-for-money implications

16.1 INTRODUCTION

Cost modelling is the symbolic representation of some observable system which exists or is proposed, and which in terms of its significant cost, features for the purposes of display, analysis, comparison or control. Cost modelling, as a term, is used when referring both to forecasting construction costs for clients and to estimating resource costs for contractors.

Calculating the costs of a proposed building or civil engineering project has traditionally consisted of applying appropriate unit rates to measured quantities and descriptions of proposed works. At the design stage of a project the measurements and descriptions may represent little more than the spatial requirements provided by the client. At the contractor's stage, these may be sufficiently detailed to describe the various components and processes of the project. In either case, quantifying the work can be reasonably precise. However, the judgement involved in allocating correct prices to these quantities and descriptions is extremely variable. Hence two distinct types of model have evolved.

The assessment of the unit rates, either for an approximate estimate or during the preparation of a tender, is usually based on some assumed standard output.

These outputs are then subjectively adjusted by the skill of the estimator or quantity surveyor. In practice little is done to verify, by reference to previous site performance, that the adjusted rates are correct. Indeed, it has been shown on several occasions that the cost code classification in use in the construction industry is too complex, incorporates considerable errors in the recording process and therefore provides for little reliability in practice. There has also been only a limited attempt to justify the principle of measuring in finished quantities. The variations in the values of these data, for example, can be as much as 200% for apparently identical items of work. The difference between contractors' tenders is typically about 10%. These figures alone cannot represent just market conditions, but reflect estimating inaccuracy. The variation in trade totals is somewhere between these percentages.

Some of the variation can be attributed to the fact that the typical bill of quantities does not indicate the location of the items on the project. The Building Research Establishment (BRE), appreciating the deficiencies of the traditional bill of quantities, developed the operational bill. This document tried to present the work in a more orderly fashion by subdividing the work into activities. Two types of operational bill were suggested. The first type allowed for measuring the work conventionally, but instead of presenting the work in trades or elements, it was presented in an operational format. The second alternative was a more radical departure from quantity surveying practice, and tried to present the information in a resource format. Neither of these approaches to bills of quantities was favoured by the profession, and there is only scant evidence of their use in practice. Contractors' estimates are, however, supposed to favour operational estimating, although they dislike single projects which require whole new databases and disrupt the work of the estimating department.

The major advances in construction price determination during recent years have resulted in improved techniques for use at the design stage. Cost planning techniques were developed during the 1960s by the then Ministry of Education. The RICS at this time organised the Building Cost Advisory Service, later to become independently run as the BCIS. Methods of early price forecasting such as the storey-enclosure method and the perimeter/floor area method were also derived in an attempt to improve the quality of price forecasting. Several research projects have sought by various means to improve the accuracy of pre-tender estimating. Figure 16.1 provides an indication of typical estimating accuracy on a construction project (see also Figure 14.4).

There have also been some attempts to improve the contractor's estimating performance, since if this can be achieved it will help to eliminate some of the variability in the price forecasting required by the client. Inevitably, they have all considered using the computer in one way or another. Two approaches have emerged. The first has sought to computerise the traditional methods already being used in the construction industry. It has been argued that successful estimating requires a huge amount of data that manual methods would find too cumbersome to handle. Methods have therefore been developed in the form of an electronic price book which can be updated quickly, efficiently and economically. Some programs allow the user to undertake price analysis from a very wide base. Often these

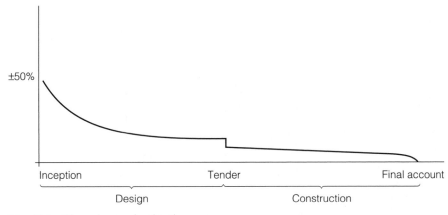

Fig. 16.1   Chronology and estimating accuracy

systems produce difficulties in use which outweigh their apparent advantages. Sometimes they have required the user to change the method of working to suit the machine. However, the modern approach seeks to fit the computer program directly to the needs of the user and allows procedures much as before but in a more efficient manner.

The second type of method has adopted a radical approach to the problem of estimating by devising an entirely new method of price forecasting. These methods use the computer to manipulate the data, often within complex mathematical formulae. There is little evidence from practice that these methods have been used at all, largely because of their radical approach, the fear of the unknown and the unfamiliarity of the mathematics and techniques used. This chapter is largely concerned with this second type of method. Cost models have equal application for the client's cost managers and the contractor's estimators.

The methods can be compared with our telephone system. Category one recognises the complexity of the system already in use, and considers that only a gradual process of refinement can take place. For economic reasons it is not possible to scrap the existing system in favour of a more efficient method of telecommunications. The second category says: here is a much improved and completely different method, let us abandon the existing system and start afresh. There are of course practical problems in doing this since it cannot be done overnight, and therefore, at least in the interim period, both systems would need to be serviced and run in parallel. Like the telephone system, some compromise has to be reached, and this is likely to be the way forward with construction estimating.

The following considers methods that can be used for cost modelling.

The first method has been in use since the beginnings of quantity surveying, and although practitioners may not recognise it as cost modelling in the sense of mathematical terms, it is nevertheless a method of modelling construction costs. The advent of the term 'cost models', however, is more generally associated with the methods described later in the chapter.

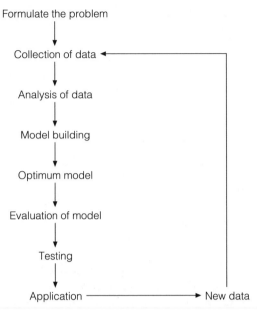

Fig. 16.2   Cost modelling

A useful distinction can be made in model building between deterministic and probabilistic models. The deterministic-type model presupposes that values can be attributed to all variables. It assumes that these are either known or can be predicted exactly. The probabilistic model, however, recognises that the values of some variables will be uncertain and can therefore only be estimated. This type of model uses concepts from probability theory. In practice, as far as the construction industry is concerned the majority of models fall into the second category. The process of model building is shown in Figure 16.2.

16.2 TYPES OF COST MODELS

16.2.1 Designers' cost models

Designers' cost models use models of previously completed buildings on which to attach estimates of future costs. The design model may be based upon an analysis of work-in-place, such as that adopted by the different rules or standard methods of measurement. At a different level the model may be based upon a number of building elements or functional units of a building. In early price models, single quantities may be adopted against which a price is then allocated.

16.2.2 Constructors' or production models

Constructors' or production models are prepared as part of the tendering process prior to construction works commencing on site. These cost models seek to model the process of construction rather than that of the finished structure. They have adopted operations or activities, often to a considerable level of detail, as a basis for determining the costs involved.

16.2.3 Mathematical models

Mathematical models have been developed by seeking to identify variables that best describe cost. They are to some extent characterised by reducing the number of priceable units in the model. These are sometimes thought to be an over-simplified representation with an insufficient number of cost centres being identified. However, Sleep (1970), as long ago as the late 1960s, claimed that such models were able to provide better forecasts of cost at much earlier dates than the traditional models.

16.3 PURPOSE OF COST MODELS

It is most important to distinguish clearly the purpose of cost modelling. Identification will influence the structure of the models and the level of quantification that will need to be provided. Cost models may be provided for several different purposes and, while there is overlap between them, their characteristics will be different.

Design optimisation models are mainly concerned with securing value-for-money aspects in building design. They are frequently used as a part of the overall design economics and cost planning process. Until the design economic consequences of construction are more fully understood then it is not possible to advise a client properly. This understanding comes from two sources: a general understanding of the principles of design economics and a particular understanding concerning the client's specific project under consideration. The strength of these models relies on their use for comparative analysis between previously completed projects and alternative proposals under consideration.

Tender prediction models are used to forecast the likely tender sum that will be accepted by a client from a contractor. In addition to identifying the probable costs of the project and the model's imperfect predictability, these models must also take into consideration the contractor's own estimating variability and those factors which affect market price. Because of such considerations, predictive models are less reliable than design-type models.

Cash flow models are prepared on behalf of both the client and the contractor. They indicate amounts and when the funds are likely to be required by the client in order to pay the contractor for the work in progress. They take into account the overall contract period and the method of construction that the contractor is likely to use.

Whole-life costing models are concerned with the whole life of a project and are thus not restricted to design and construction alone but also to use and occupation by the client. The values attributed to whole-life cost models are of much less importance than their comparison against a number of different design alternatives.

The ranking of projects, based upon the value of the model and the differences between one model and another, is of much greater significance. The use of techniques such as sensitivity analysis can assist in refining such models and in selecting the most appropriate alternative solution.

Resource-based models have been developed to assist contractors in their own estimating and forecasting process. While costs are incurred as a result of utilising resources, design cost models are not normally constructed at this level of detail. These models have sought to improve the quality of estimating accuracy. Such models also have the inherent advantage that, by reducing the contractor's own variability in estimating and tendering, they will in time improve the performance of the client's predictive models.

16.4 CLASSIFICATION OF COST MODELS

Cost modelling uses many different techniques as shown in Table 16.1. Many of these techniques have become known as single-price methods, even though in some

Table 16.1   Cost modelling of building design

Method	Description
Traditional models	
Conference	A consensus view of the team
Financial methods	Cost limits determined by the client
Unit	Used on projects having standard units of accommodation
Superficial	Total floor area of the project
Superficial perimeter	A combination of floor area and the buildings perimeter
Cube	The volume of the project
Storey-enclosure	A combination of weighted floor, wall and roof areas
Approximate quantities	An analysis of the major items of work
Elemental estimating	Used in conjunction with cost planning
Bills of quantities	Analysis prepared in accordance with detailed rules of measurement
Statistical models	
Regression analysis	Derived from the statistical analysis of variables
Causal models	Based upon algebraic expression of physical dimensions
Risk models	Monte Carlo simulation
Knowledge based	Systems such as Elsie (Brandon 1992)
Resources based	Normally a contractor's method, using schedules of labour, plant and materials
Whole-life cost models	Whole-life analysis of buildings

Source: Adapted from Ashworth (1986), Fortune and Lees (1996)

cases they use a limited number of cost descriptors or variables. The choice of a method will depend upon many different factors such as the user's familiarity and confidence with the results expected and achieved. All of the methods require access to a good source of reliable information and cost data if desired results are to be achieved.

16.5 TRENDS IN COST MODELLING

Table 16.2 suggests some of the trends in cost modelling that have occurred during the latter part of the twentieth century. The emphasis throughout has been on improving quality advice in order to allow clients to make better decisions. Coupled with these developments has been an increase in knowledge about the behaviour of costs and in the use and application of information technology. The rapid retrieval of data and the ease by which models can be updated to take into account design decisions have allowed such improved advice to be provided.

The trends have swung from a heavy reliance on the importance of experience and judgement to a rationale that construction costs can all be analysed in simple (or complex) formulae. There is now a genuine belief that cost modelling is a combination of each of these aspects. Today the emphasis is towards providing design and construction solutions that seek to resolve the economic choice while still meeting the specific needs of clients. Value for money is seen as a process of adding value to the project. Incorporated within this economic choice is the importance of the whole-life costs associated with the project.

Table 16.2   Trends in cost modelling

1940	Forecasting contractors' tenders
1940	Deterministic methods
1960	Cost planning
1965	Value for money
1970	Mathematical modelling
1975	Probabilistic methods
1975	Accuracy in estimating
1980	Simulation
1980	Whole-life costing
1985	Value analysis and value management
1990	Expert systems
1995	Added value

16.6 EMPIRICAL METHODS

Empirical types of model are based on observation, experience and intuition. They have been used and developed largely on the basis of 'right feeling'.

Within their limitations their thinking has largely been towards a commonsense method of understanding, application and presentation. Bills of quantities, for example, are an empirical model. The physical appearance of the building and the methods used for construction have been modelled in terms of descriptions and dimensions. Over a period of time the process has been continually refined, with an attempt to try to relate quantity with cost more realistically. The intention of the compilers of SMM7, for example, is to try to relate cost and the way surveyors measure building work. Their task is difficult since one of their aims is to simplify the method, while maintaining or improving the level of accuracy. The empirical models as we use them at the moment do not really take into account complex plan shapes or large numbers of storeys. The estimator has often to rely on this information being made available visually by way of drawings.

Although it is unusual to think of bills of quantities in the context of algebraic terms, it is an easy transition to see quantities and costs in these terms. For example, the price of concrete in a floor slab can be obtained from the expression:

$$L \times W \times D \times R = P$$

where L is the length on plan, W is the width on plan, D is the thickness of concrete, R is the measured rate for concrete in cubic metres in this location, and P is the price of the floor slab. The empiricist then suggests that different thicknesses of concrete will have different prices. The compilers of SMM6 classified thickness in three categories. Although this would seem a sensible approach and is not generally disprovable, it is almost solely based on opinion without any objective support and is arbitrary in its application.

The advantages of this method of cost modelling are that it is easy to understand and can be related quickly to the construction project. The majority of people at a management level in the construction industry are familiar with the documents and are able to use them for a variety of purposes.

16.7 REGRESSION ANALYSIS

Regression analysis is a technique that will find a formula or mathematical model which best describes data collected. The technique is normally used in situations where the relationship between variables is not unique, in the sense that a particular value of one variable always corresponds to the same value of the other variables. Simple linear regression analysis is a statistical technique which attempts to quantify the relationships between two variables. Multiple linear regression analysis relates three or more variables.

The idea of using regression analysis for estimating the costs of construction both at the design stage and by the contractor was developed at Loughborough University of Technology by Professor Geoffrey Trimble. Several research projects were undertaken to examine the practicalities of its use. The method was also researched in other universities in the UK. It was considered to be an appropriate method based on the following assumptions:

- Reliable estimating is based on a sound knowledge of previously achieved performance.
- The recording of performance is difficult in the construction industry owing to the variety of work undertaken by each contractor.
- The traditional method is to develop a classification system and attempt to record costs against it.
- However, a complex code system is required to cover the majority of possible items. Tests have shown that the reliability of recording substantially decreases when the number of cost codes exceeds 50. The cost code system currently used by contractors is a four-digit system.
- An alternative method of estimating is to apply regression analysis to complete projects. This method could be suitable for certain clients who are responsible for constructing similar projects, e.g. Hospital Boards and the Department for Education. It is unlikely to be of general use to the contractor, however, because it is probable that a contractor would not have a sufficient number of similar types of jobs.
- The proposed method is a compromise between detailed classification and total cost.
- It is proposed to use a limited number of cost codes to capture feedback. It is anticipated that these cost codes will represent the trades required for the construction process.

Consider the data in Table 16.3 which gives the possible sample values of bricklayer-hours and areas of brickwork from ten fictitious contracts. These data can also be plotted on a scatter diagram as shown in Figure 16.3. It can be seen that as the areas of brickwork increase, so do the bricklayer-hours required. The scatter is caused by factors other than area which affect the hours required. Bricklayer-hours required is termed the response variable, and areas of brickwork the regressor variable.

Table 16.3 Possible sample values of bricklayer-hours and areas of brickwork

Contract number n	Bricklayer-hours y	Areas of brickwork x	y^2	xy	x^2
1	800	650	640 000	520 000	422 500
2	1 000	900	1 000 000	900 000	810 000
3	1 100	1 050	1 210 000	1 155 000	1 102 500
4	1 250	1 200	1 562 000	1 500 000	1 440 000
5	1 500	1 500	2 250 000	2 250 000	2 250 000
6	1 750	1 800	3 062 500	3 150 000	3 240 000
7	2 000	2 100	4 000 000	4 200 000	4 410 000
8	2 100	2 300	4 410 000	4 830 000	5 290 000
9	2 250	2 500	5 062 500	5 625 000	6 250 000
10	2 300	2 600	290 000	5 980 000	6 760 000
10	16 050	16 600	28 487 000	30 110 000	31 975 000

Fig. 16.3   Relationship between bricklayer-hours and areas of brickwork

To avoid individual judgement in constructing the line through these points, it is necessary to calculate a 'best-fitting line'. This is derived by the method of least squares, i.e. the line is drawn in such a way that the sum of the squares of the vertical distances from the plotted points to the line is a minimum. The equations are written as follows:

$$y = an + b\Sigma x$$
$$xy = a\Sigma x + b\Sigma x^2$$

Proofs of these equations can be found in standard textbooks. Substituting values for the variables:

$$16\ 050 = 10a + 16\ 600b \tag{I}$$
$$30\ 110\ 000 = 16\ 600a + 31\ 975\ 000b \tag{II}$$

Substituting

$$a = \frac{16\ 050 - 16\ 600b}{10}$$

in equation (II) gives:

$$30\ 110\ 000 = \frac{16\ 600(16\ 050 - 16\ 600b)}{10} + 31\ 975\ 000b$$

$$b = 0.7846$$

Substituting this in equation (I):

$$16\ 050 = 10a + 16\ 600 \times 0.7846$$

$$a = \frac{16\ 050 - (16\ 500 \times 0.7846)}{10} = 310.41$$

$$y = 310.41 + 0.7846x$$

That is, when $x = 1\ 000$ m^2, $y = 1\ 087$ hours.

In fitting the regression line to a set of data we estimate several parameters which need to be tested for significance before being accepted. As an overall guide to the 'strength' of association between the two variables the correlation coefficient should be calculated:

$$r = \frac{n\Sigma xy - \Sigma x\Sigma y}{\sqrt{[n\Sigma x^2 - (\Sigma x)^2]n\Sigma y^2 - (\Sigma y)^2}}$$

$$= \frac{10 \times 30\ 110\ 000 - 16\ 600 \times 16\ 050}{\sqrt{(10 \times 31\ 975\ 000 - 16\ 600^2)(10 \times 28\ 487\ 000 - 16\ 050^2)}}$$

$$= \frac{301\ 100\ 000 - 266\ 430\ 000}{\sqrt{(319\ 750\ 000 - 275\ 560\ 000)(284\ 870\ 000 - 257\ 602\ 500)}}$$

$$= \frac{34\ 670}{\sqrt{44\ 190 \times 27\ 268}}$$

$$= \frac{34\ 670}{34\ 713} = \underline{\underline{0.998}}$$

This shows an excellent degree of correlation (for perfect correlation $r = 1$) which we would not expect to find by using one variable only. One variable would not normally explain this relationship.

Another test that we may need to make is to calculate the standard error of the estimate, i.e. the anticipated difference between the actual values and what the regression line predicts. The standard error of estimate has properties analogous to those of the standard deviation.

16.8 SIMULATION

Invariably when one hears the word 'simulation' one's mind is immediately transferred into the imitation cockpits used to train pilots of both civil and military aircraft. These £4m models allow the pilot to be trained and observed in almost lifelike conditions without removing a wheel from the runway. It is possible to simulate many of the events which occur in the real world. The dictionary describes simulation as imitation. It is not real, it is only pretending to be so, but just how good a pretence it is will depend on the skills of the model–builder. A simulation model seeks to duplicate the behaviour of the system under investigation by studying the interactions among its components. The output of a simulation model is normally presented in terms of selected measures that reflect the performance of the system.

The origins of simulation are threefold. First, there has always been a desire to avoid direct experimentation where it is possible. Direct experimentation may involve developing and testing a particular system and this may be a very costly procedure to manage. Obviously at some stage in the development of procedures this will become necessary, but not generally as a first step. The second reason stems from the solution of purely mathematical problems. Simulation, unlike mathematical problems which represent steady-state behaviour, involves observations that are subject to experimental error. This means that they must be treated as a statistical experiment and any inference regarding the performance must be subject to the tests of statistical analysis. The third reason lies in the growth area of operational research. A major difference between the subject matter of conventional scientific research and operational research is the greater variability of many of the phenomena studied in the latter.

A simulation experiment differs from a regular laboratory experiment in that it can be conducted almost totally on the computer. The relationships in the data can be gathered in very much the same way as if the real system were being observed. The nature of simulation allows much greater flexibility in representing complex systems that are normally difficult to analyse by standard mathematical models. Simulation can, however, be time-consuming, particularly where optimisation is attempted.

The method which is most properly used to solve these problems is called the Monte Carlo technique and is based on the general idea of using sampling to estimate the desired result. The sampling process requires the description of the problem under study by an appropriate probability distribution from which the samples are drawn. The present use of simulation in modelling very complex systems rests squarely on the impressive advances in the capabilities of computing power. It is unimaginable that simulation could have reached any degree of success without this.

In simulation models, sampling from any probability distribution is based on the use of random numbers. A sequence of such numbers can be found in most mathematical tables, or we can choose to use the random number generator from a computer. Randomness is not a function of the numbers themselves, but of the sequences of numbers, i.e. their interrelationships. In order to satisfy statistical conditions, each random number must have a chance of occurring and must be generated independently of previous values and be uncorrelated. The importance of random numbers in a simulation is that each value in a model's real life operation has an equal chance of occurring. The randomness is of course an imperfect representation, and hence does not itself generally have to be perfect.

The use of simulation has many possible applications within the construction industry and in the field of project cost management. The following are some examples which deserve the attention of this technique:

- Construction planning, because of the inherent risk and uncertainty associated with the project's management
- Construction estimating, particularly in the area of tender bidding and cost forecasting which are indeterminate in practice
- Whole-life costing, with the variableness in data such as life of materials and components, maintenance periods, interest rates and building life

The forecasting of construction costs can at the best of times be a hazardous occupation for both the contractor's estimators and the promoter's cost advisers. The contractor's estimator traditionally prices the work by applying unit rates to quantities in a bill of quantities. These unit rates may be composed of a number of the items shown in Figure 4.1. Much of the labour rate analysis is based on the assumed outputs of operatives which, although in theory are based on feedback, in practice generally rely on an experienced estimator's opinion. The outputs, which are so often incorrectly termed 'constants' in some estimating textbooks, will vary for very many reasons, such as the type and design of the project, the amount of repetition in the work, and the quality of the supervision provided. Contrary to popular belief, the estimator's work is not over once the pricing is complete or even when the work starts on site. Cost information data may need to be provided for a variety of purposes throughout the contract period. The following example shows, albeit in a simplified way, the use that an estimator may be able to make of Monte Carlo simulation to solve a particular problem.

Figures 16.4 and 16.5 represent the outputs and earnings of bricklayers that were assumed at the time of tender, and for the purpose of this simulation are presumed

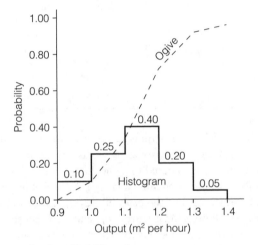

Fig. 16.4   Histogram and ogive of bricklayers' output

Fig. 16.5   Histogram and ogive of bricklayers' earnings

Table 16.4   Probability of bricklayers' output

Output (m² per hour)	Probability	Cumulative output (m² per hour)	Cumulative probability
0.9–1.0	0.10	0.9–1.0	0.10
1.0–1.1	0.25	0.9–1.1	0.35
1.1–1.2	0.40	0.9–1.2	0.75
1.2–1.3	0.20	0.9–1.3	0.95
1.3–1.4	0.05	0.9–1.4	1.00

Table 16.5   Probability of bricklayers' earnings

Earnings (£ per hour)	Probability	Cumulative earnings (£ per hour)	Cumulative probability
3.00–3.25	0.30	3.00–3.25	0.30
3.25–3.50	0.40	3.00–3.50	0.70
3.50–3.75	0.20	3.00–3.75	0.90
3.75–4.00	0.10	3.00–4.00	1.00

to be unchanged. The bricklaying outputs can be seen to vary from 0.9 m² per hour up to 1.4 m² per hour. Although there may be occasions when outputs fall beyond this range of data, these are considered to be negligible and unimportant. For each value within this range it is possible to assign the probability of occurrence. It will be observed that the sum of the probabilities equals 1, indicating that all eventualities have been allowed for. Figures 16.4 and 16.5 also show the cumulative distribution for these data in the form of an ogive. The data can also be shown in tabular form (Tables 16.4 and 16.5).

The distributions in a simulation exercise are at the centre of the technique, since it is from these that sampling will take place. These distributions can be determined only from data which have been carefully collected over a long period of time. Sampling is a simple matter of selecting a random number by using the random number generator on the computer and relating this to our distributions. For example, we may choose to select from random numbers between 1 and 100, and this can easily be done on a computer. If a random number of 40 were selected when sampling from Figure 16.4, this would indicate an output of 1.1 m² per hour. In order to determine the expected cost of bricklaying per square metre, it is necessary to repeat the simulation many times for both the outputs and the costs. This would become a very tedious operation if it had to be done manually, but by using a computer the whole process is speeded up and requires only minimum effort on the part of the user inputting the data to the computer. Table 16.6 gives the results of the simulation.

The estimator may then be presented with new data as shown in Figures 16.6 and 16.7 and asked to comment on whether this would show any change in outputs

Table 16.6 Results of simulation using estimator's original data

	Output (m² per hour)	Earnings (£ per hour)
Mean	1.076	3.305
Standard deviation	0.106	0.216

Average cost per m² = £3.071
(earning ÷ mean output)

Fig. 16.6 Histogram and ogive of bricklayers' earnings

Fig. 16.7 Histogram and ogive of bricklayers' output

Table 16.7   Results of simulation using new data

	Output (m² per hour)	Earnings (£ per hour)
Mean	1.150	3.355
Standard deviation	0.097	0.248

Average cost per m² = £2.917

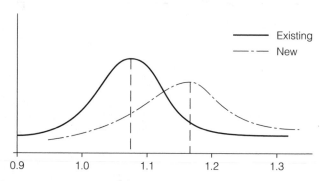

Fig. 16.8   Comparison of distribution of bricklaying outputs

or earnings for this work, which might result in an overall difference in the cost per square metre. The new data are based on a proposed incentive scheme that the site manager is anxious to introduce. The same procedure as before is adopted, and the results of this are given in Table 16.7. The results would of course be subject to statistical significance tests, but we can observe that, although both earnings and output have increased, the latter have increased to a far greater extent, with the result that the average bricklaying cost per square metre has decreased. The two distributions for the outputs are compared in Figure 16.8.

The process of building a simulation model involves formulating a set of relationships between the variables included in the model. It is necessary, therefore, to test that sufficient variables have been included and that the relationships which have been assumed are correct. This can be achieved by running the simulation results with the results obtained in practice. It is usual to start with a very simple model, but where this is too inaccurate to be of any use it must be modified by introducing additional variables. The advantages claimed for simulation are that problems can be solved which it is not possible to solve analytically, it is easier for a manager to understand, and the assumptions to be made are fewer. However, to achieve an appropriate solution it may be necessary to carry out an extensive amount of computation. Simulation is a very powerful tool for solving management problems. It can be used to select the best of a series of alternatives, to gain a deeper understanding of the behaviour of a complex system, or to determine the overall effects of a proposed change in policy. The process of applying simulation to a problem is comparable with that of other types of model building, as shown in Figure 16.2.

Simulation techniques can also be applied to the problems of approximate estimating, using, for example, the superficial area method. The method of quantification is relatively straightforward but the selection of an appropriate unit rate can be difficult, even when cost information is available. The role of simulation in this case is to model the selection of an appropriate rate. The quantity surveyor would first of all need to establish a range of rates which might be relevant. The BCIS could be used, since it provides information of this type. The possible range of values may therefore be expected to be between £320 and £400 per m². From a combination of previous data and judgement, it is then possible to establish the possibility of any of these rates occurring. Sampling from this distribution could then be performed, in a similar manner to that described previously for contractor's estimating, and this process repeated several times. The results of the simulation would then reveal a mean rate and this could be used in the approximate estimate. The simulation is therefore seen as a combination of judgement (used to establish the distribution) and luck (by use of random numbers). Approximate estimating is in reality a combination of at least these two factors in practice, although other considerations will also be taken into account.

Table 16.8 provides a comparison between approximate estimating, analytical estimating and cost modelling.

16.9 HEURISTICS

Heuristics are essentially a rule–of–thumb procedure which enables a near–optimum solution to be produced once the model has been built. A great deal of what is commonly termed trial and error is involved. The majority of the newer techniques applied to the problems of cost forecasting did not attempt to take into account the experience and skill of those involved, from either the design side or the contracting side of industry. The techniques assumed that every aspect was capable of mathematical determination. Practitioners were therefore rather suspicious of such approaches since they knew that cost forecasting was not solely a mechanical process. The practitioners insisted that good estimating was – and always would be – a combination of both objective and subjective analysis. There are of course some aspects that are quite clearly measurable and definable in mathematical terms. The professions concerned have often ignored the fact that some branches of mathematics can clearly help to produce improved accuracy in estimating, since this is the criterion to be achieved. It must be equally recognised that cost forecasting is not simply a matter of 'number crunching'. It also relies on the skill, experience and aptitude of those involved in the process. Far too little credence has been given to these factors in recent years. In the final analysis, therefore, the skills and experience of the expert cannot easily be ignored: to do so is perilous. Cost forecasting must therefore be seen partially as a value judgement. The heuristic method of solution relies on intuitive or empirical rules that have the potential to determine an improved solution relative to the current one. Heuristics are search procedures that intelligently move from one solution point to another with the

Table 16.8 Comparison of approximate estimating, analytical estimating and cost modelling

	Approximate estimating	Analytical estimating	Cost models
Accuracy	Aim is generally not to forecast actual, but contractor's tender sum. Depending on when carried out, 13% on average, depending on size of scheme, method used and luck	Claimed to be within 10% on average	Depends on the data available. Tests indicate 15%–20%. On the basis of minimum information more accurate than approximate estimating
Reliability	Quantity surveyor may intuitively have some idea of the costs of individual items	Estimator knows what rates to expect, therefore incorrectly calculated rates should be recognised	Users largely unfamiliar with what results might be expected
Usage	Methods used by generations of surveyors. Easily understood and applied	Estimators very conversant with techniques used, and whatever the disadvantages these do not outweigh those of an unknown system where control may be ceded	Methods largely either unknown or not understood. Reluctance to change is therefore considerable
Calculations	Single quantity × all-in measured rate, or elemental analysis	Individual measured items × appropriate analysed rates	Model's formulae × rate generated by the computer
Cost control	If prepared in an elemental format, elements must be complete to allow true cost comparisons	Usually required to be done for operations to work rather than individual bill items	Depending on the method used for model building, will influence the methods used for cost control
Dealing with variations	Correct the estimate on the basis of the new information	Remeasure the work and value in accordance with the contract provisions	Amend the data and rerun the program
Updating the process	Use of new schemes if published, or calculated by the office concerned	Outputs rarely adjusted, and only where a number of schemes indicate consistent results	New schemes added to the model and the coefficients are then revised
Use of feedback	Performance data are used but these are substantially adjusted, largely by intuition alone	Because of the complex cost code classification, performance data are of little value. Experience and hunch used to a large extent	Methods rely on the use of actual data for model-building purposes.
Speed	With a single-price method, costs can be obtained within the hour. A more detailed estimate may require a few days	In practice estimators generally require about one month before calculations can be completed	The model's data are already stored on the computer. The new data will need to be input with the results in a hour
Cost of calculations	The use of an experienced quantity surveyor	An estimator together with assistants	Estimator or surveyor to complete data input sheets, computer operator and computer time

objective of improving the value of the model. When no further improvements can be achieved, the best attained solution is the approximate solution to the model. In machine intelligence developments, a heuristic is a rule which dictates a course of action depending on the state of information available at the time. Such decision rules are seldom mathematically or statistically based.

16.10 EXPERT SYSTEMS

'Artificial intelligence has always been a proper subject for science fiction writers. For some it has also been a proper subject for research. But a short while ago it tried to make its entrance into what the practical men and women who manage our commerce and society would call, and with some justification, the real world. Now that was a mistake. The computer got away with a similar infiltration some thirty years ago, since although it masqueraded as a giant brain, it was quite clearly stupid. It did what it was told for one thing. And, society being short on obedience at the time, a rule follower was quite welcome. The computer knew its place, so to speak. But as for artificial intelligence – well, that was taking management head on. Once this pretentious gaffe was spotted, the infant prodigy's name was changed to expert systems and was promptly relaunched. This time, we are told, with greater success' (Price Waterhouse, *Information Technology Review*).

Expert systems are defined as computer systems which behave like experts. They do not break new ground, but pick the brains of someone who already has the know-how to solve a problem and carefully store all the person's rules of thumb in the computer's memory. The scope is enormous. We call on experts continually, in private life and business, but generally, they are not there, too busy or too expensive; often all three. How much better to have that ache in your arm or fault in your factory diagnosed by a computer ready and waiting on your desk. Not according to the Price Waterhouse computer panel. Despite the success claims of expert systems' vendors, it seems that very few UK companies admit to using them. Table 16.9 lists the top ten reasons for not using expert systems at the present time.

The construction industry has few expert systems that have ever been up and running in practice. Until 1990 quantity surveying was one of the few professions in the UK which could lay claim to a knowledge-based computer system, tailor-made to suit specific business requirements. The Lead Consultant (LC) system, better known as 'Elsie', was launched in 1988 after development as part of the Alvey Project by a research team at Salford University headed by Professor Peter Brandon. Knowledge from experts in practice has been built into the system for use on a number of different building types, originally commercial buildings (offices, shops, hotels etc.) but then light industrial units, such as factories and warehouses. The difficulty of deriving the expert knowledge arises from the fact that not all experts agree. This is especially the case where the expertise relies on human judgement and opinion rather than scientific fact for its decision-making.

The Elsie knowledge-based software estimating system is built around the four models of budget, procurement, time and development appraisal (Table 16.10).

Table 16.9   The top ten reasons for not using expert systems

Reason	Mentions (%)
Corporate awareness	66
Finding suitable applications	53
Cost-justifying applications	37
Availability of technical skills	31
Integration with existing systems	29
Acceptance by users	25
Delivered practical systems (as opposed to developed theoretical systems)	24
Capturing expert's knowledge	22
Maintaining captured knowledge	14

Source: PriceWaterhouseCooper

Table 16.10   Elsie modules

Module	Characteristics
Budget	Use of minimal or no drawn information Caters for the complexities of the various elements' qualities and combinations Expressed in BCIS format Detailed assumptions report Report includes advice and a breakdown of all elements
Procurement	Evaluates all the major project characteristics and client priorities Tests and ranks the five main methods of procurement Report contains detailed explanations and reasoning
Time	Forecast of likely project duration at the concept stage Considers client's need and job constraints Identifies and reports the critical activities in three phases: feasibility, design and procurement, and construction and commissioning Variations in duration can be explored by changing constraints and key dates Report contains breakdown of events and reasoning
Development appraisal	Used to give advice on the viability of a development project Derives and tests five residual values: building cost, profit, rent or sale, market yield and land value Construction cash flow is generated

Source: Imaginor Systems

It claims that it is able to produce estimates from no design, saves feasibility costs and is adaptable based on a 'what if?' feature. A series of question-sets about buildings are displayed on the computer's monitor and are answered by the user. The nature of the series depends on the module invoked, the project being considered and the answers given to earlier question-sets. Elsie is constantly using its expert knowledge to deduce the next question, and at the same time it is constructing a virtual model of the desired building and its key attributes.

16.11 VALUE-FOR-MONEY CONSIDERATIONS

An integral part of all cost modelling is an attempt to offer a client improved value for money in design and construction. Traditionally the early forms of cost models had but a single objective of attempting to forecast the contractors' tender sums. This was frequently carried out for the client's own budgeting purposes and to obtain formal approval of the scheme from a board or committee. The cost model forecasts were also sometimes done within the constraint of a cost limit on the design.

The importance of providing early cost advice to clients was to some extent limited to budget forecasts. During the post-war building boom in the UK in the early 1960s it became apparent that much more information could be made available to clients, particularly in the area of value for money. Throughout this period the importance of value for money had become a popular theme, not just in areas of building design.

The early forms of traditional cost models could not clearly identify this aspect other than in the very broadest sense that more expensive buildings probably added value in some way. The development of elemental cost models claimed as one of their objectives that of adding value for money in building design. These models claimed this because of their ability to examine the individual elements of the building and their relationship to each other. However, this technique achieved its major impact only when outline drawn information had been prepared. It was argued that to attempt to cost model a building in this level of detail before shape, spatial layout and specification had been suggested was of very limited use to the design team.

Early design investigations help to focus a design team on achieving value for money. Once a design has been formulated on the drawing board the further efforts involved in achieving value for money become much more restricted. Value for money in building design is about seeking to do more for less. It aims to turn the ordinary into something that is out of the ordinary. Issues that early price models seek to address include:

- Lowest initial costs
- Lowest whole-life costs
- Balanced distribution of design costs
- Highest value for lowest cost

16.12 DEVELOPMENT OF MODELS

A good understanding of the behaviour of construction costs is required in order to model them adequately. This knowledge has increased considerably during the past 25 years, but is still lacking in many respects. Researchers have too often been content to accept the status quo or believed in the perceived wisdom of practitioners who are involved in quantification and analysis. There are still many myths that need to be exploded. What is not easily accepted is that, given the nature of cost forecasting, modelling accuracy is perhaps now almost at its limits.

The history, development and appraisal of cost models has revealed that cost models can be described as traditional and manual or mathematical and computer-assisted. All cost models rely upon an adequacy of historic cost data. In many cases this adequacy is to be questioned, particularly when one takes into account the vagaries of tendering. The traditional models also place a great deal of emphasis on judgement. Models that have been developed that represent a numerical analysis have only a limited chance of acceptance in practice. They are flawed because they misunderstand the nature of construction costs and the inherent desire to incorporate some form of human interaction. Practitioners are also likely to be suspicious of such attempts and unlikely to adopt such approaches in practice. The development of expert systems recognised that progress in cost modelling relies not only on utilising techniques and information technology, but also on incorporating what is already best practice from those employed in commerce and industry.

Until this time the professions involved and their clients chose to receive these forecasts on the basis of single lump sum amounts. Early forms of cost modelling copied this procedure. Deterministic models were therefore produced. It was not until the early 1980s that uncertainty in designs became accepted as a fact of life, and the later models recognising this became probabilistic in nature. Ironically about 70 years had elapsed since uncertainty had been embraced as a fact of the future in many other disciplines.

Later models have encompassed expert systems. Many of the former models had been dismissed, often because of their supposed complexity, but in truth because they provided results that were no better than the traditional models. Expert systems sought to build models that would forecast price, evaluate alternative designs, prepare cash flows and carry out whole-life costs in a manner akin to the way that those in professional practice did this work. These types of cost model have perhaps the best chance of success at the present time.

CONCLUSIONS

The poor quality and reliability of current cost forecasting practice have caused some surveyors to look towards radical methods of price prediction in the construction industry. Cost models are just some of the methods that may provide the results we are all looking for. The use of models has caused the pendulum to swing from the traditional highly subjective procedures to those allowing for little use of experience once the appropriate model has been constructed. The pendulum is now beginning to swing a little way back to allow mathematical models to be tempered with the very valuable experience gained by surveyors and estimators over the very many years of working in practice. Expert systems seek to capture the best of good practice. Sometimes a logically calculated answer to a problem is not necessarily the correct solution to a given set of circumstances. This can be due to the importance of additional factors which were not or could not be quantified. Because construction work and outputs are so variable, the total mathematical approach had little chance of success. Had they been able to achieve consistent results they would still have been rejected because of the protectionist views of surveyors and estimators. This never was, and is unlikely ever to be, a contest in those terms. The usefulness of models is to assist those who are responsible for forecasting building costs in some attempt to improve their performance.

If the full potential of the computer is to be properly harnessed for the benefit of the construction industry, it is likely that cost models will have some part to play in the not too distant future. The time-consuming task of calculating algebraic formulae has in the past discouraged this approach. The computer, with its appetite to perform repetitive and complex arithmetic without effort, is particularly suited to this task. The development of cost models and their wider application to aspects of construction pricing do have the following advantages:

- More information can be generated so that better-informed decisions can be made.
- This information will be more reliable, introducing greater confidence into the decision-making process.
- The cost information can be provided more quickly.
- Suitable cost information can be produced at an earlier stage within the design process.

It was Lord Denning who said, 'a professional man cannot properly advise his client unless he is in possession of the full facts'. Much of the cost advice, particularly at the pre-tender stage, often by its very nature lacks the full facts. Cost models do, however, allow us to provide advice in a more reliable and informed manner by taking account of other variable factors.

Finally, a good cost model should incorporate the following criteria:

- The data requirements for the model should be freely available in the appropriate form and amount. Many models have suffered because of a lack of adequate data.
- The model should allow for continuous updating by incorporating new data that become available.
- The model should be capable of evolving to suit the needs of the changing situation that is prevalent in the construction industry.
- The entire process of cost model management should be able to be done quickly, cheaply and efficiently.
- The model should accurately and reliably represent that which it is attempting to predict.

The only valid question for the use and adoption of cost models for price forecasting is 'Do they work?' There are three possible answers to this question:

1. Yes. In this case the construction industry will need to be a little less conservative and protectionist about their introduction into practice.
2. No, and nothing further need be done. Research should be channelled in other directions.
3. No, but it appears from the work so far that, given time and the necessary expertise, the required results will be achieved.

SELF ASSESSMENT QUESTIONS

1. Identify the different kinds of cost models used in the construction industry and comment upon their effectiveness in practice.

2. Cost modelling has always relied upon a combination of analysis and judgement. Explain the reasons why this is so and whether this approach is likely to continue into the future.

3. Through investigating other industries' methods of forecasting or controlling costs, describe other techniques or models that may be appropriate to meet the needs of the construction industry.

BIBLIOGRAPHY

Ashworth, A., 'Regression analysis: an assessment of its potential', unpublished MSc Thesis, Loughborough University of Technology 1977.
Ashworth, A., 'Cost modelling for the construction industry', *The Quantity Surveyor*, 1981.
Ashworth, A., 'Simulation and construction estimating', *Consulting Engineer*, October 1985.
Ashworth, A., *Cost Models: their History, Development and Appraisal*. CIOB Technical Information Service 1986.
Ashworth, A. and Skitmore, R.M., *Accuracy in Estimating*. CIOB 1982.
Ashworth, A., Neale, R.H. and Trimble, E.G., 'An analysis of the accuracy of some builders' estimating', *The Quantity Surveyor*, April 1980.
Beeston, D., *One Statistician's View of Estimating*. PSA 1973.
Beeston, D.T., 'Cost models', *Chartered Surveyor*, B & QS Quarterly, Summer 1978.
Beeston, D.T., *Modelling and Judgement*. BCIS Cost Study 29a, December 1982.
Beeston, D.T., *Statistical Methods for Building Price Data*. Spon 1983.
Bowen, P., 'An alternative estimating approach', *Chartered Quantity Surveyor*, February 1982.
Bowen, P., 'Problems in econometric cost modelling', *The Quantity Surveyor*, May 1982.
Brandon, P.S., 'A framework for cost exploration and strategic cost planning in design', *Chartered Surveyor*, B & QS Quarterly, Summer 1978.
Brandon, P.S. (ed.), *Building Cost Techniques – New Directions*. Spon 1982.
Brandon, P.S. (ed.), *Building Cost Modelling and Computers*. Spon 1987.
Brandon, P.S., *Quantity Surveying Techniques: New Directions*. Blackwell Scientific 1992.
Buchanan, J.S., *Cost Models for Estimating*. RICS 1972.
Fortune, C.J. and Hinks, A.J., 'The selection of building price forecasting models' in *Proceedings of Economic Management of Innovation, Productivity and Quality in Construction*, Zagreb 1996.

Fortune, C. and Lees, M., 'The relative performance of new and traditional cost models in strategic advice to clients', *Royal Institution of Chartered Surveyors Research Papers*, 1996.

Mathur, K., 'Resource and performance optimisation', *Chartered Quantity Surveyor*, June 1982.

McCaffer, R., 'Some examples of the use of regression analysis as an estimating tool', *The Quantity Surveyor*, 1975.

Raftery, J., *Models for Construction Costs and Price Forecasting*. Surveyors Publications 1991.

Raftery, J. and Newton, S., *Building Cost Modelling*. Blackwell 1993.

Reynolds, J.G., 'Predicting building costs', *Chartered Quantity Surveyor*, May 1980.

Skitmore, M., 'Heuristics and building price forecasting', *Second South East Asian Survey Congress, Hong Kong*, 1983.

Skitmore, M. and Marston, V., *Cost Modelling*. Spon 1999.

Skoyles, E.R., *Introducing Bills of Quantities (Operational Format)*. BRE 1968.

Sleep, R.P., *Topographical Cost Models*. Highways Economics Unit, Department of the Environment 1970.

Southwell, J., *Building Cost Forecasting*. RICS 1971.

Wilson, A.J., 'Cost modelling building design', *Chartered Surveyor*, B & QS Quarterly, Summer 1978.

CHAPTER 17

WHOLE-LIFE COSTING 1: FACTORS TO CONSIDER

LEARNING OUTCOMES

After reading this chapter, you should have an understanding about the factors that need to be considered when carrying out whole-life costing in the construction industry. You should be able to:

- Understand the principles that affect a building's life
- Identify the factors that affect the physical deterioration of buildings
- Consider the different forms of obsolescence that affect property
- Recognise the variability in the lives of building components
- Identify the problems that are inherent with component life data
- Understand the relationships between inflation, interest rates and discount rates
- Recognise the significance of taxation on whole-life costing calculations

17.1 INTRODUCTION

It has long been recognised that to evaluate the costs of buildings on the basis of their initial costs alone is unsatisfactory. Some consideration must also be given to the costs-in-use that will be necessary during the lifetime of the building. The latter factor will be influenced by the type of client and will be a more important consideration to some than to others. For example, developers who construct buildings for sale will be concerned only with future costs-in-use items that may make the project an acceptable proposition for ownership by intending purchasers. Different degrees of importance will therefore be attributed to costs-in-use factors depending on whether the project is to be constructed for sale, lease or owner-occupation.

Whole-life costing is a trivially obvious idea, in that all costs arising from an investment decision are relevant to that decision. The image of a whole life is one of progression through a number of phases, and it also implies renewal as the project undergoes changes throughout its existence. The pursuit of economic whole-life costs is the central theme of the whole evaluation. The method of application

Table 17.1   Whole-life phases

Whole-life phase	Description	Associated costs
1. Specification 2. Design	The formulation of the client's requirements and translating these into an acceptable design	*Initial costs* connected with land purchase, professional fees and construction
3. Installation 4. Commissioning	The construction process up to completion and the handing-over of the project to the client	*Recurring costs* necessary for occupational charges such as rates, insurance, repairs, improvements, fuel, cleaning and estate control
5. Maintenance	The use of the project for its intended purpose	
6. Modification	Alterations necessary to keep the project in a good standard of repair or to improve to current-day standards	*Recurring costs* required for major changes to building in respect of refurbishment and redevelopment
7. Replacement	The evaluation of the project for major refurbishment, or the site for redevelopment	

incorporates the combination of managerial, financial and technical skills in all the phases of the whole life of the building. The proper consideration of the costs-in-use aspects of a project during the design stage is likely to result in a building offering better value for money.

The sequence of the different phases is described appropriately in British Standard 3811 : 1974, and although this adopts engineering terminology the definition of physical assets includes that of buildings. These are described in Table 17.1.

The primary use of whole-life costing is in the evaluation of alternative solutions to specific design problems. For example, a choice may be available for roofing a new project. It would be important to consider not just the initial cost alone, but also maintenance and repair costs, thermal insulation properties, life expectancy, appearance and the possible effect on value arising from the choices which are available. Although appearance is an aesthetic consideration, and therefore largely subjective, it cannot be ignored in the total evaluation of such alternatives. Whole-life costing is therefore a combination of calculation and judgement. The component with the lowest whole-life cost will not automatically be selected. For example, certain types of flat roof are less expensive in their lifetime than are pitched roofs, but preference and bad reports of flat roofs may result in the choice of a pitched roof. It is necessary, however, even in these circumstances to carry out the calculation, since it is important to know the full financial implications of a design.

The application of whole-life costing to capital works projects in the construction industry may result in the commissioning of totally different buildings and structures. A problem occurs in practice, however, that while initial construction costs are relatively clear and predictable at the design stage, costs-in-use are not. They are subject to considerable errors in their assessment. A factor which to some extent mitigates these errors is the fact that all future costs need to be discounted in order to bring them into the same timescale as the initial costs. Also, the comparative evaluation is the main purpose of its application.

Costs-in-use   These are the costs associated with a building or structure while the project is in commission by the owner or occupier. Costs-in-use are therefore recurring costs which may be either annual or periodic in nature.

Whole-life costs   The whole-life costs of a building or structure incorporate the total costs associated with it from inception through to eventual demolition. In addition to costs-in-use described above, therefore, they include all the costs associated with initial construction and the costs of final clearance of the site when the building or structure is no longer required. Whole-life costs therefore include all costs arising from the project, and these can include land charges and professional fees.

17.2 THE IMPORTANCE OF LONG-TERM FORECASTING

The importance of counting the cost before you build was recognised at least 2,000 years ago in St Luke's Gospel (14:28): 'Suppose that one of you wants to build a tower. Will he not first sit down and estimate the cost to see if he has enough money to complete it?'. The emphasis in this example is also on the whole-life cost! Forecasting is required for a variety of purposes such as early price estimating, the setting of budgets, invitation of tenders, cash flow analysis, final account predictions and whole-life costing. While it is recognised that there are confidence and reliability problems associated with initial cost estimating, these are not of the same magnitude as those associated with whole-life costing. A large amount of research has been undertaken in an attempt to improve the forecasting reliability of the former. By comparison the acquisition of whole-life costing knowledge and skills through research and application is still in its infancy, with a considerable gap between theory and practice. It is also difficult to provide confidence criteria, due largely to an absence of historical perspectives, professional judgement and a feeling for a correct solution. The fundamental problem associated with the application of whole-life costing in practice is the requirement to be able to forecast a long time ahead. While this is not in absolute terms, it must be done with sufficient reliability to allow the selection of project options which offer the lowest whole-life economic solutions. The major difficulties facing the application of whole-life costing in practice are therefore related to predicting future events. While some of these events can at least be considered, analysed and evaluated, there are other aspects that cannot even be imagined today. These therefore remain outside the scope of prediction and probability, and cannot even be considered, let alone assessed in the analysis. The key criterion, however, for whole-life costing is not so much in the accuracy of the forecast as in allowing the correct economic solution to be made.

17.3   A BUILDING'S LIFE

Over a period of time, existing buildings decay and become obsolete and require maintenance, repair, adaptation and modernisation. The lifespans of buildings are diverse from their inception to construction, use, renewal and demolition. There is also a varied pattern of existence, where buildings are subject to periods of occupancy, vacancy, modification and extension.

As soon as buildings are erected, deterioration and obsolescence commence. During the 1960s, at a time of rapid expansion and growth in construction activities, there were those who thought that buildings should be designed with short lives and be disposable after a life of about twenty years. Society would require modern buildings to reflect the rapid advances in the age of the *white heat of technology*. Others have suggested that building designs need to be as flexible and adaptable as possible with theories promulgated by the architect Alex Gordon, based upon long life, loose fit and low energy. This would help to assist in delaying obsolescence as long as possible.

A building structure may be designed using materials, components and technology that may last for about a hundred years or more depending upon the quality and standards expected from users. However, the engineering services components in buildings have a much shorter life, with an expectancy of at the most about fifteen years, and the life expectancies of finishes and fittings are now frequently less than ten years. By comparison, information technology hardware and software systems are becoming outdated even after a period of only three years.

The useful life of any building is governed by several different factors and their coincidence. These include the sufficiency of the design, its constructional details and the methods used for construction on and off site. It is also dependent upon the way that the building is used and the maintenance policies and practice undertaken during its life by its owners.

The forecasting of a building's life expectancy is a fundamental prerequisite for whole-life cost calculations. However, the prediction of future events, some of which cannot even be imagined at the outset of a forecast, is fraught with problems. While to some extent a building's life relies upon the lives of the individual building components, this may be less crucial than first imagined, since the major structural elements usually have a life far beyond those of the replaceable elements alone.

There is a general shortage of data on the life expectancy of buildings and structures. There is also evidence that owners and users are unaware of either the total lifespan or the life expectancy up to renewal. They will have theoretical assumptions but these are unlikely to mirror actual practice. For example, many temporary structures such as additional classrooms on schools often achieve a life beyond what is normally understood by their description. The prefabricated concrete bungalows constructed at the end of the Second World War as emergency dwellings with an expected life of about ten years were in many locations still in use after 30 or more years. There are other examples of buildings constructed for a 'normal' lifespan that have been demolished early to make way for newer developments. This occurs in those locations where relative land prices are high and there is a commercial need to gain as much as possible from the land and buildings.

The conundrum of predicting building life or life up to renewal remains unresolved. It has been recommended that for whole-life costing purposes the timescale should be the lesser of physical, functional and economic life. Sensitivity analysis can then be usefully applied to test the validity of lifespans selected. Where the physical lifespan is the shortest then this will be used as the basis. However, in practice this is rarely the case, with one of the different forms of obsolescence being of overriding importance. Physical repair is possible in the majority of cases. It is more likely that one of the forms of obsolescence triggers the need for building renewal.

17.4 DETERIORATION AND OBSOLESCENCE

A distinction needs to be made between obsolescence and the deterioration of buildings as shown in Figure 17.1. The physical deterioration of buildings is largely a function of time and use. While it can be controlled to some extent by selecting the appropriate materials and components at the design stage and through correct maintenance while in use, deterioration is inevitable as an ageing process. Obsolescence is much more difficult to control since it is concerned with uncertain events such as the prediction of changes in fashion, technological development and innovation in the design and use of buildings. Deterioration eventually results in an absolute loss of use of a facility, whereas buildings that become obsolete accept that better facilities are available elsewhere. While deterioration in buildings can be remedied at a price, obsolescence is much less easy to resolve. Obsolescence can be defined as value decline that is not caused directly by use or passage of time.

Because of the large investment that is required in a building, demolition due to obsolescence is a last resort and will only take place where the building either is not capable of renewal or is in a wrong or decaying environment. In other cases the site on which the building stands may be required for other purposes.

The word obsolescence, which has been in use since the middle of the sixteenth century, has the following meanings: *That which is no longer practised or used, discarded, out of date, worn out, effaced through wearing down, atrophy or degeneration.* Such a definition relates to the decay of tangible and intangible things. All human products have an irresistible tendency to become old, but the speed of ageing is different for different objects and circumstances. Obsolescence is largely to do with changing requirements which the object is no longer able to fulfil. For example, when existing standards of performance are replaced by new ones, functional obsolescence takes place.

Obsolescence is an inevitable consequence of rapid economic and technological change. As a result of rapid innovation and development, buildings in the future are likely to enjoy a shorter useful life as a result of early obsolescence. This has important implications for the design and management of property and the

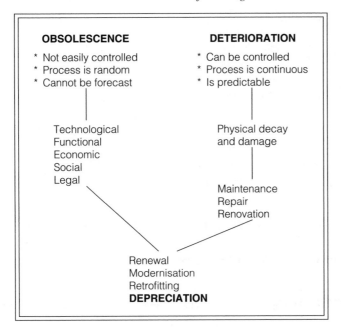

Fig. 17.1 Obsolescence, deterioration and depreciation
Source: Adapted from Flanagan *et al.* (1989)

allocation of financial resources. Initial building design decisions always contain cost implications for the future. The risks associated with physical deterioration can to some extent be controlled at the design stage. The long-term costs arising from this can be minimised through the appropriate application of whole-life costing techniques. However, the problems associated with obsolescence are less easily allowed for, since their impact is unpredictable, as shown in Figure 17.2. An important criterion to delay early obsolescence is to design flexible and adaptable buildings. Western culture is marked by an irresistible acceleration of obsolescence. According to some authors, buildings can only truly be defined as obsolete when they have become completely useless with respect to all possible uses that they have been called upon to support.

   Physical deterioration occurs more slowly than the various forms of functional and other types of obsolescence. The blame for a great deal of obsolescence lies with inflexible planning and designing buildings that were unsuitable for adaptation should their original function cease. However, a majority of clients commissioning buildings require bespoke design solutions to meet their individual needs. Some of these solutions, perhaps decades later when the property is no longer required, are difficult to adapt to changing circumstances. Obsolescence is also to some extent coupled with population relocation, which may make even the most adaptable structure obsolete if it is located in an area of declining desirability or usefulness.

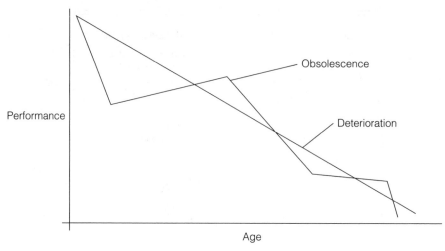

Fig. 17.2   Relationship between deterioration and obsolescence

17.5 PHYSICAL DETERIORATION

Buildings wear out at different rates depending upon the type and quality of materials used and the standards and methods that were adopted for their construction. Ultimate physical deterioration is reached when a building is likely to collapse due to structural failure. However, in practice buildings rarely reach this stage before they are demolished, normally for one of the reasons of obsolescence. The various components used within buildings each in themselves have different lifespans and these are capable of life extension or reduction depending upon the user's needs and the care exercised over their use. Where a building has been carefully designed and constructed and properly maintained its physical life can be almost indefinite. Physical obsolescence is the deterioration of the physical structure of the building. It is not simply a factor of age but a combination of age, use and scale of maintenance.

With heavy use and the passing of time, the costs of maintaining the physical fabric of a building will rise and ultimately buildings and their component parts will reach the end of their physical lives. The factors affecting the physical deterioration of building components are shown in Figure 17.3.

Figure 17.4 illustrates the relationship between physical deterioration and maintenance. Initially a building is designed to meet a maximum level of performance, but from then on throughout its life this is not again achieved as deterioration of different types continues its course. It is even difficult to maintain adequate optimum performance, since the criteria selected at inception will have changed and evolved. Adequate building maintenance will seek to secure a minimum level of building performance below which the premises will fail to meet the essential objectives of the owner and users. However, in practice many buildings fall into disrepair and fail to meet even this level adequately. Without suitable plans for repair and maintenance, the building will reach a state of early demolition.

Weathering factors:	Radiation
	Temperature
	Water
	Normal air constituents
	Freeze–thaw
	Wind
Biological factors:	Fungi
	Bacteria
Stress factors	
Incompatibility factors:	Chemical
	Physical
Use factors:	Design
	Manufacture and construction
	Use
	Maintenance

Fig. 17.3   Factors affecting the deterioration of materials and components in buildings
Source: Adapted from Sneck (1984)

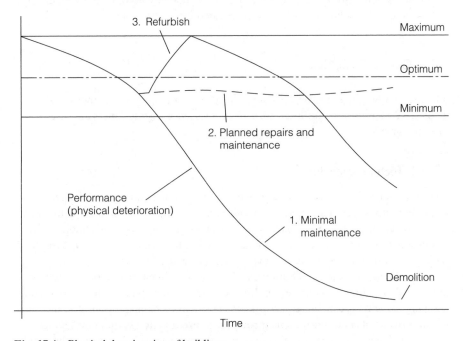

Fig. 17.4   Physical deterioration of buildings

The rates of physical deterioration can be forecast within tolerable levels of accuracy using the lives of the respective building components. However, it must be remembered that considerable variation exists in the lives of even the same building component depending upon a wide range of different circumstances.

The rapid deterioration of buildings and their components can be attributed to many different causes:

- An emphasis upon initial building costs without considering the consequences of costs-in-use
- Inappropriate design and detailing of buildings and their components
- Use of materials and components that have insufficient data on their longevity
- Constructional practices on site that were poorly managed, supervised and inspected
- A lack of understanding of the various mechanisms of deterioration
- Insufficient attention given to maintenance of the building stock
- Inappropriate use by owners and occupiers

The rate of physical deterioration can to some extent be controlled by the designer through the correct choice of materials, methods of construction and appropriate standards of maintenance. Obsolescence is more difficult to control other than through providing a flexible and adaptable design solution to facilitate easier adaptation and renewal at some later date. Also physical deterioration is a continuous process, in contrast to obsolescence which is irregular and unpredictable. Fashionable buildings in high demand in one era suddenly become in less demand unless an age-free solution has been incorporated into their design.

17.6 OBSOLESCENCE

The life of a building may be considered in several different and distinct ways. Table 17.2 identifies the different sorts of obsolescence that designers and users need to consider.

17.6.1 Technological life

The technological form of obsolescence occurs when the building is no longer technologically superior to alternatives and replacement is undertaken to achieve lower operating costs or greater efficiency. A building may become technologically obsolete before half of its physical life has passed. The speed of change in current society suggests that in the future this life will be reduced even faster.

Some components used in buildings will require frequent maintenance attention, others will last almost indefinitely and be described as maintenance-free, while other components will need to be replaced because of their technological obsolescence and the improvements and innovations through new product design and development. This is evidenced in many building components and is especially acute in the rapid development associated with engineering services.

If an office building, for example, is unable to accommodate modern information technology systems because of physical constraints, e.g. insufficient vertical space for the introduction of new office technology such as computers and communication networks, this results in the building reaching the end of its technological life, unless other uses can be found for it. Warehouses that are unable to accommodate modern methods of material handling, such as new generations of fork lift trucks, also suffer

Table 17.2   Building life and obsolescence

Condition	Definition	Examples
Deterioration		
Physical	Deterioration beyond normal repair	Structural decay of building components
Obsolescence		
Technological	Advances in sciences and engineering result in outdated building	Office buildings unable to accommodate modern information and communications technology
Functional	Original designed use of the building is no longer required	Cotton mills converted into shopping units
Economic	Cost objectives can be achieved in a better way	Chapels converted into warehouses Site value is worth more than the value of the current activities
Social	Changes in the needs of society result in a lack of use of certain types of buildings	Multi-storey flats unsuitable for family accommodation in Britain
Legal	Legislation resulting in the prohibitive use of buildings unless major changes are introduced	Asbestos materials, fire regulations
Aesthetic	Style of architecture is no longer fashionable	Office building designs of the 1960s

Source: Adapted from RICS (1986)

the same fate. This occurs, for example, where the trucks cannot be operated owing to insufficient headroom. New industrial technologies developed during the 1950s relied upon the conveyor belt approach; buildings that were not capable of this adaptation had to be demolished or used for other purposes. In manufacturing industry, the shape and composition of the building are mostly determined by the manufacturing process being used. Technological advances in manufacturing technology including the use of robotics have resulted in the need to construct new buildings to replace those that could not be adapted or redeveloped for such purposes.

17.6.2 Functional life

The life of a building sometimes comes to an end because the original function or purpose of the building has ceased. This may have arisen because of changes in, for example, technological or social developments. In some circumstances it may be appropriate to adapt a building for other purposes and change its functional use entirely. Sometimes, in order to achieve this, radical redesign, modification and renewal may be required. There are many buildings that at some time during their lives will be subject to major upgrading, modernisation or renewal in this way.

The reordering of spatial layouts rarely accommodates the changes as easily as a solution that has been developed from new.

In some cases the only alternative for a building is a change in use. In these circumstances the building is no longer suitable or even adaptable for the function for which it was originally designed. In other cases the demand for a particular type of building no longer exists. This is so in connection with many mills, chapels and churches, cinemas and even the local corner shop where lifestyles and social patterns have changed.

Functional obsolescence does not normally result in demolition since such buildings are frequently in a good enough structural condition to allow a form of renewal to take place so that they can be used to serve some other purpose.

17.6.3 Economic life

Economic obsolescence can be defined as when the benefits less the costs of continuing to use the building in its present state are less than the benefits less the costs (including renewal costs) of using the building or site for some alternative purpose. Economic obsolescence is therefore concerned with the least-cost alternative for meeting a particular objective. This might occur where the land value of the building is worth more for potential development than any equivalent rental income derived from the letting of the building in its present state. It is concerned with the highest and best use for the land or the property that stands upon it.

A useful indicator of economic life is to compare the costs of maintaining and using an existing building against its replacement with a new building. The economic life is heavily influenced by the value of the site on which the building is standing. There are several examples around the world where buildings have been demolished a long time before the end of either their physical or functional life because a better economic use was available for the site. For example, in Hong Kong a prestige hotel was recently demolished to make way for new commercial premises. In other cases buildings have been engulfed by new developments resulting in their demolition.

Where the use of the site will result in a higher value than its present use then demolition and redevelopment will occur. A high-value city centre location is therefore likely to favour buildings where rents can be at their highest.

It should be noted that economic obsolescence is largely a function of appreciation rather than depreciation. A building becomes economically obsolete not as a result of depreciation of the existing structure, but through the enhancement of the developing potential of the underlying nature of the site and its environs.

17.6.4 Social life

Buildings are fixed assets in immovable locations. Unlike many other goods and services they cannot normally be moved to more suitable positions. One of the initial failures of the Canary Wharf venture was that it failed to move the City financial institutions to its new location and thus failed to attract those companies that would act as a magnet for other firms. A building may therefore become socially obsolete because, while it is suitable for the process envisaged, it is situated in the wrong location and therefore of only limited practical use. Multi-storey housing in the UK, while developed on a huge scale in every town and city after the War, is now recognised as an unsuitable form of dwelling because of changes in social expectations and behaviour.

During the late 1940s many high-rise developments were planned. Probably as a result of experiences with tenement-type dwellings many local councils had wished to avoid high-rise construction wherever possible. However, incentives offered by central government during the next two decades and the apparent possibilities of new and innovative designs and technology led to a high-rise boom in almost every town and city throughout the country. By the late 1960s, evidence of problems associated with high- and medium-rise dwellings was all too apparent. Coinciding with the partial collapse of the Ronan Point building in 1968 many local authorities therefore turned their backs on high-rise dwellings.

17.6.5 Legal life

New legislation is constantly being introduced to make buildings safer to use and to reduce hazardous or harmful materials that may formerly, without adequate knowledge and understanding, have been acceptable in a building. Fire regulations are revised, often after a disaster, in order to avoid the same problem occurring again in the future. Asbestos and other hazardous materials to health are now prohibited in new buildings and where they occur in existing buildings they need to be either removed or provided with sealed protection systems. The general condition of a building may in some cases make this financially prohibitive, even where grants for their removal are available, and the building will need to be demolished. Legal obsolescence occurs where a building fails to meet current legislation requirements and the costs involved in bringing the building up to the required standards are prohibitive for the type of building concerned. In these cases legislation will advance demolition before the end of the building's physical life.

Both social and legal obsolescence occur when human desires change and dictate replacement for non-economic reasons.

17.6.6 Aesthetic life

Everyone wants beautiful buildings. Sometimes costs and other factors inhibit the possibilities of architectural design. Also modern architectural design does not satisfy everyone. The same buildings that are acclaimed by architectural reviewers as being of excellence are frequently dismissed by the public at large as being poor examples. There are also the designs that sometimes receive general praise only to then appear to be outdated even within a few years and to be wanted by no one in the medium term. Only a relatively few buildings survive over time and eventually find their way onto lists resulting in their preservation and protection. Changes in fashion and architectural style provoke an adverse reaction against those buildings characterised by a preceding era. This effect of changes in fashion and aesthetics often results in building design solutions that are conservative and a compromise.

17.7 COMPONENT LIFE

The lifespans of the individual materials and components have a contributory effect upon the lifespan of the building. However, data from practice suggest widely varying life expectancies, even for common building components. It is also not so much a question of how long a component will last, but of how long a component will be retained. The particular circumstances of each case will have a significant influence upon component longevity. These will include the original specification of the component, its appropriate installation within the building, interaction with adjacent materials, use and abuse, frequency and standards of maintenance, local conditions and the acceptable level of actual performance by the user. The management policies used by owners or occupiers are perhaps the most crucial factors in determining the length of component lives. There is a general absence of such characteristics in retrieved maintenance data.

The design must also recognise the difference between those parts of the building with long, stable life and those parts where constant change, wide variation in aesthetic character and short life are the principal characteristics. There seems to be little merit in including building components with long lives in situations where rapid change and modernisation are to be expected.

There is a variety of sources of published information and data on the life expectancy of different building components. These include:

- Component and building materials manufacturers
- Building Maintenance Information
- The former National Building Agency
- Housing and Property Manual
- The former Property Services Agency
- Royal Institution of Chartered Surveyors
- Building Research Establishment

Some of these sources emphasise the longevity of some building components. It can be argued that if a building is properly designed and constructed then it can be maintained almost indefinitely. There are many examples of buildings where the original components remain in use for hundreds of years. All components have widely different life expectancies depending upon whether physical, economic, functional, technological or social and legal obsolescence is the paramount factor influencing their life. For example, the rapid advancements in fuel technology can mean that it is economically (and environmentally) sensible to replace a central heating boiler, even though it is capable of providing good service for a much longer life. Its life expectancy is curtailed owing to its having reached technical obsolescence.

An important and useful source of data for those involved in whole-life costing is their own accumulated research and expertise. In the absence of this, one or more of

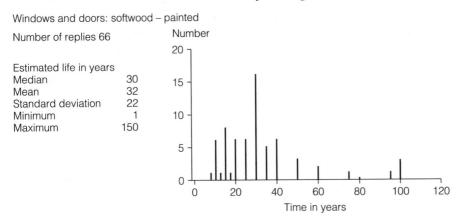

Windows and doors: softwood – painted

Number of replies 66

Estimated life in years
Median 30
Mean 32
Standard deviation 22
Minimum 1
Maximum 150

Fig. 17.5 Life expectancies of softwood windows and doors

the sources of information identified above could be used. The data included in *Life Expectancies of Building Components* (RICS/BRE 1992) represent the findings from a questionnaire issued to a number of building surveyors. The information is typically represented in the format shown in Figure 17.5. It provides an indication of the sample size and the estimated component life in years using a variety of statistical measures.

It can be observed from these data that the life expectancy of softwood windows and doors can vary between one and 150 years. Typically the data show a life expectancy of about 30 years. Furthermore it would be foolish, for example, to prepare a whole-life cost based upon 150 years, even where guaranteed maintenance is promised, owing to the possibility of advancing obsolescence in buildings as identified above. Changes in use, the implications of fashion and the development of new technologies will also have some impact on life expectancies. The important message from investment analysts that 'past performance is no guarantee of future projections' can so easily be applied to building lifespans and the forecast of building components' lives.

A further analysis of the data shown in Figure 17.5 is provided in Table 17.3. This combines three separate charts from the RICS/BRE (1992) survey. It indicates that 14% of the survey respondents expected softwood windows to last (be retained) less than ten years, almost 40% expected them to last for less than twenty years and over 70% for no longer than 30 years.

The survey does not provide an indication of the possible reasons for the predicted different life expectancies. The replacement of the windows may be due to general decay, vandalism, fashion, the installation of double glazing, in order to reduce long-term maintenance, development of new technologies etc. These and other data characteristics are not provided. If this information was included, then the range of values under a particular set of circumstances would be reduced. This would then allow reuse of the values in new situations to be made with greater confidence. On the basis of this and other information alone, it is not possible to

Table 17.3   Life expectancies of softwood windows: percentage distribution of respondents

Life expectancy (years)	Painted (%)	Microporous painted (%)	Stained and varnished (%)	Totals (%)
0–10	11	17	15	14
11–20	24	25	28	25
21–30	33	28	30	31
31–40	17	15	10	14
40+	15	15	17	16
	100%	100%	100%	100%

Source: Adapted from RICS/BRE (1992)

select a precise life expectancy for a particular building component. Different techniques, such as sensitivity analysis or simulation, can be used to test the effects of best, worst, typical and any other scenario in terms of assessing the life expectancy.

The *Housing and Property Manual* (*HAPM*) *Technical Note Number 6* (1995), *Lifespan of Building Components*, provides yet further evidence of the variability of lives of building components data. These are described more fully in the *HAPM Component Life Manual* (1992). The component lives indicated in this manual are based upon some general assumptions of good practice that include:

- Installation in accordance with the manufacturers' directions, relevant Codes of Practice and British Standards
- The use of appropriate design details
- The compliance with the conditions of any relevant third party assurance certificates
- The carrying out of a basic level of maintenance as indicated alongside each benchmark in the *Component Life Manual*

In assessing the expected component lives a number of different factors are considered.

Sources of lifespan data   The sources include the British Standards Institution (BSI), building research organisations such as the Building Research Establishment (BRE), Building Services Research and Information Association (BSIRA), Construction Industry Research and Information Association (CIRIA), Timber Research and Development Association (TRADA) and the British Cement Association (BCA).

However, an important point that should not be missed is that the prediction of building lives, for whole-life costing purposes, is not so much concerned with how long a component will physically last, but how long it will be retained. The scientific data are almost solely concerned with component longevity and not with obsolescence.

While manufacturers and trade associations offer a valuable source of information it needs to be remembered that the component life of a product may be described under ideal or perfect circumstances that rarely occur in practice.

Modes of component failure　The reasons identified for component failure are a combination and coincidence of several different factors. Researchers have recognised for a long time the vagaries associated with costs-in-use data. Even under similar circumstances, identical components frequently have different lifespans in practice.

Risk of component failure　The assessment of the life expectancy of individual components carries an element of risk. This is partially controlled through the long-term use of a component and the independent quality control and assurance procedures developed by, for example, the British Board of Agrément.

Practical experience　The use of the *Component Life Manual* is a good starting point and guide for those needing to assess the lives of building components. The information is likely to be modified in practice by its users to suit particular circumstances of particular buildings. It is also subject to revision on the basis of any past recorded data and the experiences of individual users.

17.8　PROBLEMS WITH COMPONENT LIFE DATA

17.8.1　Maintenance policies

It needs to be accepted that while some companies and organisations have preventative and planned maintenance policies, many do not. Even some of the large central and local government offices, while having maintenance policies, recognise that these are budget-orientated rather than needs-driven. Such budgets are generally insufficient to meet anything like total need. In other words the work carried out occurs where the needs for maintenance and the adequacy in the provision of funding coexist. Thus historic maintenance cost data, in terms of time and cost, represent only that which was affordable, by stretching the available funds to meet as wide a range of maintenance needs as possible.

17.8.2　Data classification

The examination of maintenance records indicates that they are prepared for accounting purposes rather than for future budgeting needs. The sorts of data recorded fall into broad classification systems that are too coarse to be used for any other purposes. Since the retrieval of useful cost information is not a priority of the historic recording of maintenance data, the adequacy for this function has been at least temporarily lost. It is generally not possible therefore to retrieve costs in the manner required for whole-life costing purposes. For example, accounting headings of 'General building' or 'Repairs and maintenance' disclose too limited information for those involved in whole-life costing.

17.8.3 Causes of component failure

Whilst the causes of component failures should enable the designers of future buildings to ensure that such faults are not repeated in future projects, their better understanding should also help to improve the assumptions on component life expectancies. The causes of failure have been broadly classified as bad design and detailing, inappropriate specification, poor standards of workmanship or a failure in supervision and inspection. However, the causes of component failure can be attributed to a wide range of different issues. In some cases a component will be replaced because of the failure of some other aspect of the building or the replacement of another component. Also, rather than undertake maintenance in a piecemeal fashion, often at high cost, delays in repairs may occur until a sufficient quantity of work can be completed. This has the effect of distorting the maintenance information.

17.8.4 Non-identical replacements

The rapid developments in design and technology should mean that building components are likely to be produced with a higher degree of cost efficiency and improved quality and standards. These changes in the specification of a component are likely to result in differences in costs, not necessarily of the same order of magnitude as for the item being replaced. This will distort any cost retrieval system. Such changes in design and technology, or fashion, will have the effect of distorting any whole-life cost predictions that may already have been made.

17.8.5 Time lag delays

Building failures and defects are generally not known until they actually affect the inhabitants of the building or until some routine inspection or maintenance work is being carried out. For example, the decay in external woodwork, e.g. fascia boards or windows, is frequently only discovered at the time of their repainting. Had the decay been discovered earlier then the cost of remedying the defect might have been less and the life expectancy of the component longer. The following represents a typical scenario:

- Problem may be undetected for some time
- Problem might not be notified when detected
- Initial response to the problem might be delayed
- Budget may have already been spent so that work cannot be done in the present reporting period
- When authorised, the remedial work may be in a queuing system
- This may cause further deterioration in the component

17.8.6 Hidden costs

Some maintenance and repair items are concealed in major refurbishment programmes. Dual objectives are therefore met, although the cost information that might have been provided then becomes further distorted.

17.8.7 Potential distortion in the timing of maintenance systems

Since maintenance work is frequently delayed because of a lack of adequate finance, the corresponding life expectancies and costs of the components involved may not necessarily be reliable indicators for a particular component.

17.8.8 Delayed work

The delayed action in carrying out required maintenance work may have a possible knock-on effect upon other components. For example, a leaking roof, if left unattended, may create other maintenance problems elsewhere in, for example, decorations. Also the delay in maintenance work to save costs may result in actual expenditure that is out of all proportion to the costs of the initial maintenance problem and may distort the true life expectancies that would otherwise have been recorded.

17.8.9 Predicting future maintenance costs

For the prediction of future component lifespans and their consequent maintenance costs three assessments need to be made:

1. Those components that fail due to ageing and predictable wear and tear (replacement predictable)
2. The risk of accidental damage such as storm, vandalism and misuse where prediction is only possible on the basis of probability
3. The effect of delaying maintenance which may be indicated by a partial failure of components

Using historical data for these purposes can be very misleading for some of the reasons suggested above. The former Property Services Agency (PSA) has produced a useful set of *Costs in Use Tables* to assist in the assessment of more reliable information on building component lifespans. This is very much like the standard outputs used for capital cost estimating. However, the tables provide guide information only. Even so, the information may bear little resemblance to actual performance and lifespans in practice and must be adjusted to suit local practices and maintenance policies.

17.9 TARGETING THE MAJOR ELEMENTS OF COSTS-IN-USE

The following represents a breakdown of operational costs for four major building types. These represent those areas of costs-in-use where savings can be made. This

Table 17.4   Targeting the major elements of costs-in-use

Spending category	Commercial	Recreation	Educational	Residential
Utilities and energy	35	22	32	22
Overheads	24	19	23	13
Administration	12	29	14	20
Cleaning	12	13	16	25
Fabric maintenance	9	6	5	6
Services maintenance	5	7	7	6
Decorations	3	4	3	8
Totals	100	100	100	100

Source: The *Construction Best Practice Programme*

information is prepared annually by the Building Maintenance Information (BMI) Service. However, such analyses of historic cost data can hide significant variations in practice between the best and worst performers. In the most efficiently managed projects, cleaning costs, for example, can be up to 50% below those in the least efficient. Too often such costs are seen as fixed overheads rather than variable costs that can be managed to provide long-term cost savings.

Table 17.4 confirms that engineering services represent the highest cost-in-use element. Typically, when the maintenance of services is included this represents at least 30% of the ongoing costs on an annual basis. Anecdotal rule-of-thumb calculations suggest that for every unit of capital cost of construction, over a 30-year period there will be ten units spent on maintaining the building and 100 units on staffing the business activity within it. Improvements obtained through adopting a whole-life cost approach can make a significant difference, particularly if expressed in terms of core business activity. This is an important feature in the work of a facilities manager.

Options that are identified during the design stage of the project can result in reduced initial construction costs and costs-in-use. Irregular plan layouts can result in an additional 30% heat loss compared with a more compact design.

There are now large amounts of data that are available to indicate strategic ways of reducing energy consumption and improving the continuing efficiency gains of plant and equipment. As indicated in Table 17.4, energy and utilities are areas where major savings can be achieved without reducing the overall level of comfort that is now expected from modern buildings. Within the Construction Best Practice programme there is an Energy Efficiency Best Practice programme that provides guidance on how organisations in almost every sector can achieve future energy savings, often within a payback period of between three and five years. The savings become greater, often with commensurate reductions in capital costs, if the options are considered at the design stage and are more fully integrated within the main building design. In summary:

- The earlier that whole-life costing is considered, the greater are the potential benefits
- Lower capital costs, reflecting a more efficient design, can result in lower whole-life costs
- Buildings with low in-built running costs will have a lower impact on the building owner's or occupier's business
- Buildings whose performance can be readily and economically maintained will benefit the occupants and help to improve their productivity.

17.10 INFLATION

Throughout almost the whole of the twentieth century the UK has experienced erosion in the purchasing power of the pound. Much has been written on the causes and the possible cure. The effects of inflation and the problem that it causes in capital investment decisions need to be taken into account in a whole-life costing comparison. The following are some of the characteristics of inflation:

- Inflation refers to the way that the price of goods and services tend to change over time.
- Inflation causes money to lose its purchasing power because the same amount buys less.
- The most commonly used measure of inflation in the UK is the retail prices index (RPI). This is supposed to measure the costs of goods and services of a typical family's spending.
- The nominal rate of return on an asset or investment is the amount you get back; the real rate of return is the return after inflation has been taken into account.
- Cash deposits such as savings accounts, although secure, do not keep pace with inflation.
- Interest rates are used to control inflation. By raising interest rates, governments can dampen consumer spending which results in reducing economic activity.
- Low inflation is supposed to be a good thing because it leads to price stability.
- The opposite threat of deflation is considered to be just as much a threat as inflation.
- Zero inflation is rarely desirable. The level of interest rates needed to achieve this would discourage economic activity.
- Europe measures inflation using a harmonised index. If this were adopted in the UK, the inflation record would look much better than is currently reported.

Even with relatively low levels of inflation (say, less than 5%), prices will be substantially affected over long periods of time. An item costing £100.00 today would cost £127.60 after five years at a rate of 5% per annum. The UK tends to regard such a rate of inflation as modest. During the 1970s, the inflation rate increased to almost 30% and today this would be unacceptable in a developed country. However, in developing countries or countries which lack political stability, inflation can reach very high rates. In 1997, Bulgaria had an inflation rate of 70%, compared with 3% for the USA and 1.5% for Japan. However, even Bulgaria's rate looks attractive compared with Brazil in the early 1990s where it averaged 1,270%!

The principal problem facing the decision-maker is whether to forecast future cash flows associated with an investment project in real terms or in money terms. Real terms here means in terms of today's (the date of decision) price levels. Money terms refers to the actual price levels which are forecast to obtain at the date of the future cash flow.

Two different approaches may therefore be used to deal with the problem of inflation. First, inflation could be ignored on the assumption that it is impossible to forecast future inflation levels with any reasonable degree of accuracy. The argument is reinforced in that there is often only a small change in the relative values of the various items in a whole-life cost plan. Thus, a future increase in the values of the cost of building components is likely to be matched by a similar increase in terms of other goods and commodities. There is therefore some argument for working with today's costs and values. Also, since we are attempting to measure comparative values, real costs can perhaps be ignored.

However, changes in costs and prices and their interaction with each other are not uniform over time. Also property values tend to move in booms or surges whereas changes in building costs are much more gradual. It should be noted that building costs do not necessarily increase in line with inflation. Reference to a range of different material or component costs over a period of time will show that these do not follow a uniform trend or pattern. Even similar materials, such as plumbing goods, can show wide differences even over a ten-year cycle of comparisons. To ignore such differences will at least create minor discrepancies in the calculations.

It needs to be remembered that the main purpose of whole-life costing is to correctly inform on the evaluation of options. When such evaluations are economically comparable, it would be unwise to make the selection on the basis of a minor cost advantage to one particular system or commodity.

The alternative approach in whole-life costing is to attempt to make some allowance for inflation within the calculations. This may be done, with some apprehension, using evidence of market expectations, published short- and long-term forecasts and intuitive judgements relating to the prevailing economic conditions. It is worth remembering that in common with all forecasting there will be a degree of error. The forecasting of inflation is a science and art in its own right. Mathematical models are constructed using a wide range of data to assist in their predictions. The models can only consider future events that may occur. In reality events occur that could not be predicted even a few years earlier.

In practice, over the longer periods, the market cost of capital is frequently higher than inflation in order to provide investors with an adequate return on their funds. Thus where inflation is forecast at 10% and if the market cost of capital is 15%, then a discount rate of 5% can be applied. The use of these percentages may not necessarily be uniformly applied to all parts of the calculation. Recent analyses have indicated that where inflation is low and stable then a good relationship will exist between inflation and a bank's preferred lending rate. It needs to be reinforced that whole-life costing is attempting to compare and contrast the worth of alternative systems and not their costs.

17.11 DISCOUNT RATE

The selection of an appropriate discount rate to be used in the whole-life cost calculations will depend upon the financial status of each individual client. The public sector is frequently able to obtain preferential rates for borrowing over those achieved by private companies. The discount rate to be selected will be influenced by the sources of capital that are available to be used for the investment. The client may intend to use retained profits or to borrow from one of a number of commercial lenders. This to some extent can be tested and adjusted by the use of sensitivity analysis (see p. 458).

It can be argued that the choice of a discount rate is one of the more crucial variables to be used in the analysis. The decision to build or to proceed with an investment may be influenced by the discount rate that is chosen. The selection of a suitable discount rate is generally inferred to mean the opportunity cost of capital. This is defined as meaning the real rate of return available on the best alternative use of the funds to be devoted to the proposed project. In practice the discount rate that is selected often represents the costs of borrowing, whether from the firm's own funds (loss of interest) or at a higher rate from a commercial bank or other financial institution. The discount rate that is proposed should then be adjusted by the expected rate of inflation or the time that the project remains live.

For simplicity, it is acceptable to arrive at a discount rate by deducting the expected rate of inflation from the cost of capital percentage. It is more correct to calculate the discount rate by using the formula

$$r = \left(\frac{1+d}{1+i} - 1 \right) \times 100$$

where r is the net of inflation discount rate (real discount rate), d is the interest rate (cost of capital) and i is the rate of inflation percentage.

For example, the rate of inflation is 4% and the cost of capital is 8%:

$$r = \left(\frac{1+0.08}{1+0.04} - 1 \right) \times 100 = 3.846\%$$

Figure 17.6 illustrates the effect of differing interest rates and their effects on present value costs. The higher the discount rate that is used and the further in the future that this is applied then the less impact this will have on present whole-life cost values.

Table 17.5 illustrates this using the three percentage values that are used in Figure 17.6. For example a one-off item costing £10,000 in year 5 using a discount rate of 3% would yield a present value of £8,626.60 (£10,000 × 0.86260). If the same item was replaced in year 25 with a discount rate of 10%, then the present value would be worth only £922.90 (£10,000 × 0.09229).

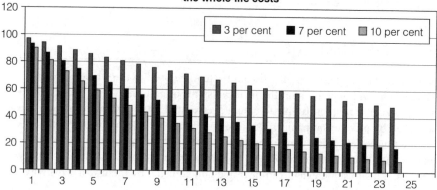

Fig. 17.6   Discount rates and their effect

Table 17.5   Varying present values depending on the discount rate used

Present value of £1 for the different years stated	Year 5	Year 10	Year 25
Discount rate 3%	0.86260	0.74409	0.47760
Discount rate 7%	0.81629	0.50834	0.18424
Discount rate 10%	0.75131	0.38554	0.09229

It is very important that for each option that is being considered the respective cash flows are calculated on exactly the same basis. If cash flows are to be estimated in nominal terms, i.e. include an estimate for inflation, they should be discounted at a nominal discount rate. This should then be applied to all of the options under consideration. It is difficult to be definitive regarding which approach to adopt. Where cost estimates are assumed to inflate at the same rate then it is preferable to perform all calculations in current prices and to apply a real discount rate. However, where inflation is expected to operate differentially then the calculations should be done in nominal terms with explicit account then being taken of the differential rates of inflation.

To select a discount rate that is too low will favour or bias decisions towards short-term, low-capital-cost options. A discount rate that is too high will give an undue bias towards future cost savings at the expense of higher initial outlays. The most correct discount rate should reflect the particular circumstances of the project, the client and prevailing market conditions. It is all too easy to tamper with the discount rate to make the calculation reflect the desired outcome. It is a matter of judgement but one that is done within the context of best professional practice and ethics.

17.12 RATE OF INTEREST

The rate of interest, being the price of loans, is determined by the demand for loans and the supply of funds that are available. The natural rate or equilibrium rate is that which equates demand with supply. The extent to which individuals prefer to retain their assets in cash is termed their liquidity preference. According to economics the reasons for holding cash are threefold:

1. The transactions motive: money is required for everyday needs
2. The precautionary motive: money is needed as a reserve against unforeseen contingencies
3. The speculative motive: money is held in the expectation that prices will fall or interest rates will rise

The need to hold cash will therefore vary with the rate of interest. The calculation of interest rates includes the following:

- Pure interest
- Payment for risk
- Management charges

The period of the loan will also affect the rate charged. Shorter-term loans will generally require higher rates of interest, which partially reflect higher management charges. The interest rate charges are largely influenced by:

- Liquidity
- Yield
- Hedge against inflation

Interest rates severely affect investment by companies and this in turn affects the workload of the construction industry. Government policy is in practice largely responsible for determining interest rates that are set by the Bank of England. Other lenders use this as their yardstick for the setting of their own interest rates to borrowers and lenders. The bank base interest rate is adjusted for the following reasons:

- To reduce the amount of borrowing, a factor that is claimed to affect inflation
- To make investment in a country more attractive than elsewhere
- To create a strong currency and therefore make a country's currency attractive to others
- To provide a suitable return on investments

17.13 TAXATION

Cash flows associated with taxation must be brought into the calculation during the assessment of the project. Most projects will cause differences to corporation tax. This may be due to capital expenditure attracting relief through capital allowances, profits from the project resulting in additional taxes or losses attracting tax relief. Tax is not assessed by the Inland Revenue project by project but for the company as a whole. Cash flows must therefore be considered in this context and calculated on whether the project is carried out, delayed or abandoned. The matter is further complicated since the project may be spread over one or more tax years. Careful accounting may result in beneficial effects through tax avoidance measures. Capital allowances are set against taxable profits in order to relieve the expenditure on fixed assets. There are several categories of asset into which statute has placed the various types of business fixed assets. Each of these have their own rules and basis for granting the allowance. In practice they are a combination of writing-down allowances and balancing charges.

Relief varies, sometimes depending upon the type of building and in some cases in order to encourage development of certain types of buildings. In the case of industrial buildings, for example, companies are able to deduct 4% of the cost of the building from the taxable profit in each year of its ownership and use. On disposal, the proceeds will cause a claw-back of excess allowances or an additional allowance if the difference between cost and the disposal proceeds has not already been fully relieved. Aspects of taxation are considered further in Chapter 11.

17.14 ACCOUNTING FOR UNCERTAINTY

Whole-life costing is intrinsically linked with events in the future and since the future is unknown, predictions can never be made with certainty. Ashworth (1987), as long ago as 1987, posed the question, Can it really work in practice? In this article he presented a number of different scenarios, that are still valid today, suggesting that our lack of knowledge about the future can easily allow us to make incorrect predictions based upon our current understanding. The future includes those events that are both unknown and in some cases even unimaginable. Undoubtedly, and particularly as practice wants to make use of whole-life costing, there is now a feeling of a greater sense of confidence and security. Kishk et al. (2002) noted that uncertainty is endemic to whole-life costing. They also stated that there is a need to be able to forecast a long way ahead for life cycles, future operating and maintenance costs, and discount and inflation rates. The difficulty of doing so is exacerbated not only by a lack of adequate data but also by the fact that when such data are available they often produce conflicting results. Figure 17.7 illustrates that the treatment of uncertainty in information and data remains crucial to a successful implementation of whole-life costing. A number of risk assessment techniques can be used but these alone will never eliminate the uncertainty that should be anticipated.

17.14.1 Sensitivity analysis

This is a modelling technique that is used to identify the impact of change in variables such as building life, component life, interest rates, etc. Whole-life costing is a technique that is able to assist users in the best or most economic use of the alternative assumptions that are available. Different values are therefore attributed

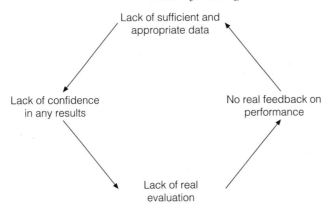

Fig. 17.7   The vicious circle of whole-life cost implementation
Source: Kishk and Al-Hajj (1999)

to these different variables and comparative models developed to test out the different assumptions. It needs to be noted that only assumptions can be tested but these might not provide an optimum solution, since other as yet unknown factors cannot be included in these models. As the technique of whole-life costing is developed then smarter solutions will undoubtedly be made possible. Kirk and Dell'Isola (1995) suggest that the objective is usually to determine the breakeven point defined as the value of the input-data element that causes the least-cost alternative to be better than the next-lowest-cost alternative. A major advantage of this technique is that it explicitly shows the robustness of the ranking of alternative solutions (Flanagan and Norman 1983).

However, sensitivity analysis can only really be interpreted by changing the value of one variable at a time. Second, it does not aim to quantify risk but rather to identify factors that are risk-sensitive. Thus, it does not provide a definitive method of making the decision.

17.14.2 Probability techniques

In the probabilistic approach to risk analysis, all uncertainties are assumed to follow the characteristics of random uncertainty. A random process is one in which the outcomes of any particular realisation of the process are strictly a matter of chance. The following are two suitable probability-based techniques.

Confidence index approach

The confidence index technique (Kirk and Dell'Isola 1995) is a simplified probabilistic approach. It is based on two assumptions:

■ Uncertainties in all cost data are normally distributed
■ High and low 90% estimates for each cost do in fact correspond to the true 90% points of the normal probability distribution for that cost

Monte Carlo simulation

This is a means of examining problems for which unique solutions cannot be obtained. It has been used in whole-life cost modelling by many authors such as Flanagan et al. (1987). In a typical simulation, uncertain variables are treated as random variables, usually but not necessarily uniformly distributed. In this probabilistic framework, the whole-life costing variables, usually the net present values, also become random variables. In the last phase of evaluation, various alternatives are ranked in order of ascendant magnitude and the best alternative is selected such that it has the highest probability of being first. The decision-maker must weigh the implied trade-off between the lower expected cost of one alternative and the higher risk that this cost will be exceeded by an amount sufficient to justify choice of an alternative. Although the technique provides the decision-maker with a wider view in the final choice between alternatives, this will not remove the need for the decision-maker to apply judgement and there will be, inevitably, a degree of subjectivity in this judgement.

17.14.3 Fuzzy set theory

The above techniques recognise shortcomings suggesting that other alternative approaches might be more appropriate. There has been a growing interest in many science domains in the idea of using the fuzzy set theory (FST) to model uncertainty. Fuzzy set theory is an appropriate technique where human reasoning, human perception and human decision-making are inextricably involved. In addition, it is easier to define fuzzy variables than random variables when no or limited information is available. Mathematical concepts and operations within the framework of FST are much simpler than those within the probability theory, especially when dealing with several variables.

Byrne (1997) carried out a critical assessment of the fuzzy methodology as a potentially useful tool in discounted cash flow modelling. However, this work was mainly to investigate the fuzzy approach as a potential substitute for probabilistic simulation models. Kaufmann and Gupta (1988) described how to manipulate fuzzy numbers in the discounting problem. They introduced an approximate method to simplify the mathematical calculations with fuzzy numbers.

17.14.4 Integrated approach

Kishk and Al-Hajj (1999) proposed an integrated framework to handle uncertainty in whole-life costing. It is based on the simple idea that a complex problem may be deconstructed into simpler tasks. The appropriate tools are assigned a subset of tasks that match their capabilities.

17.14.5 Summary

Although the sensitivity analysis approach is simple, it is effective only when the uncertainty in one input-data element is predominant. Furthermore, it does not provide a definitive method of making the decision. The confidence index method is a simplified probabilistic method that has been found to lack the generality of application. Simulation techniques are more powerful but they have been criticised for their complexity and expense in terms of the time and expertise required. Probability theory can only deal with random uncertainty. The fuzzy algorithms are superior due to their ability to deal with judgemental assessments of all state variables. The integrated approach can handle both statistically significant data and expert assessments within the same whole-life cost model.

17.15 WHOLE-LIFE VALUE

This is a term that has been developed by the Building Research Establishment (BRE) to describe the various aspects of sustainability in the design, construction, use, demolition and, where appropriate, the reuse of a built asset. It involves achieving compromise and synergy between the following three different sets of values:

- **Economic value.** This is focused on economically sound sustainability: it is growth-oriented and seeks to safeguard the opportunities of future generations.
- **Social value.** This is concerned with aspects of a societal nature and covers a wider scope of social, cultural, ethical and juridical impacts.
- **Environmental value.** This focuses on environmental aspects of development such as pollution, waste and CO_2 emissions. These issues involve the initial manufacture of construction materials, the construction of the project, its use and eventual replacement. In this context value is maximised when environmental pressures are minimised to the level of the carrying capacity of ecological systems while using natural resources effectively and safeguarding natural capital and its productivity

The Whole Life Value Framework established by BRE is a search facility that points users to a variety of design tools that deal with the issue of sustainability and are appropriate to the particular user group and stages in the procurement process and life of an intended facility. It enables designers and their clients to take account of the most significant aspects of sustainability and to predict the whole-life value of their projects. The Framework has been populated with tools, guidance and procedures identified during a BRE review of material in the public domain, as well as being further populated with tools identified in the questionnaire and by the user group to ensure that the WLV Framework meets industry's requirements.

The WLV Framework has also been populated with 'embedded knowledge', a term used to describe information that the designer is likely to find useful when trying to design a built asset that takes into consideration the whole-life value ethos. For further information refer to www.bre.co.uk.

Additional information can also be obtained from the Association for Environment Conscious Building (AECB). This was established in 1989 to increase awareness within the construction industry of the need to respect, protect, preserve and enhance the environment, principles that have now become known collectively as sustainability. The Association numbers local authorities, housing associations, builders, architects, designers, consultants and manufacturers among its membership. Further information can be found at www.aecb.net.

17.16 WHOLE-LIFE COST FORUM (WLCF)

The WLCF was formed in 1999 by a cross-section of construction industry representatives. The Forum organisation, which is financially supported by its members, includes property clients, property owners and operators, construction consultants, contractors, PFI contractors, suppliers and material manufacturers. Whilst the government does not offer any direct financial support, several government departments are members.

The remit of the WLCF was to develop a set of definitions and methodologies that could be accepted by the construction industry as a whole that would provide a degree of certainty when talking about WLC or providing quotations on a WLC basis.

The WLCF has developed a web-based WLC Comparator Tool which uses the WLCF methodologies and provides a secure process that does not allow users to access the calculations. Therefore any output or results from the WLCF Comparator Tool can be considered as authenticated as there is no possibility of manipulation of the results. Further information can be obtained from: www.wlcf.org.uk.

17.17 WHOLE-LIFE COST MANAGEMENT

Recognising the importance of considering whole-life costs rather than initial costs alone, why do some clients focus only on the former? Speculative developers, for example, who have no interest in the project after it has been constructed and sold to a client will only consider the effects of longer-term costs where these might have an important influence on the initial sale of a property. In some cases excessive consideration of whole-life costs may result in a property that is difficult to dispose of because its costs may compare unfavourably with those of a competitor. Also, some clients may be forced to keep the initial costs as low as possible otherwise the project would not be constructed since funding may not be available to cover the higher whole-life costs.

However, for clients who will own a property for all of the foreseeable future, a proper consideration of not just the initial construction costs but also the costs of owning and operating a building should be encouraged.

Cost management is a process of planning, estimating, co-ordination, control and reporting of all cost-related aspects from project initiation up to the time of an asset's eventual disposal. It involves identifying all of the costs that are involved in order to make informed choices about delivering best-value solutions and in managing these costs throughout the life of the project.

Long-term costs over the life of a project are more reliable indicators of added value than the initial costs alone used as a yardstick. Whilst the concept of whole-life costs has been around for a good number of years, new considerations of sustainability, design quality and its management, waste reduction techniques and the emphasis on reducing recurring costs (repairs, running and replacement costs) have become important to a great many clients in more recent years.

Wherever possible a client should be encouraged to think about these ideas and concepts and to construct a building that results in the lowest possible whole-life costs. It is essential to consider the long-term consequences of initial design decisions and to develop best practices in adding value. Promoting excellence in design does not necessarily mean high initial costs, but these costs need to be viewed in the context of whole-life costs, as indicated in Tables 17.6 and 17.7. Table 17.6 is concerned with the costs of ownership, and clearly shows that the cost of operations, and especially salaries, is the largest cost over a period of time. Table 17.7 limits the breakdown to the proportions of construction-specific costs. Each of these tables emphasises the relative, although not insignificant, construction costs.

In order to properly cost-manage a construction project there are clear links between cost management and time management. Table 17.8 suggests ways of avoiding cost overruns.

All of the parties who are involved in the supply chain process must have reliable data and information on the operational costs and life expectancies of their products. Whilst historic costs and values should be considered these need to be set in a context of future trends. Value-management techniques should be used to reduce

Table 17.6   Relative costs of owning an office building for 30 years

Construction costs	1.00
Costs-in-use	5.00
Office costs, including salaries	200.00
Consultancy fees	0.15

Source: Adapted from Evans *et al.* (1998)

Table 17.7   Capturing whole-life building costs

Costs	%
Design	3
Construction	17
Running and maintenance	40
Repair	30
Periodic replacement refurbishment	10
Demolition	?
Costs of ownership	100

Source: Evans *et al.* (1998)

Table 17.8   Avoiding cost overruns

- Clear objectives at the outset
- Keeping change orders to a minimum
- Clear and reliable estimates at the outset (not being unduly optimistic)
- Project brief that is complete, clear and consistent
- Design that meets planning and statutory requirements
- Design that incorporates sustainability, buildability and maintainability
- Risk allocation that is clear and fair
- Sound management and leadership
- Simple payment mechanisms that are honoured

Source: Adapted from Office of Government Commerce (2007)

the potential for waste and inefficiency. Wherever possible the data should be benchmarked against other similar facilities using, for example, data from the BCIS. Clients are frequently too optimistic about what their project should cost and how well they are capable of managing the risk that is involved. Suggestions have been made about a socially acceptable price. This encourages clients, especially governments, to then develop a project that is based upon such an unrealistic budget. Without the concept of a socially acceptable price many government-funded projects would probably never be constructed.

It is at the design stage that the greatest added-value gains can be achieved. Once a scheme has been fully developed, designers are reluctant to change their schemes to deliver improved added value over the whole life cycle. Poorly designed facilities are frequently expensive to both construct and maintain. Specifications should be output-based, allowing constructors as much freedom as possible to achieve the best solutions.

SELF ASSESSMENT QUESTIONS

1. Building life can be described and measured in many different ways. Explain which you consider to be the most appropriate descriptor in the context of whole-life costing.

2. In determining the life expectancies of building components describe the various factors that should be considered.

3. Explain the relationship between inflation, discount rates and interest rates.

BIBLIOGRAPHY

Ashworth, A., 'Life Cycle Costing – Can it really work in practice?', in *Building Cost Modelling and Computers* (edited by P.S. Brandon). E. & F.N. Spon 1987.

Ashworth, A., 'Data difficulties of building components for use in life cycle costing', *Journal of Structural Survey*, Vol. 14, No. 3, 1996.

Ashworth, A., 'Estimating the life expectancies of building components in life cycle costing calculations', *The Surveyor. Journal of the Institution of Surveyors, Malaysia*, Fourth Quarter 1996.

Ashworth, A., 'Life cycle costing: predicting the unknown', *Journal of the Association of Building Engineers*, 1996.

Best, R. and De Valance, G., *Building in Value*. Arnold 1999.

BRE, *Whole-Life Costing: A Clients Guide*. Construction Clients' Forum 1999.

Browne, R., 'Building deteriology, the study and prediction of building life and performance', *Chemistry and Industry*, 1986.

Building Maintenance Information, *Occupancy Cost Planning*. Building Maintenance Information Service 1992.

Byrne, P., 'Fuzzy D C F: a contradiction in terms, or a way to better investment appraisal?', *Proceedings of RICS Cutting Edge Conference* 1997.

Corus, *Life Cycle Assessment in Steel Construction*. Corus Construction Centre 2001.

Cowan, P., *Depreciation, Obsolescence and Ageing*. The Architect's Journal Information Library 1965.

CRISP, 'Whole life cost forum'. www.wlcf.org.uk.

Evans, R., Horyott, R. and Haste, N. *The Long-term Costs of Owning and Operating Buildings*. Royal Academy of Engineering 1998.

Flanagan, R., *Whole Life Costing*. Blackwell Publishing 2004.

Flanagan, R. and Jewel, C., *Whole Life Appraisal for Construction*. Blackwell 2005.

Flanagan, R. and Norman, G., *Life Cycle Costing for Construction*. RICS 1983.

Flanagan, R., Kendell, A., Norman, G. and Robinson, G., 'Life cycle costing and risk management', *Construction Management and Economics*, Vol. 5, 1987.

Flanagan, R., Norman, G., Meadows, J. and Robinson, G., *Life Cycle Costing: Theory and Practice*. Blackwell Scientific 1989.

Gordon, A., 'The three "Ls" principle: long life, loose fit, low energy', *Chartered Surveyor*, B & QS Quarterly 1977.

Housing and Property Manual. Component Life Manual. Spon 1992.

Housing and Property Manual: Technical Note Number 6, Lifespans of Building Components. Spon 1995.

Kaufmann, A. and Gupta, M.M., *Fuzzy Mathematical Models in Engineering and Management Science*. Elsevier Science Publishers 1988.

Kirk, S.J. and Dell'Isola, A.J., *Life cycle costing for design professionals*. McGrew-Hill Book Company 1995.

Kirwin, R. and Martin, D.B., *The Economics of Urban Residential Renewal and Improvement*. Working Paper No. 77, Centre for Environmental Studies, London 1972.

Kishk, M. and Al-Hajj, A., 'An integrated framework for life cycle costing in buildings', *Proceedings of RICS COBRA* 1999.

Kishk, M., Al-Hajj, A., Pollock, R., Aouad, G., Bakis, N., and Sun, M., *Whole Life Costing in Construction: A State of the Art Review*. RICS Foundation 2002.

National Building Agency, *Maintenance Cycles Life Expectancies*. National Building Agency 1985.

National Union of Teachers, *Crumbling Schools*. National Union of Teachers 1994.

Nicoletti, M., 'Obsolescence', *Architectural Review*, Vol. 143, 1968.

Nutt, B. *et al.*, *Obsolescence in Housing*. Saxon House Publications 1976.

Office of Government Commerce. *Whole life Costing and Cost Management*. HMSO 2007.

Property Services Agency, *Costs in Use Tables*. HMSO 1991.

Richardson, B., 'A dedication to quality', *Building Owner and Manager*, February, pp. 29–30, 1993.

Royal Institution of Chartered Surveyors and the Building Research Establishment, *Life Expectancies of Building Components*. Preliminary results from a survey of building surveyors' views. Royal Institution of Chartered Surveyors Research Paper No. 11, 1992.

Royal Institution of Chartered Surveyors, *A Guide to Life Cycle Costing for Construction*. Surveyors Publications, 1986.

Sneck, T., *General Report No. 1. Proceedings of the Third International Conference on the Durability of Building Materials and Components*, Espoo, 1984.

CHAPTER 18

WHOLE-LIFE COSTING 2: PRACTICE

LEARNING OUTCOMES

After reading this chapter, you should have an understanding about the practice of whole-life costing as applied to projects in the construction industry. You should be able to:

- Identify where whole-life costing can be applied in the project's life
- Distinguish between whole-life costing and costs-in-use
- Apply the principles of whole-life costing to new situations
- Understand the application of sensitivity analysis
- Recognise the difficulties in forecasting
- Recognise the limitations of whole-life costing

18.1 WHOLE-LIFE COSTING APPLICATIONS

The following are some of the advantages of whole-life costing associated with capital works projects:

- The emphasis is on a whole- or total-cost approach undertaken in the acquisition of any capital cost project or asset, rather than concentrating on the initial capital costs alone.
- It allows a more effective choice to be made between competing proposals of a stated objective. The method will take into account the capital, repairs, running and replacement costs and express these in consistent and comparable terms. It can allow for different solutions for the different variables involved and set up hypotheses to test the confidence in the results achieved.
- It is an asset management tool that will allow the operating costs of premises to be evaluated at frequent intervals.
- It will enable these costs to be correlated with changes in working practices, such as hours of operation, the introduction of new plant or machinery and use of maintenance analysis.

Whole-life costing can be used during the following phases.

At inception

Whole-life costing can be used as a component part of an investment appraisal. This is the systematic approach to capital investment decisions regarding proposed projects. The technique is used to balance the associated costs of construction and maintenance with rental values and needs expectancies. It is a necessary part of property portfolio management. It recognises that many projects are built for investment purposes. The way that future costs-in-use are dealt with therefore depends largely on the expected ownership criteria of occupation, lease or sale, or indeed a combination of these alternatives.

At the design stage

A major use of whole-life costing is at the design stage or pre-contract phase of a project. Whole-life costing can be used to evaluate the various options in the design in order to assess their economic impact throughout the project's life. It is unrealistic to attempt to assess all the items concerned; indeed the cost of undertaking such an exercise might well rule out any possible overall cost savings. The sensible approach is to target the areas where financial benefits can be more easily achieved. As familiarity with the technique increases, it becomes easier to carry out the analysis, and this may prompt a more in-depth study of other components or elements of construction. While some of the areas of importance will occur on every project, others will depend on the type of project being planned. For example, roofing is probably an important area for whole-life costing on most projects, whereas drainage work is perhaps not. However, on a major highway scheme, where repeatability in the design of the drainage work occurs, then the small savings which might be achieved through whole-life costing can be magnified to such an extent as to make the analysis worthwhile. The important criterion to adopt is that of cost sensitivity in respect of the whole project costs.

Whole-life costing is perhaps most effective at this stage in terms of the overall cost consequences of construction. It can be particularly effective at the conceptual and preliminary design stage, where changes can be made more easily and resistance to such changes is less likely. When a design is nearing completion, the designer may be reluctant to redesign part of the project even though long-term cost savings could be realised.

In selecting a design from a possible choice of options, the one with the lowest whole-life cost will usually be the first choice, provided other performance measures or criteria have been met. Using whole-life costing with other techniques such as value engineering should enable the scheme to be designed within a framework that is more cost-effective without the loss of any of the design's desirable attributes.

At procurement

The concept of the lowest tender bid price should be modified in the context of whole-life costing. Under the present contractual and procurement arrangements, both manufacturers and suppliers are encouraged to supply goods, materials and components which ensure their lowest initial cost irrespective of their future costs-in-use. In order to operate a whole-life cost programme in the procurement of capital works projects, greater emphasis should be placed on the economic performance in the longer term, in order to reduce future maintenance and associated costs. The inconvenience that often arises during maintenance and the other associated replacement costs, which may be out of all proportion to the costs of the part that has failed, also need to be examined. The different methods of procurement which are available may make it easier and beneficial for the contractor to consider the effects of whole-life costing on a design.

At the construction stage

While the major input of whole-life costing is at the design stage, since its correct application here is likely to achieve the best overall long-term economic savings, it should not be assumed that this is where the use of the technique ceases. At the construction phase, three broad applications should be considered.

The first of these concerns the contractor's method of construction, which unless prescribed by the designer is left to the contractor to determine. In some instances the contractor may be allowed to choose materials or components that comply with the specification but will nevertheless have an impact on the whole-life costs of the project. The method of construction the contractor chooses to employ can have a major influence on the timing of cash flows and hence the time value of such payments. This is perhaps more pertinent to works of major civil engineering construction, where the methods available are more diverse. Buildability aspects which might enable the project to be constructed more efficiently, and hence more economically, may also have a knock-on effect in the longer term and hence have an influence on the related costs-in-use.

Second, the contractor can benefit from adopting a whole-life costing approach to the purchase, lease or hire of the construction plant and equipment. The probable savings resulting from this evaluation may then have an impact on future tendering and estimating strategy and project costs.

Third, the construction managers can provide a professional input to the scrutiny of the design, if involved sufficiently early in the project's life. They may be able to identify whole-life cost implications of the design in the context of manufacture and construction and in the way that the project will be assembled on site.

During the project's use and occupation

Whole-life costing has an important part to play in physical asset maintenance management. The costs attributable to maintenance do not remain uniform or static throughout a project's life, and therefore need to be reviewed at frequent intervals to assess their implications within the management of costs-in-use. Taxation rates and allowances will change, and can have an impact on the maintenance policies being used. Grants may become available for building repairs or to address specific issues such as energy usage or environmental considerations. The changes in the way the project is used and the hours of occupancy, for example, need to be monitored to maintain an economic whole-life cost, as the project evolves to meet new demands placed on it.

When a project nears the end of its useful economic life, careful judgement needs to be exercised before further expenditure is apportioned. The criterion for replacing a component is a comparison of the rising running costs with the costs of its replacement and the associated running costs. Additional non-economic benefits should also be considered, and will need to be accounted for in some way in the analysis. For example, the improved efficiency of central heating boilers and their systems suggest that these, on economic terms alone, should be replaced every ten to twelve years irrespective of their working condition. A simple whole-life cost analysis can show that the improved efficiency of the burners and the better environmental controls will outweigh the replacement costs within this period of time.

18.1.1 For energy conservation

A major goal of the developed nations is a reduction in the use of energy in all its costly and harmful forms. This is true for the governments concerned, who have introduced taxation penalties, and for private industry, which is seeking ways of reducing its own energy consumption and hence the associated costs. Whole-life costing is an appropriate technique in the energy audit of premises. A reduction in energy usage has been encouraged due to the rising costs of foreign oil supplies, the finite availability of fossil fuels and the rising urgency to combat the effect of climate change. The energy audit requires a detailed study and investigation of the premises, recording of outputs and other data, tariff documentation and an appropriate monitoring system. The way that the premises is used, plus typical or likely expectations of energy usage and sound professional judgements, are important criteria for such an analysis. The recommendations may include, for example, providing additional insulation in walls and roofs, replacing obsolete equipment, and suggesting values for temperature gauges, thermostats and other control equipment. An energy audit is not a one-off calculation, but one that needs to be repeated frequently in order to monitor the changes in the variables which affect the overall financial implications.

18.2 WHOLE-LIFE COST PLAN

A whole-life cost plan is a plan of the proposed expenditure of a construction project over its entire lifespan. Table 18.1 is a summary of the items which may need to be considered for inclusion. The total information is shown as either a net present value (NPV) or an annual equivalent (AE). Capital cost is the estimate of the initial cost, and this is already a PV amount. It should be remembered that all estimates include errors of prediction, and this will be especially true of the estimate of costs-in-use. Maintenance costs would be estimated on an annual basis using historical

Table 18.1 Whole-life cost plan

		Project:	
		Date:	
		Discount rate:	
	Summary		

Description	Discount factor	Estimated cost	Present value
1. Capital cost			
2. Maintenance (per annum)			
3. Redecoration (intervals)			
4. Minor new works (year)			
5. Energy (per annum)			
Heating			
Lighting			
Power			
6. Cleaning (per annum)			
7. General rates (per annum)			
8. Insurance (per annum)			
9. Estate management (annual)			
10. Additional tax allowances (per annum)			
Total net present value (NPV)			
Annual equivalent (AE)			

information coupled with current knowledge. The summary may be supported with more detailed schedules, as shown in Table 18.2. The £4,178 is the NPV of the decorations discounted at 5% over a twelve-year period. The AE can be obtained by dividing this amount by 8.863 (PV of £1 per annum), and this equals £471.39 per annum on decorations. Alternatively, for costing purposes, it could be assumed that full redecoration will occur every three years. The NPV of £1,660 on this basis is £4,667, or for four-yearly intervals £3,414. Clearly, the longer the time between these decoration periods, the lower will be the NPV.

18.3 COSTS-IN-USE

The costs-in-use of a building project depend on the owners and the users of the building, and the designer's efforts to minimise recurring costs. It should be recognised that some costs-in-use are largely beyond the scope and control of the designer, e.g. rateable values and some other occupancy costs. The design may encourage users to be thrifty, and schemes such as a recent government's 'Save It campaign' have had at least a limited success. In order to obtain the best benefits from a scheme, the project should be maintained in accordance with the designer's instructions. Unfortunately, in-use manuals of such a major asset as buildings are

Table 18.2   Costs-in-use

Project: Smith's House
Date: January 1993
Discount rate: 5%

Decoration schedule

Room	Amount (£)	1	2	3	4	5	6	7	8	9	10	11	12
External	350		*		*		*		*		*		*
Lounge	240			*			*			*			*
Kitchen	150		*			*			*			*	
Hall	250				*				*				*
Dining	160					*					*		
Bathroom	80			*			*			*			*
Study	90						*						*
Bed 1	100					*					*		
Bed 2	140						*						*
Bed 3	100					*					*		
	1 600												

Year	Amount (£)	DCF	NPV
1	0	0.952	–
2	500	0.907	454
3	320	0.864	276
4	600	0.823	494
5	510	0.784	400
6	900	0.746	671
7	0	0.711	–
8	750	0.677	508
9	320	0.645	206
10	710	0.614	436
11	150	0.585	88
12	1 160	0.556	645

NPV of decoration = £4 178

AE over 12 years at 5%

$$= \frac{£4\,178}{8.863} = £471.39 \text{ or approximately } £4.28 \text{ per m}^2 \text{ per annum (based on a GIFA of 110 m}^2)$$

seldom prepared. Little attempt is made to recommend how the building should be maintained even in the case of engineering services.

Costs-in-use data cover the three 'R's: running, repairs and replacements. These costs may usefully be classified as follows.

18.3.1 Maintenance costs

The expenditure of money in time and materials on building maintenance is high and growing because of the necessity of having to maintain an ever-increasing stock of older properties. There is generally some relationship between maintenance costs and the age of a building.

The major factors which make buildings inefficient or expensive to maintain are:

- Incorrect specification of materials used either initially or during subsequent repairs
- Incorrect use of spaces
- Poor constructional detailing, resulting in inadequate weather resistance, rapid corrosion and deterioration
- Inadequate care in use by owners and occupiers

18.3.2 Redecorations

Redecorations are somewhat different in character from other maintenance costs. They normally follow a predetermined cycle and can therefore be anticipated. An estimate of cost can therefore be obtained in advance. They are also works which in time of recession are often deferred. During cuts in local government expenditure in the early 1980s, the internal decoration of school buildings in some local authorities was delayed from a nine-year cycle to an eleven-year cycle. The cycle of redecorations depends on many factors, such as user requirements, type of use, finish and exposure. During the early years of a building's life, no costs will be allocated to this element of costs-in-use. Delays in external redecoration can sometimes result in costs out of all proportion to the apparent saving, particularly on joinery and other items adversely affected by the weather.

18.3.3 Minor new works

Many buildings often incur expenditure which cannot be truly categorised as repair and maintenance in the context of fair wear and tear. Refurbishment and alterations will occur in varying degrees, some of which may be necessary due to changes in use of premises or for modernisation purposes.

Minor new works in some circumstances may be classified as capital works, whereas in other circumstances the funding may be from a maintenance and repairs account. The distinction in practice is often not easy to make, and misallocation of monies from one fund to the other can therefore be expected. Only the larger items of minor works expenditure will require authorisation from a separate fund. The size of the company or organisation concerned will usually determine the line of demarcation. No general guide can be given for predicting this expenditure, since in some buildings no minor new works are ever carried out, whereas in others they may become necessary during the first few years of life.

18.3.4 Energy

The term 'energy' is used in the context of providing all the required fuel for heating, lighting, cooling and power requirements of a building. Although the individual costs of the above items are important, it is necessary to measure the total energy consumption when comparing this element between buildings. The cost of any attendant labour for stoking boilers and adjusting controls, or for servicing and maintenance, must also be included under this heading. The Chartered Institute of Building Services (CIBSE) has produced several guides aimed at making energy usage more efficient. A wide variation in annual costs may be expected depending on the type and amount of services provided, the building's design and insulation provision and the control and use of equipment.

18.3.5 Cleaning

The cleaning costs of buildings, which are generally very labour intensive, depend on the function of the spaces to be cleaned, the type of finishes used and the cleaning interval. Offices tend to have a higher cleaning cost than most other types of building.

18.3.6 General rates and insurance

Designers have little control over the rateable value of the buildings they design, since this is largely determined by location, size and the amenities provided. The client can be advised to build in a less expensive area, provide a building of a smaller plan size or reduce the amenities within the building. Insurance costs of the building in use are also largely self-determinable by the nature of the project. Factors such as type of structure, method of construction, materials used, class of trade, materials stored and number of employees will influence the premium to be charged. For example, timber-frame houses generally require the payment of a higher premium than the more traditionally constructed types. Increases in costs in line with inflation may be expected for both rates and insurance, the latter being determined on the increased insurable sum of the property. In some circumstances, due to a high-spending local authority, rates may outstrip inflation. Where insurance claims are excessively high, a review by the insurance company of premiums generally may become necessary.

18.3.7 Estate management

Large construction projects when in commission may require some form of expenditure on estate management, security, porterage etc. Such sums can be determined in advance by the owners, but a good design and method of construction can help to reduce the required amounts. There will be times when a greater input of estate management is required, particularly prior to extensive refurbishment or alteration.

18.4 CALCULATIONS

One of the apparent difficulties of using whole-life costing in practice is the mathematics associated with the evaluation. An understanding of the principles involved in discounting the value of future receipts and payments is an essential feature of such an evaluation. Although the arithmetic associated with discounting may appear complicated, the concept is simple. This is that the capital in the hand today is worth more than the capital at some time in the future. Even ignoring inflation, it would be more profitable to choose to receive a given sum today than the same sum next year.

If the current rate of interest is 5% per annum, then £100 invested today will yield £105 in twelve months' time. Conversely, £100 to be received next year has a PV of:

$$£100 \times \frac{100}{105} = £95.24$$

Money is generally invested at compound interest; discounting can therefore be viewed as the reverse of this process. When discounting we know the future sum and wish to find the PV. In order to use the technique in practice it is not necessary to understand the mathematical basis of the subject, since the relevant information is contained in discounting tables. The most commonly occurring calculations when discounting are:

- Calculating the PV of a lump sum to be paid in the future
- Calculating the PV of a regular annual payment for a number of years
- Calculating the AE of a lump sum to be paid now

A typical scenario requiring the use of discounted cash flow (DCF) techniques is as follows. It is already presumed that the window selection will be made not on the basis of initial costs alone but on their whole-life costs. Factors regarding the window type and size etc. will have been previously determined.

Example 1

A choice is available between the use of softwood, hardwood or aluminium windows for a detached house. Before the calculation can be undertaken it is necessary at the outset to predict two factors: expected lives and the discount rate to be used. Both the life of the building and the life of the components will need to be determined from previous historic data (if any), coupled with future predictions based on fashion, trends and personal judgement. The economic life for buildings is assumed to be 60 years. The whole-life calculation may be carried out as follows.

1 Softwood windows
   10 No. softwood double-glazed standard pattern windows,
      fixed to brickwork including decoration and window furniture   =   2 500

Renewal (A) every 15 years:
  Year 15 0.31524
     30 0.09938
     45 <u>0.03133</u>
      0.44595 × £2 650 (B)   =   1 182

Redecoration every 5 years:
  Year 5 0.68058 35 0.06763
     10 0.46319 40 0.04603
     15 (C)   45 (C)
     20 0.21455 50 0.02132
     25 0.14602 55 0.01451
     30 (C)   60 (End of supposed building life)
          1.65383 × £100   = 165

Cleaning (D) per annum
£50 × 12.377   = 619
  NPV   = £4 466

This calculation is based on a 60-year life and a discount rate of 8%.

A. Renewal periods in practice will be different and will depend on initial quality, location, aspect, use and maintenance. The argument is often used that once softwood windows start to rot the only course of action is their renewal. Some authorities may attempt to splice in new sections of timber in order to prolong their life and the period to renewal. Opinions will vary, but fifteen to twenty years is assumed to be a good life expectancy (see Figure 17.5).

B. Inflation is generally ignored in whole-life cost calculations, because the amounts are a comparative value. The renewal costs include a sum of £15 per window for taking out the existing window, clearing away off-site, preparing the opening to receive the new window and any making good required.

C. On average, external decoration may take place every three to seven years. Often when companies or individuals have to reduce their expenditure this is one of the items of work to have its maintenance period extended. The decoration of the renewed windows in years 15, 30 and 45 is included with renewal costs. It is assumed that because economic life is 60 years nothing will be done in year 60. There is an allowance of £10 per window for redecoration i.e. 10 windows × £10 = £100.

D. The cost of the cleaning of windows depends very much on the frequency of this work. Assuming they are tradesman-cleaned, an annual sum of about £50 may be expended.

One method of reducing each of these sums to a common timescale is to calculate their NPVs. The initial costs are already in this form and can therefore be entered directly. In order to convert the renewal costs to PVs we discount their amounts from the 'PV of £1 table' or from the equation $(1 + i)^n$, where i is the interest rate and n is the number of years. The same table is used for redecoration. We could also use the same table for discounting the cleaning costs, but this would be a very cumbersome method of calculation. In this case we refer to an alternative table, 'the PV of £1 per annum', sometimes referred to as the 'YP single-rate table'. The equation is a little more complicated, but the terms mean the same.

$$\frac{1 - 1/(1 + i)^n}{i}$$

The alternative approach is to compare the annual equivalents of the three different types of windows. In order to convert the NPV to an AE we divide by the NPV of £1 per annum for the length of years at the appropriate discount rate.

Softwood windows $\dfrac{£4\,466}{12.377} = £360.83 = AE$

Either way, on a comparative basis the results will be the same.

II Hardwood windows

10 No. sapele double-glazed windows, fixing, decoration and furniture	=	4 500
Renewal every 30 years £4 500 + 150 × 0.09938		462
Redecoration every 5 years £50 × 2.0004		
(This assumes £5 per window and redecoration every 5 years except renewal year 30)	=	100
Cleaning – costs as previous	=	619
		£5 681

III Aluminium windows

10 No. aluminium double-glazed windows, fixing, decoration and furniture	=	5 500
Renewal. Life expectancy assumed 60 years (E)	=	–
Redecoration – none	=	–
Cleaning – costs as previous	=	619
		£6 119

E. Although the life expectancy of the aluminium windows is thought to be at least 60 years, there is little evidence to support this in practice. Should replacement be necessary, using these data, aluminium windows would have an immense cost disadvantage.

Too much emphasis should not be placed on the final figures. These will be variable, depending, for example, on the quality of each of the windows offered. They will also be sensitive to changes in interest rate and life expectancy achieved in practice. No attempt has been made to measure any thermal differences. The method used is the important consideration.

Summary of costs

	Softwood	Hardwood	Aluminium
Initial	2 500	4 500	5 500
Renewal	1 182	462	–
Redecoration	165	100	–
Cleaning	619	619	619
Total (NPV)	£4 466	£5 681	£6 119
AE	£360.83	£459.00	£464.38
Initial	56%	79%	90%
Renewal	26%	8%	–
Redecoration	4%	2%	–
Cleaning	14%	11%	10%
	100%	100%	100%

18.4.1 Effects of changes in lives of components and buildings

Life of components (softwood windows 25 years and redecoration every seven years)

Initial cost	=	2 500

Renewal every 25 years:

Year 25 0.14602

50 <u>0.02132</u>

<u>0.16734</u> × 2 650   =   443

Redecorate every 7 years:

Year   7 0.58349

14 0.34046

21 0.19866

(25 + 7) 32 0.08520

39 0.04971

46 0.02901

(50 + 7) 57 <u>0.01244</u>

1.29897 × £100   =   130

Cleaning – as previous   =   <u> 619</u>

NPV   = <u>£3 692</u>

There is a significant difference in the NPV for renewal costs. The difference in redecoration, however, is only marginal. The extended life of these softwood windows increases their cost advantage against either of the other two choices available. Due to the life expectancy of the aluminium windows, already at a maximum, any reduction in this will only increase their cost disadvantage.

Life of buildings

The effect of extending the life of the building will be to increase its NPV, but often by only a minimal amount. Where components are used in buildings that are demolished before the end of their life expectancy, some form of bonus will be added to their evaluation. The one factor militating against this is that costs so far into the future, i.e. beyond 60 years, are of minimal value when they are discounted in terms of either NPVs or AEs. The 60-year life is therefore taken as being realistic in practice, and also for whole-life cost evaluation purposes.

18.4.2 Changes in interest rates

The interest rates used for discounting purposes tend to have the following importance. Higher interest rates have the effect of reducing the importance of future payments, and the further into the future such payments occur, the less importance will be attached to them. Low interest rates mean that future sums are of much greater importance when evaluated against their initial costs, which are NPV amounts.

Example 2

A client is considering replacing his heating system. System A is the standard scheme whereas system B relies on additional insulation being provided. Evaluate the alternatives (insulation costs are not considered).

Initial costs	System A	System B
Boiler	160 000	145 000
Pipework and units	48 000	42 000
Insulation	12 000	32 000

Recurring costs		
Repairs	3 000 per annum	2 800 per annum
Replacement	40 000 (every 20 years)	32 000 (every 30 years)
Overhaul	15 000 (every 5 years)	15 000 (every 10 years)
Fuel	15 000 per annum	11 000 per annum

The expected life of each building is 60 years and the discount rate to be used is 8%.

System A – NPV 8%/60 years

Initial costs (16 000 + 48 000 + 12 000)	=	220 000
Repairs £3 000 × 12.377	=	37 131
Replacement: Year 20 £40 000 × 0.21455	=	8 582
40 £40 000 × 0.04603	=	1 841

Overhaul: Year   5   0.68058
           10   0.46319
           15   0.31524
           20   Replace
           25   0.14602
           30   0.09938

                        35 0.06763
                        40 Replace
                        45 0.03133
                        50 0.02132
                        55 0.01451
                        60 Demolish

£15 000 × 1.83920   =   27 588
Fuel £15 000 × 12.377   =   185 655
                                                                      £480 797

System B

Initial cost (145 000 + 42 000 + 32 000)   =   219 000
Repairs £2 800 × 12.377   =   34 656
Replacement: Year 30 32 000 × 0.09938   =   3 180
Overhaul: Year 10 0.46319
                20 0.21455
                30 Replace
                40 0.04603
                50 0.02132
                60 Demolish

£15 000 × 0.74509   =   11 176
Fuel £11 000 × 12.377   =   136 147
                                                                      £404 159

On the basis of the above solutions, system B would be selected as the more
economic proposition. This argument might be reinforced by the fact that
more of system A's costs are in the future and are therefore more susceptible
to prediction errors. System B will require fewer shutdowns for major overhauls
and replacements. Table 18.3 gives the comparative amounts and percentage
proportions for each system.

Table 18.3 Cost comparisons of alternative heating systems

	System A		System B	
	£	%	£	%
Initial costs	220 000	46	219 000	54
Repairs	37 131	8	34 656	9
Replacement	10 423	2	3 180	1
Overhaul	27 588	6	11 176	3
Fuel	185 655	38	136 147	33
Total (NPV)	480 797	100	404 159	100

Example 3

Calculate the comparative whole-life costs of the following two buildings, using only the data provided. The opportunity cost of capital is assumed to be 6%. Both buildings provide a similar size of accommodation.

C. High initial cost building with low costs-in-use

Initial cost	£75 000
Repairs	£500 every 10 years
Maintenance	£300 per annum
Heating, lighting etc.	£1 200 per annum
Demolition and disposal	£3 000

D. Low initial cost building with high costs-in-use

Initial cost	£45 000
Repairs	£1 500 every 5 years
Maintenance	£900 per annum
Heating, lighting etc.	£2 100 per annum
Major modifications every 15 years	£10 000
Demolition and disposal	£4 500

The question sometimes arises in practice of whether high initial cost buildings are a less expensive proposition in terms of the whole-life cost. A way of achieving a solution to such a question is to use DCF on projected data (i.e. future data based on past costs). The generally held view is that buildings which are to be constructed for only a short lifespan should have low initial costs. This will invariably result in high maintenance costs. Evidence exists in practice to support this opinion in the form of buildings such as temporary classrooms to schools, buildings in locations which are shortly to be redeveloped, temporary accommodation prior to the building of permanent structures and short-life housing. Calculations of DCF will also generally support this viewpoint. However, it needs to be realised that many temporary buildings erected in the past have far exceeded their design life, sometimes with huge penalties in terms of cost. The prefabricated concrete bungalows constructed in the late 1940s were intended for only a ten-year use but were retained as housing stock for over 40 years.

High initial cost buildings do not automatically result in future cost savings. Improved thermal insulation, the use of self-finished materials and deterioration-free construction will help to reduce costs-in-use expenditure. Some high-quality construction, however, will result in increased expenditure in the future in order to maintain them at this standard. The increased cost content of engineering services has increased the costs-in-use of this element.

In the private sector the question under consideration becomes distorted because of taxation requirements. This tends to encourage low initial cost type buildings, since taxation relief is generally available only against repairs and maintenance items. This also militates against the government's 'Save It campaign' in connection with energy consumption. Because many energy-saving measures have to be paid for in full, without taxation relief, this reduces their attractiveness. In connection with buildings C and D, after taking taxation into account, building D would always be shown to be the best economic choice.

High interest rates also tend to favour low initial costs. High interest rates on high capital cost expenditure result in large mortgage repayments over a long term of years. High interest rates also favour future expenditure, since these sums can be discounted in a smaller proportion.

Both buildings are functionally comparable. In addition, building C is of high-quality construction incorporating many costs-in-use saving measures to reduce future expenditure. High-value thermal insulation has resulted in savings in the running costs of the heating installation. Building D is intended to be more of a prefabricated design, of inferior construction and quality resulting in more extensive repairs and maintenance and high running costs, and necessitating major modifications throughout. Although building D is of a more flimsy construction, demolition is likely to be less expensive because of the possible salvage and scrap value of some materials. On the basis of the data provided, a shorter-life building with higher maintenance costs is the best economic choice. For a longer-life building the result is not conclusive on the basis of these data.

Building C (High initial cost)		10-year life	60-year life
Initial cost		75 000	75 000
Repairs: Year 10 0.55839			
20 0.31180			
30 0.17411			
40 0.09722			
50 0.05429			
60 Nil			
1.19581 × £500		Nil	598
Maintenance £300 × 7.360		2 208	
Maintenance £300 × 16.161			4 848
Running costs £1 200 × 7.360		8 832	
£1 200 × 16.161			19 393
Demolition: Year 10 £3 000 × 0.55839		1 675	
60 £3 000 × 0.03031			91
NPV		£87 715	£99 930
AE		£11 918	£6 183

Compare first of all the shorter life buildings. If building C is to become a serious cost competitor over this lifespan, then either some of its costs are too high or building D's are too low. On examining building C it is not possible to reduce the initial cost, since to do so is likely to have adverse repercussions on the costs-in-use and it is the intention to keep these to a minimum. The question of course must be asked, 'Can building D be constructed for the price suggested?' It should be remembered that prefabricated buildings, although quick to erect, are often not an inexpensive alternative. There is the real possibility, however, that after ten years' usage a buyer will be found and the prefabricated parts of the structure can be transported reasonably economically to a new site. This will increase the attractiveness of the building D proposal. There is also the possibility of adopting an inadequate maintenance schedule so that rapid deterioration will make any resale impossible. It will be observed that maintenance is three times as high on building D, with running costs 75% higher than on building C. This may appear to be on the high side, and this is also in favour of the building D option.

Building D (Low initial cost)	*10-year life*	*60-year life*
Initial cost	45 000	45 000
Repairs: Year 5 0.74726 × £1 500	1 121	
10 0.55839		
15 0.41727		
20 0.31180		
25 0.23300		
30 0.17411		
35 0.13011		
40 0.09722		
45 0.07265		
50 0.05429		
55 0.04057		
60 Nil		
2.83667 × £1 500		4 255
Maintenance £900 × 7.360	6 624	
Maintenance £900 × 16.161		14 544
Running costs £2 100 × 7.360	15 456	
£2 100 × 16.161		33 938
Major modifications:		
Year 15 0.41727		
30 0.17411		
45 0.07265		
0.66403 × £10 000	Nil	6 640
Demolition: Year 10 £4 500 × 0.55839	2 513	
60 £4 500 × 0.03031		136
NPV	£70 714	£104 513
AE	£9 608	£6 667

An alternative method of evaluation is to take the whole-life costs of building C (£87 715) and to deduct from this the total costs-in-use of building D (£25 714), leaving a maximum of £62 001 to be spent on the initial cost of building D. If a tender sum for this building can be obtained below this amount then building D will be the economic choice for the short-life building.

Perhaps the main argument against the possibility of using building D for the longer lifespan is that it may not be capable of lasting for that length of time. The argument becomes somewhat distorted, since the higher quality construction used in building C will provide a better appearance, will be less inconvenient in terms of

Table 18.4   Percentage comparisons

	Building C		Building D	
	10-year	60-year	10-year	60-year
Initial cost	85.50	75.05	63.64	43.05
Repairs	–	0.60	1.59	4.07
Maintenance	2.52	4.85	9.37	13.92
Running	10.07	19.41	21.85	32.47
Major modifications	–	–	–	6.36
Demolition	1.91	0.09	3.55	0.13
	100%	100%	100%	100%

repair and maintenance, and differs very little in terms of the whole-life cost. Where the amounts are so comparable it is clearly preferable to choose the better quality building.

Table 18.4 shows in percentage terms the comparison between the four different projects. Even in the lowest case over 40% of the whole-life costs can be attributed to the initial costs, and typically this represents about 70% using the above assumptions.

Example 4

A contractor is considering purchasing some computing equipment at head office, which will cost £60 000. The installation of this equipment should enable a saving of £9 000, mainly on the salaries of clerical staff. The machine is expected to have a useful life of six years, after which it will have a residual value of £12 000. The rate of writing-down allowances has been agreed at 25% and the tax payable by the firm is 52%.

This information is shown in Table 18.5 as yearly cash flows and NPVs.

The contractor has decided to write down this equipment at an annual rate of 25%, until its eventual disposal by sale. The writing-down allowance is the amount which can be offset against taxation liabilities. On disposal an adjustment will be made by the Inland Revenue to balance the written-down value with the actual value. In the above, £49 321 has been allowed as depreciation whereas the actual sum is only £48 000. The difference between these two amounts then forms a balancing charge which is subject to tax. If the written-down value had been less than the actual depreciation then a further balancing allowance could have been used to adjust against any tax payable.

Cost savings of £9 000 are made in respect of clerical staff employed for each of the six years considered. This cost is assumed to take into account all the costs associated with employing staff for which the employer is responsible. The cost saving is theoretically transferred to taxable profits, but since the tax payable by

Table 18.5 Annual and residual cash flows

Year	Cash saving	Tax on cost saving	Capital allowances*	Tax saving	Cash flow	PV factor for 8%	Present value
1	9 000	–	15 000	–	9 000	0.92593	8 333
2	9 000	–4 680	11 250	+7 800	12 120	0.85734	10 391
3	9 000	–4 680	8 437	+5 850	10 170	0.79383	8 074
4	9 000	–4 680	6 328	+4 387	8 707	0.73503	6 400
5	9 000	–4 680	4 746	+3 291	7 611	0.68058	5 180
6	9 000	–4 680	3 560	+2 468	6 788	0.63017	4 278
7	–	–4 680	(1 321)	+1 851	–2 829	0.58349	–1 651
8	–	–	–	–687	–687	0.54027	–371
							NPV £40 634

*See Table 18.6.

NPV, taking into account capital outlay = £60 000 – £40 634 = £19 366

This amount represents the net cost of the equivalent in today's values.

Table 18.6 Calculation of capital allowances

Capital cost of equipment	=	60 000
Writing-down allowance 25% – Year 1	=	15 000
		45 000
Writing-down allowance 25% – Year 2	=	11 250
		33 750
Writing-down allowance 25% – Year 3	=	8 437
		25 313
Writing-down allowance 25% – Year 4	=	6 328
		18 985
Writing-down allowance 25% – Year 5	=	4 746
		14 239
Writing-down allowance 25% – Year 6	=	3 560
Written-down value		£10 679
Realised on sale		12 000
Balancing charge		£1 321

companies is generally at least twelve months in arrears, the tax will be paid one year later. The capital allowances are as given in Table 18.6, and the same principle applies regarding any tax payment. The cash flow is calculated as cost saving – tax on cost saving + tax saving on capital allowances. Since the balancing amount is a charge, the tax is a negative amount and is thus a tax payment. The calculated cash flows can then be discounted to a present value amount by using the firm's opportunity cost of capital.

The information can be shown for accounting purposes in the 'books' and compared with the cash flow (Table 18.7).

Table 18.7   Comparison of book values and cash flows

Total used in cash flows			Total recorded in accounts		
Increase in net profit before			Profit:		
depreciation £9 000 × 6	=	54 000	Saving	=	54 000
Tax at 52%	=	28 080	Depreciation:		
		25 920	£60 000 − £12 000	=	48 000
				=	6 000
Capital allowances:					
£48 000 × 52%	=	24 960	Tax at 52%	=	3 120
		50 880	After-tax profit	=	2 880
Cash from sale of	=	12 000			
equipment			Cash flows shown by:		
			After-tax profit		2 880
			Depreciation added back		48 000
			Proceeds of sale of equipment		12 000
		£62 880			£62 880

18.4.3 Sensitivity analysis

During a whole-life cost analysis, a large number of assumptions need to be made, for example the discount rates to be applied and the life of the project and its various components. It is necessary to test the sensitivity of these assumptions in order to avoid any possible error in the overall analysis. There is the need to provide the decision-maker with all relevant information which may influence the final outcome of the solution. A way of testing whether the results of the whole-life cost analysis are sound is to repeat the calculations in a methodical manner by changing the values that have been attributed to the above assumptions. This would be a tedious process, but the use of a suitable computer program makes the task simple. The less the final outcome is altered by these changes, the less sensitive is the project to changes in costs and the more reliable the analysis will be.

In making total cost evaluations, four variables need to be tested by sensitivity analysis. These are:

1. Building life
2. Components' life
3. Discount rate
4. Estimate of initial cost

Example 5

Capital cost on investment of new machinery £150 000.
Revenue accruing from above for 5 years £40 000 per annum.
Opportunity cost of capital 5%. Taxation factors have been ignored.

Capital cost	−150 000
Revenue income £40 000 for 5 years at 5% × 4.329	173 160
NPV of scheme	= +£23 160

Sensitivity to changes in costs and revenue:

	+10% on capital cost	−10% on capital cost	+10% on revenue income	−10% on revenue income
Cost	165 000	135 000	150 000	150 000
Revenue	173 160	173 160	190 476	155 844
	+£8 160	+£38 160	+£40 476	+£5 844

Using the above data, only if the capital cost increased by 10% and revenue decreased by 10% would the project fail to show a positive NPV. It would of course be necessary to compare the results against other options which might be available, and a scheme would not necessarily be viable simply because it showed a positive NPV.

Alternatively it may be preferable to calculate the internal rate of return (IRR) for the project. This may necessitate two trial discounts, with the true IRR then being determined by interpolation. It should be remembered that the IRR reduces the NPV of a project to zero (see Chapter 13).

	Trial discount of 10%	Trial discount of 12%
Capital cost	−150 000	−150 000
Revenue income £40 000 for 5 years	+151 640	+144 200
	+1 640	−5 800

The IRR in the above example can be determined by either calculation or graphical methods (see Figure 18.1).

$$\frac{1\ 640}{1\ 640 + 5\ 800} \times (12 - 10) = 0.44$$

The IRR is therefore 10.44%, and this would then be compared with the costs of finance. If the latter were below the IRR then the project would yield a positive rate of return, or if above then a negative value. It is also possible to measure the sensitivity to changes in costs and revenue, and these can be summarised as follows:

+10% on capital costs = −3.6% on return
−10% on capital costs = +4.3% on return
+10% on revenues = +3.9% on return
−10% on revenues = −4.0% on return

Example 6

The following calculations compare two projects with three different interest rates. This example illustrates the fact that high interest rates favour low-cost expenditures, which results in a lower whole-life cost for the project. A comparison of the effect of these interest rates on the net present values (NPVs) of projects X and Y is shown graphically in Figure 18.1.

Fig. 18.1 NPV versus interest rates for projects X and Y

Project X
1. Initial cost   £200 000
2. Annual costs   £20 000 per annum
3. Periodic costs   £30 000 every 35 years
4. Regular costs   £10 000 every 10 years

Life of building 60 years

	5%	10%	14%
1.	200 000	200 000	200 000
2. 20 000 × 18.929	= 378 580		
20 000 × 9.967		= 199 340	
20 000 × 7.140			= 142 800
3. 30 000 × 0.2953	= 8 859		
30 000 × 0.09230		= 2 769	
30 000 × 0.03779			= 1 134

4. 10 000 × 1.45143   =   14 514
   10 000 × 0.62210   =   6 221
   10 000 × 0.36885   =   3 689

	5%	10%	14%
NPV	£601 953	£408 330	£347 623

Project Y
1. Initial cost   £270 000
2. Annual costs   £15 000 per annum
3. Periodic costs   £25 000 every 40 years
4. Regular costs   £7 000 every 15 years

Life of building 60 years

		5%	10%	14%
1.		270 000	270 000	270 000
2.	15 000 × 18.929 =	283 935		
	15 000 × 9.967 =		149 505	
	15 000 × 7.140 =			107 100
3.	25 000 × 0.14205 =	3 551		
	25 000 × 0.02209 =		552	
	25 000 × 0.00529 =			132
4.	7 000 × 0.82370 =	5 766		
	7 000 × 0.31042 =		2 173	
	7 000 × 0.16248 =			1 137
NPV		£563 252	£422 230	£378 369

Example 7

The following example illustrates how interest rates and different amounts of inflation can be considered using the simplified calculation of:

discount rate = inflation − cost of capital (interest rate)

	Project A	Project B
Construction costs	100 000	140 000
Heating (per annum)	3 000	2 500
Lighting (per annum)	3 400	2 200
Cleaning (per annum)	1 800	3 100
Maintenance (every 5 years)	16 000	7 000
Resale value	100 000	140 000

The cost of capital is 18%.
The inflation rates are:
   Heating   11%
   Lighting   9%

Cleaning   8%
Maintenance   10%
Resale value   13%
Expected life 25 years.

Project A – taking inflation into account
  Construction costs   =   100 000
  Heating 3 000 × 11.654
    (PV of £1 per annum for 25 years at 18% – 11%
      = 7% discount rate)   =   34 962
  Lighting £3 400 × 9.823
    (18% – 9% = 9%)   =   33 398
  Cleaning £1 800 × 9.077
    (18% – 8% = 10%)   =   16 339
  Maintenance: Year   5 0.68058
    (18% – 10% = 8%)   10 0.46319
                                15 0.31524
                                20 0.21455
                £16 000 × 1.67356   =   26 777
  Resale 100 000 × 0.29530   = –29 530
    (18% – 13% = 5%)   £181 946

Project B – taking inflation into account
  Construction costs   =   140 000
  Heating £2 500 × 11.654   =   29 135
  Lighting £2 200 × 9.823   =   21 611
  Cleaning £3 100 × 9.077   =   28 138
  Maintenance 7 000 × 1.67356   =   11 715
  Resale 140 000 × 0.29530   = –41 342
                                             £189 257

It should also be noted in calculations of this type that different projects may, for example, have different types of heating systems with fuel costs which inflate at different rates.

Example 8

Your client has investments earning 14%. Last year the client's fuel bill was £6 000 and it is likely that this will rise by 10% per annum. The client intends to remain in these present premises for the next five years and has decided to withdraw money to spend on fuel-saving improvements. The following likely possibilities have been identified:

	Initial cost	Fuel saving
	£	%
Double glazing	20 000	unknown
Roof insulation	1 500	5.0
Thermostatic heating valves	1 000	7.5
Pipe lagging	2 000	2.5
Cavity insulation	to be agreed	10.0

Evaluate each item and advise the client on the action to be taken. What fuel saving would need to be made to justify the expenditure if the double glazing was selected? Recommend a cost limit for the installation of the cavity insulation.

Fuel costs
Last year the fuel costs were £6 000 and these are rising at 10% per annum. These increases may be due to a variety of different reasons, which might include also inflationary factors. For each of the next five years these are as follows:

	£
Year 1	6 600
Year 2	7 260
Year 3	7 986
Year 4	8 785
Year 5	9 663
	40 294

These values can be discounted to a net present value (NPV) by multiplying each year by the respective discount factor from the present value of £1 table, which represents the opportunity cost of capital. The following thus represent the above fuel costs, but discounted to today's amounts:

Year 1	6 600	×	0.87719	=	5 789
Year 2	7 260	×	0.76947	=	5 586
Year 3	7 986	×	0.67497	=	5 390
Year 4	8 785	×	0.59208	=	5 201
Year 5	9 663	×	0.51937	=	5 019
					£26 985

The above therefore represents the NPV for the fuel costs, allowing for both interest and inflation. This sum can easily be converted to an annual equivalent by dividing this total by the factor from the present value of £1 per annum table for five years, as follows:

$$\frac{26\,985}{2.914} = £9\,260$$

It would be incorrect to divide £26 985 by five years, since this will not take into account the time value of money. It is equally incorrect to use £40 294, since this does not allow for the various sums of money being paid at different times. Consider, for example, a single payment of £40 294 in year five with the various payments made in years 1–5. The time values are evidently different.

Double glazing
Initial cost = £20 000
NPV of fuel costs = £26 985

The saving to be achieved on fuel if double glazing is used is:

$$\frac{20\ 000}{26\ 985} = 74\%$$

Since double glazing normally results in a saving on fuel expenditure of between 5% and 10%, this improvement is not considered to be in any sense cost-effective. However, the installation of double glazing will probably assist the client to sell the property and for a higher price in five years' time, and this may need to be considered.

Roof insulation
Initial cost = £1 500
Saving on fuel of 5% of NPV £26 985 = £1 349

Within the five-year period this would not be cost-effective and should therefore not be carried out. This is marginal. This is also unlikely to affect significantly the selling price of the property after five years. There may be some possibility of obtaining a local authority grant for this work, if the property is a private dwelling. This might have the effect of making such an investment worthwhile.

Thermostatic heating valves
Initial cost £1 000
Saving on fuel of 7.5% of NPV £26 985 = £2 024

This is a cost-effective solution and should be carried out.

Pipe lagging
Initial cost £2 000
Saving on fuel of 2.5% of NPV £26 985 = £674

This is not a cost-effective solution.

Cavity insulation
Saving on fuel of 10% of NPV £26 985 = £2 699

The insulation costs are to be negotiated with an installer. If a price of below £2 631 can be agreed then this will be a worthwhile measure to undertake. It should be remembered that predicting future costs is always difficult and uncertain. Where the amounts calculated for fuel savings are close to the initial cost of the fuel-saving improvements, such as in the case of the roof insulation, it may be desirable to test the results using the technique of sensitivity analysis.

Example 9

A local church is currently considering modernising and extending its premises to provide for its increased congregation for both social and worship purposes. It anticipates an increase in membership from 340 to 600 and assesses that this

will provide an extra weekly income of £10 per week per new member.
The following costs are associated with this proposal:

	£
Modernisation	300 000
Additional repairs allowances	3 000 per annum
Renewals (every 7 years)	10 000
Extension heating costs	10% of capital building costs per annum
Additional cleaning/insurances	1 000 per annum
Furniture and fittings	50 000
Refurbishment (every 14 years)	30 000

What is the maximum amount that can be made available for building a proposed extension?

Income
With questions of this type two important factors need to be immediately considered: the expected life of the project and the discount rate to be applied. Each of these can of course be amended later using sensitivity analysis techniques. One would assume that the extension would have a life expectancy of at least 60 years, but this time horizon is too far ahead, since the funding will probably have to be borrowed against the above projections. Financial lenders might also be more cautious about the future and the ability to make future payments. A term of probably no longer than 20–25 years would seem to be appropriate. The other important consideration is the discount rate to be applied. Whilst rates, in the UK, are currently running at a level not seen for over 50 years, these might not be expected to remain at these levels for even the immediate future. The following calculation can easily be repeated, varying these values as required.

The following calculation is based on 25 years' life and a discount rate of 8%.

Additional income 600–340 = 260 × £10 × 52	=	135 200

Discounted to a net present value

8% for 25 years = 10.675 × £135 200	=	£1 443 260

This amount must therefore cover the following costs:

Modernisation		=	300 000
Refurbishment at year 14 £30 000 × 0.34046		=	10 214
Renewals at year	7 0.58349		
	14 refurbish*		
	21 0.19866		
	0.78215 × £10 000	=	7 822
Repairs (annual) £3 000 × 10.675		=	32 025
Cleaning and insurances £1 000 × 10.675		=	10 675
Furniture and fittings		=	50 000
			£ 410 736

* This always causes some debate. It depends upon what the different costs are meant to cover. For this client, any refurbishment would include renewal costs.

Thus £1 032 524 (1 443 260 – 410 736) is the amount that is available for construction costs and future heating bills.

The construction costs would include professional fees for the architect, the structural engineer and the quantity surveyor if the church wanted an independent design approach. Otherwise these costs could be absorbed within a design-and-build arrangement. Different advice and preferences would determine which method was the most appropriate.

Assuming that a traditional single-stage tendering approach was to be used then 10% might be required to cover the various professional fees that are involved.

If BC = the building costs, then the heating costs are 1/10 of this and the professional fees are also 1/10 of the building costs. However, the heating costs will need to be discounted over the 25-year period at 8%. Therefore:

$$BC + (1/10BC \times 10.675) + 1/10BC = £1\ 032\ 524$$

$$BC + 1.0675BC + 0.1BC = 2.1675 = £1\ 032\ 524$$

$$BC = £476\ 366 = \text{the amount that is available for the actual construction costs.}$$

Check

Building cost	=	476 366
Professional fees 10%	=	47 637
Heating £47 637 × 10.675	=	508 525
		£1 032 528

(Note: difference due to rounding errors.)

18.5 FORECASTING CHANGE

18.5.1 Technological change

It is difficult to forecast with any degree of accuracy the changes in technology, materials and construction methods that may occur even up to the end of the first decade of the 21st century. The construction industry, its process and its product are under a purposeful change and evolution. There is a constant striving to develop excellence in both design and manufacture and to introduce new materials having the desired characteristics of quality and reliability in use. The changes in technology can often be sudden and unexpected, and sometimes prototypes which fail initially when used in practice are eventually refined and improved to produce a worthwhile product. The introduction of new technology and good solutions to age-old problems can have a major impact on the whole-life cost forecasts and the pursuit of whole-life construction economy.

18.5.2 Fashion changes

These changes are less gradual and more unpredictable than changes in technology, and are also subject to a degree of speculation. Themes, within the construction industry, have been developed in different eras, such as 'built to last', 'inexpensive initial cost', 'industrialisation', 'long life, loose fit, low energy', and the present attitudes towards refurbishment. Changes, for example, in the type and standards of provision, the use of space or the level of quality expectations can be observed from the historical study of buildings. Changes in the way that buildings might be used in the future are already predicted. Some of these are hopelessly fanciful. Others reflect an attitude to work and leisure, changes in an individual's personal expectations, demographic trends and developments generally in society. A whole-life cost analysis must, however, attempt to anticipate future trends and their future effect on the overall economic solution. Fashion changes are the result of the desire to provide something new, sometimes solely to address a reason for change. In other cases they arise because of our social awareness and perception of human development and advancement. A whole-life cost analysis which considers only the status quo is of very limited value in practice.

18.5.3 Cost and value changes

The erratic pattern of inflation throughout the past 50 years could not have been predicted even a decade earlier. The high inflation percentages experienced during the 1970s would not have been thought a possibility in the 1950s. An examination of building tender prices throughout the 1960s and 1970s indicates a general upward trend in values. This pattern has existed since the end of the depression of the 1930s. In the early 1980s, however, tender price levels showed a downturn, which at the time was an unusual and unexpected phenomenon, since the preceding years had been financially difficult times for builders and contractors. The more recent variability of oil price levels illustrates how volatile the market place really is. Inflation rates and interest rates are intertwined and influenced by such factors. Slumps follow booms and vice versa, but seem to be beyond the scope of present indicators and predictors. Costs and values do not move in tandem; indices for the different materials, products or components do not follow similar patterns but are subject to wide degrees of fluctuation (see Tables 7.3 and 7.4). Economists have indicated that costs and prices cannot be assumed to rise indefinitely, and that there may be a future lapse or even a reversal of the traditional historical patterns.

18.5.4 Policy-making and decision-making changes

One of the most important whole-life costing variables is the future use and maintenance policy of the project by the owner. This factor is generally absent from the sparse historic data sources available. It is now widely recognised, for example, that maintenance work is not needs-orientated but budget-led. The maintenance work carried out is thus largely determined by the amount of funds available, and once these have been expended no further amounts are available until the following year's budget allocations have been determined. There may therefore be only limited value in comparing the whole-life costs of, say, wall tiling with those of repainting, in the absence of such a policy. The tiling may be shown to be the economic choice, but if the owner, due to a shortage of available funds, does not repaint the walls at the intervals stipulated in the whole-life cost plan then the economic comparison may prove to have been at best optimistic or even a false assumption. The policy of the owner and the use by the occupants are likely to be at least as important as the theoretical design and construct values in the determination of the relevant maintenance costs.

The way in which owners and occupiers use and care for their buildings or other structures also needs to be considered. The desire for proper maintenance of the physical asset is influenced by the costs and inconvenience involved. Different owners will set differing priorities. Proper maintenance cannot be assumed on the historical precedents of apportionments of other buildings, unless it is certain that the uses and priorities are compatible. Studies emphasise that maintenance cycles and their associated costs must first be set properly within the maintenance objectives of the particular organisation concerned and the policies employed for planned and responsive maintenance.

18.6 HISTORICAL PERSPECTIVES

It is often presumed that whole-life costing will assist in the selection of the most economic solution for a design, taking into account all the costs associated with that project. A brief precursory glance at the past suggests that this may not always be the case. Consider, for example, a simple exercise concerned with the evaluation of timber and cast-iron rainwater gutters which might have been made at the beginning of the twentieth century. Cast iron would have been selected as the economic solution largely because of its durability and low costs-in-use when compared with timber. However, within a few years of such a decision being taken a new material now known as PVC had been discovered for use in gutters. The correct economic solution based on hindsight and historical fact would have been to install the timber gutters and when replacement became necessary to renew them with PVC.

Flat roofs are out of fashion today, primarily because of their apparent short life, high repair cost when compared with pitched roofs, and low reliability. Whole-life cost calculations do not generally favour them, even under the most optimistic conditions, when compared with an inexpensive pitched roof construction. The recommendation today, therefore, after all the economic considerations have been examined, is to choose the latter. However, it is possible that within a few years material scientists may discover or invent a material for flat roofing that is inexpensive, highly durable and reliable and has a higher life expectancy and lower costs-in-use than those of even moderately priced pitched roofs. The correct economic choice may therefore be to install the cheaper alternative flat roof construction, and then replace it after its expected short life with this yet-to-be-discovered material.

The provision of insulation in buildings is a reflection of the relationship between the annual cost of fuel for heating purposes and the initial cost of the insulation. The search in recent years for alternative and less expensive forms of fuel has yet to be realised. When these are discovered, many of the present levels of insulation in buildings may become redundant in terms of their cost-effectiveness. The real reduction in the price of fossil fuels and other energy sources in recent years, together with the added efficiency of mechanical heating plant and equipment, gives this argument some validity.

The illustrations above, which range from fact to fiction, indicate that it is possible to use whole-life costing to help us select the wrong economic option in a total-cost appraisal. It is worth noting that such a choice could have been made even in ignorance of this technique! The question, however, that needs to be answered is 'if the technique is applied to projects constructed tomorrow, will the long-term desired objectives be achieved?' The technique does not remove from the user the responsibility to apply judgements and to make decisions, but it needs to offer a reliable analysis on which to base those decisions.

CONCLUSIONS

The importance of attempting to account for future costs-in-use in an economic appraisal of any construction project has already been established in theory. The question of whether whole-life costing works in practice is of crucial importance. The general belief is that whole-life costing, when applied to capital works projects, will enable the selection of the most economic solution over the project's whole life. This might not be so. If it can be shown, for example, that whole-life costing might have encouraged the choice of the least economic alternative, then its continued use in practice and its further development become of questionable worth for clients and their practitioners. If the forecasts are unreliable because of an absence of appropriate data, this should be a problem that can be remedied, at least in the long term, by properly assembling the data sets with the appropriate characteristics. If estimates are misleading because they rely on the myth of being able to forecast the future, then the efforts in evaluating alternative designs and methods of construction might be better spent in considering other more suitable techniques. The world is now undergoing a rapid change, with new technologies affecting all aspects of society. Values in society are also under constant scrutiny and evolution, and it is virtually impossible to predict how these factors might influence the future. What is certain is that these aspects do affect whole-life cost predictions. In the past some of these would have been at best misleading.

Whole-life costing does offer some potential. Its philosophy of whole-cost appraisal is certainly preferable to the somewhat narrow initial cost estimating approach. The widespread efforts so far expended in its research and development are a positive move, but more research is necessary to sharpen up the realities of the problems encountered. Also, there is sometimes an eagerness to introduce a new method of evaluation without being fully aware of all the facts. Improving the education of those responsible for the design of capital works projects and encouraging them to consider the future effects of their design and constructional details is an urgent priority. Educating owners and users in how to obtain the best from their buildings is another useful course of action to follow. The implementation of maintenance manuals or building owners' handbooks might also provide an improvement in the performance of buildings in use. At this stage it is doubtful whether a large amount of emphasis or importance should be placed on the actual numerical results, due to the vagaries within the calculations. Although the use of this technique in practice will, it is hoped, continue to increase for the reasons described above, this must be done with some caution until results achieved in practice can be verified.

Whole-life costing is at best a snapshot in time, in the light of present-day knowledge and practice and anticipated future applications. Some of the factors involved are of a crucial nature and can only be tested over a range of known values. Others are currently beyond our expectations, may not even be regarded as important today, and may not come to light until observed in practice at some time in the future. Some of the assumptions may also be untenable in practice. Are we asking too much of the technique?

SELF ASSESSMENT QUESTIONS

1. Describe the relationship between whole-life costing and costs-in-use.

2. Explain the importance of applying sensitivity analysis to whole-life costing calculations.

3. Recognising that the future cannot really be predicted, is whole-life costing a technique that can usefully be applied to the future costs of buildings?

BIBLIOGRAPHY

AIA, *Life Cycle Cost Analysis: A Guide for Architects*. American Institute of Architects 1977.

Ashworth, A., 'Life cycle costing – can it really work in practice?', *Proceedings of Building Cost Modelling and Computers, University of Salford*. Spon 1987.

Ashworth, A., 'How life-cycle costing could have improved existing costing', in *Life Cycle Costing for Construction* (ed. J.W. Bull). Blackie Academic and Professional 1993.

Ashworth, A. and Au-Yeung, P., 'The evaluation of life cycle costing as a practical tool during building design', *CIB Proceedings of the Fourth International Symposium on Building Economics*, Vol. A, 1987.

Best, R. and De Valance, G., *Building in Value*. Arnold 1999.

Dell'Isola, P.E. and Kirk, S.J., *Life Cycle Costing for Design Professionals*. McGraw-Hill 1981.

Flanagan, R., *Whole Life Costing*. Blackwell Publishing 2004.

Flanagan, R. and Norman, G., *Life Cycle Costing: Theory and Practice*. BSP Professional Books 1989.

Flanagan, R., Norman, G. and Furbur, D., *Life Cycle Costing for Construction*. Surveyors Publications 1983.

Hardy, S.C., Norman, A. and Perry, J.G., 'Evaluation bids for construction contracts using discounted cash flow techniques', *Proceedings of the Institution of Civil Engineers*, February 1981.

Jelen, F.C. and Black, J.H., *Cost and Optimisation Engineering*. McGraw-Hill 1983.

Ministry of Public Building and Works, *The Decision to Build*. HMSO 1970.

Orshan, O., 'Life cycle cost: a tool for comparing building alternatives', *Proceedings Symposium on Quality and Cost in Building, Lausanne*, Vol. 1, 1980.

Royal Institute of British Architects, 'First costs or life cycle costs', *Proceedings Conference at the RIBA*, 1983.

Royal Institute of British Architects, *Life Cycle Costs for Architects, a Draft Design Manual*. College of Estate Management 1986.

Royal Institution of Chartered Surveyors, *A Guide to Life Cycle Costing for Construction*. Surveyors Publications 1986.

Royal Institution of Chartered Surveyors, *Life Cycle Costing: a Worked Example*. Surveyors Publications 1987.

Smith, G. *et al.*, *Life Cycle Cost Planning*. Society of Chief Quantity Surveyors in Local Government 1984.

Southwell, J., *Total Building Cost Appraisal*. Royal Institution of Chartered Surveyors 1967.

Southwell, J., *Total Building Cost Appraisal*. Research and Information Group of the Royal Institution of Chartered Surveyors 1972.

CHAPTER 19

VALUE MANAGEMENT

LEARNING OUTCOMES

After reading this chapter, you should have an understanding about the principles and practice of value management as it is used in the construction industry. You should be able to:

- Outline the origins and expectations of value management
- Distinguish between value engineering, value analysis and value management
- Understand the methodology used in value management
- Realise the importance of cost–value reductions in building in eliminating unnecessary costs
- Identify the different techniques that are used in value management studies
- Recognise the wider implications of securing value for money

19.1 INTRODUCTION

Everyone involved in the construction industry should be concerned with value for money. One of the principal objectives of any design solution for a proposed building or civil engineering project is to provide value for money for the client or owners of the project. This is of course only one of the criteria by which the project will be evaluated but it does encompass many of the project's required attributes. The search for value for money is nothing new. During the 1960s one of the aims in the development of the techniques associated with cost planning (see Chapter 15) was to help to secure value for money. It can be reasonably argued that the introduction of cost planning was partially successful in this respect. Project designs were more carefully examined and alternative design solutions were developed to meet the client's objective of value for money. The introduction of cost limits for public sector projects also helped to eliminate wastage in design and construction and to some extent to reduce construction costs. It was able to do this, while still maintaining standards of quality, through more carefully considered design solutions that fully met client objectives. Cost planning as originally envisaged and cost limits paid only scant regard to long-term, whole-life costs. However, while the use of such techniques helped to introduce changes in design methodology and culture they only partially succeeded in improving value for money. The importing of value (engineering and analysis) techniques from the USA in the 1980s would bring further developments aimed at improving value for money.

The classical definition of value is: to obtain a required level of quality for the least cost or the highest level of quality for a given cost. This definition implies that there must be a more cost-efficient way of achieving the same objective. While one speaks of optimum design solutions this can only ever be within the limited knowledge and capabilities that currently exist. When an optimum solution is found it is never very long before someone is able to improve it! Value is about doing more for less, an attitude and practice that will continue to gain momentum into the early part of the next century. It is now common in all walks of life.

19.2 ORIGINS OF VALUE ENGINEERING

The origins of value engineering are generally attributed to Lawrence Miles of the General Electric Company (GEC) in the USA during the Second World War. The technique has since been used extensively in a variety of industries and situations. While it was originally applied to the purchasing function within GEC, it soon became a part of the manufacturing and production processes that were employed. Value engineering was an innovation resulting from a shortage of materials and other resources due to wartime activity. Out of necessity the company began to search for alternative materials and to substitute these wherever possible in their designs and processes. Surprisingly, they found that many of these materials not only did the job, but in a great many cases they offered superior performance for a lower cost. Because of this the application of the technique was extended, developed and made more formal. In order to improve product efficiency they began to intentionally develop substitute materials and methods of manufacture to replace the hitherto more expensive materials and components that had been used in their manufacturing business.

The application of value engineering in the construction industry is supposed to have started in the USA in 1963. The use of the technique in that industry spread rapidly and by 1972 the US General Services Administration required that a clause on value engineering was to be included in all public sector construction contracts (see Table 19.1). The estimated cost savings reportedly ran into millions of dollars. Value engineering in the UK construction industry followed much later and did not really gain any momentum until the late 1980s. The Royal Institution of Chartered Surveyors published a report of a study of *Value Engineering and Quantity Surveying Practice* in 1987.

19.3 TERMINOLOGY

Like many embryonic subjects it takes time for a body of knowledge to be developed and during this time the terminology that is used often adapts and changes in

Table 19.1   Value engineering activity introduced by US government agencies

Date	Government department
1963	Department of Defense: Navy Facilities Engineering Command
1965	Department of Defense: Army Corps of Engineers
1968	Facilities Division: National Aeronautics and Space Agency
1973	General Services Administration: introduction of value engineering service contract clauses
1974	General Accounting Office: identified need for increased use of value engineering
1976	Environment Protection Agency: value engineering mandatory on projects over $10 million

Source: Adapted from Kelly and Male (1993)

meaning as new directions are considered. The following are some of the definitions that have been used for value engineering by key writers on the subject:

- *Crum, 1971*: A disciplined procedure directed towards the achievement of necessary function for minimum cost without detriment to quality, reliability, performance or delivery
- *Macedo* et al., *1978*: The systematic review and control of costs associated with acquiring and owning a facility or system
- *Dell'Isola, 1982*: The creative organised approach whose objective is to optimise cost and/or performance of a facility or system
- *Kelly and Male, 1988*: An organised effort to attain optimum value in a product, system or service by providing the necessary functions at the lowest cost
- *Green, 1992*: A systematic approach to delivering the required functions at the lowest cost without detriment to quality, performance or reliability

The debate about the subject used to centre around the distinctions made between value engineering and value analysis. The definition given for value analysis is:

- *Miles, 1972*: Value analysis is an organised approach to providing the necessary functions at the lowest cost

The subject is now more correctly referred to as value management, where value engineering and value analysis are incorporated as a part of the overall process. Value management is a strategy for identifying the project that provides the best value for money through the best use of the limited resources that are available. A definition for value management is:

- *Connaughton and Green, 1996*: A structured approach to defining what value means to a client in meeting a perceived need by clearly defining and agreeing project objectives and how they can be achieved

Table 19.2 The job plan (after Miles)

Phase	Title	Description
1	Orientation	What is to be accomplished
2	Information	Provision of drawings, specifications, quantities, costs, methods etc.
3	Speculation	Consider alternative solutions
4	Analysis/evaluation	Analyse costs of alternative solutions
5	Development	Accepted ideas are considered in further detail
6	Selection	Refined ideas are further developed
7	Conclusion	Proposals are presented to the client

Source: Miles (1972)

19.4 METHODOLOGY

A standard methodology for the application of value management has become widely established and is referred to as the job plan following the outline described by its originator, Miles (1972), as shown in Table 19.2.

19.4.1 Phase 1 Orientation

Phase 1 is the introductory phase when it is being decided just what is to be accomplished. It will seek to separate the client's needs from the client's wants and to establish the objectives and the desirable characteristics of the proposed project. It allows everyone who is involved in the project to understand the issues and constraints.

19.4.2 Phase 2 Information

In Phase 2 as much as possible of the information appertaining to the project is collected together. The objective of this is to identify the functions of the whole or parts of the project, as seen by the client. As much factual evidence as possible should be collected. The quality of the decision-making is based upon the reliability of this information. It will seek to separate client wants from client needs, the project constraints, budget limits and the time available for both design and construction.

19.4.3 Phase 3 Speculation

Phase 3 is the creative phase of value management and engineering. The team along with the value management consultants seek to develop ideas for the project. Research has indicated that good and original ideas are just as likely to come from any member of the team as from the individual experts. Some methods used to aid the group ideas are considered later in this chapter.

19.4.4 Phase 4 Analysis and evaluation

Phase 4 forms a crude filter for reducing the ideas that have been generated to a manageable set of propositions. The value engineering and management team will analyse what has been suggested and use appropriate evaluative techniques to eliminate the various options.

19.4.5 Phase 5 Development

The ideas that have not been eliminated during Phase 4 are now examined in detail. Outline designs will be prepared and technical and economic feasibility and viability studies will be carried out. At the end of this phase the team will consider what has been achieved and other ideas that may have been introduced during the developmental phase.

19.4.6 Phase 6 Selection

The refined ideas will now be carried forward towards the final proposal along with working drawings, calculations and costs.

19.4.7 Phase 7 Conclusion

The final phase involves presenting the findings to the client. Throughout the whole process the client will have been kept informed about progress and possible solutions. The final result should not therefore present the client with any unwanted surprises.

19.5 VALUE MANAGEMENT WORKSHOPS

The 40-hour workshop is the most commonly adopted approach used for a value engineering study. The 40-hour workshop is usually carried out at a point when 35% of the design has been completed. There will be a considerable amount of information to assimilate within this time period in terms of the overall concepts, design, specification and costs. In addition, the group dynamics involved, unless the team are familiar with each other, will take some time to get used to. It can be argued that the appropriate duration of a value engineering workshop is really dependent upon the scope of the study. Case studies of value engineering in practice indicate that the period of time required is frequently in excess of the supposed 40-hour workshop. Familiarity with the value engineering process could result in a shorter period of time. Other workshops have been arranged on a 3+2+2 day basis, a total of 56 hours. In this case a three-day intensive study is proposed followed by a break of about one week, two days' study followed by another week's break and finishing off with a further two days of study.

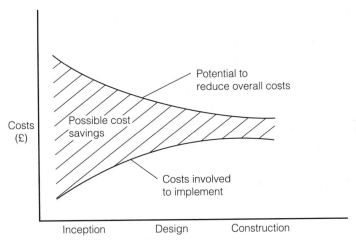

Fig. 19.1   Opportunity to change a design

19.6 COST–VALUE REDUCTIONS

Cost–value savings can be made on a project at any time from inception to completion. However, it is easier to make such changes during the earlier stages of a project than when the project design or the construction phase is nearing completion. Designers are reluctant to become involved in abortive design work, because it is demotivating and is often done at their own expense. Changes to design are more difficult because of the possible knock-on effects with other aspects of the project. It is also not a good policy to spend more on preparing design changes to achieve cost reductions that are less than the administrative costs involved, unless this also achieves some longer-term cost savings. Figure 19.1 illustrates the likely opportunity to revise or make changes to the design of a project. Figure 19.2 suggests that, as the project develops, the ability to change costs decreases rapidly during the design stage.

19.7 THE 'WHAT?' SCENARIOS

19.7.1 What is it?

A prerequisite of functional analysis is the selection of the function to be analysed. It is the consensus view of the specialists that the most productive areas to be selected for the analysis are the ones that are the most expensive. Some knowledge of the costs of the components in a project are therefore important. Those parts of a project that are described as cost-sensitive are likely to be the most fruitful areas for study. In a typical bill of quantities, for example, 80% of the cost is frequently contained in only 20% of the items. The value managers should therefore quickly concentrate on these sorts of items in order to be the most effective.

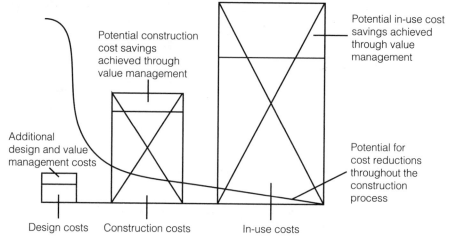

Potential construction
cost savings
achieved through
value management

Potential in-use cost
savings achieved
through value
management

Additional
design and value
management costs

Potential for
cost reductions
throughout the
construction
process

Design costs   Construction costs   In-use costs

Fig. 19.2   Declining influence on costs

19.7.2 What does it do?

This is the key value management question. It requires the definition of the
function under study. The form of this definition is prescribed by established
value analysis procedures as two words, a noun and a verb. The simplicity of
the approach is deceptive. Selection of the proper two-word description is quite
often difficult, requiring a comprehensive understanding of the items under study.
The most obvious functional analysis is not necessarily correct, and therein lies the
potential for successful analysis. The description is not limited to a single two-word
description; a series of descriptions can apply to the same item. For example, it can
be suggested that the doorway rather than the door is the true basic reasoning for
positioning the door in a given location. The basic purpose of the doorway will be
different in different locations. In an office building it may be to provide access;
in a prison its function may be to control access; the purpose of a fire door may
be to control the spread of fire.

For example, the function of a door in its normal form might be as follows:

Permit access
Prevent access
Insulate from noise

In the case of an external glazed door we might want to add:

Provide security
Allow view
Stop draughts

The many different kinds of doors such as garage doors, revolving doors, fire doors,
automatic doors etc. would allow these lists to be extended. The logical analysis of

the function of a component or element will provide new information to the design team and help to clarify the client's own objectives. There are a number of techniques available to elicit such information, such as the functional analysis system techniques or FAST diagramming.

19.7.3 What is it worth?

Worth is a measure of value and represents the least expenditure required to provide the defined function. This evaluation can be quite specific, but it does include elements of subjective judgement. A method of measuring this worth is the identification of the cost-comparable items that provide a similar requirement. If a building or engineering structure is to be evaluated on its value, then the worth will be in that building or structure. However, the human interaction with buildings requires a subtle and significant consideration. Emphasis will be placed not only on the initial construction costs but also on the project's whole-life costs.

19.7.4 What does it cost?

When the item has been identified sufficiently so that quantities of work can be measured, then it is possible to calculate their cost and compare them with their worth. A value index can be found by dividing cost into worth. The construction industry in the UK is fortunate in that it has a comprehensive databank of construction costs and prices. Comparison figures are therefore readily available, and information on inflation and cost escalation is easily accessible to convert these costs and prices to different timescales.

19.7.5 What else will do?

Once the functional definition has been established and the worth and cost have been assessed, the next key stage is to assess the alternatives which would also perform the value function.

19.7.6 What does that cost?

These basic value questions are asked many times during the course of the value management of a project. Each improvement is then analysed in a similar way, in order to select the option that provides the best value for money.

19.8 UNNECESSARY COSTS

The main purpose of value management is the reduction of unnecessary costs. A well-developed understanding of the nature of construction costs is therefore required prior to embarking on a value management study of a particular project. Some of the costs involved in construction projects are unnecessary and these are the costs that are targeted through value management. They have been defined in various ways. They may be costs that do not make any meaningful contribution to the project, costs that add no intrinsic value to the project or costs that do not add any of the attributes of quality, function, appearance, life expectancy or client requirements.

Projects can be designed and constructed in many different ways. Each different design attracts particular costs. Where two different designs satisfy the main client requirements then the difference between the costs of these designs can be described as an unnecessary cost.

The most obvious occurrence of unnecessary cost is when a particular building component included within a project serves no real function, e.g. the provision of tile floor finishes in buildings where users would subsequently cover them with carpets. The purpose of value management techniques is not to reduce the aesthetic appeal of a design, since this appeal would in the designer's opinion add intrinsic value to the project.

The second category is where costs are expended on unnecessary materials or where an inexpensive material could be used in their place. For example, an external wall of a building is to be constructed in external facing bricks and then rendered. There seems to be no point in paying the additional costs for the facing bricks where they are subsequently to be covered with the render. A cheaper material could therefore be used.

The importance of buildability in scheme designs is now well understood in theory but not always completely in practice. The extent to which the design of a building facilitates ease of construction can result in the project being constructed more quickly and economically on site. Where buildability is not considered at the design stage then this may result in a third category of unnecessary costs.

The emphasis of controlling building costs in the past has always been directed towards initial building costs alone. Too little emphasis has been placed on the longer-term application of cost control through whole-life costing. Any design today that ignores the expenditure on future costs of whatever kind will be guilty of introducing a further category of unnecessary costs.

An investment appraisal of a construction project is often based on a consideration of costs and revenues. For example, when considering the costs of two designs for a new office project it is necessary to balance the costs against the possible revenue that the project will provide. If the design of one of the schemes provides a greater amount of lettable floor area for the same cost as the alternative design, then, all other things being equal, this project should be selected since it adds the greater value. These costs are sometimes termed opportunity costs and represent a fifth unnecessary cost when they are not considered.

19.9 CREATIVITY

Once a workable solution to a particular design has been achieved then it becomes extremely difficult to persuade designers to change their minds. They become closed to other solutions and defend their original concepts against most arguments that are put forward. Once a scheme has been committed to paper, changes proposed by others are resisted. Some designers claim that the principles involved in value management should already have been considered as a normal part of the design process. Clearly as more designers become fully aware of the practice and implications of value management then these will begin to influence the shape of the design. However, no matter how good the designer is who works on a scheme, the involvement of others will often bring positive aspects to the design. Sometimes it takes a third party to point out and identify weaknesses in a design, such as unnecessary costs. Also the continuing success of value engineering studies and applications with the identification of solutions that were not previously considered provides considerable evidence in support of value management.

Edward De Bono has made significant contributions towards the area of creativity by identifying two different kinds of thought processes. Vertical thinking is described as the traditional thought process which results in progressing from one step to another. Lateral thinking seeks to break out of that progression by making a conceptual step sideways. De Bono compared vertical thinking with the process of digging a hole deeper and lateral thinking was seen as finding somewhere else to dig the hole.

The search for creativity can also benefit from the traditional value management approach which requires teams to work together through intensive group activity. In this way ideas can be developed that were beyond the knowledge or experience of the original designers. Creativity can also be further enhanced through the synergy of the group where participants not only contribute their own ideas but are also able to build on the ideas of others.

In order for group activity to work effectively it is necessary to develop actions that encourage creativity rather than those that discourage it. The following are some of the actions that encourage:

- Protect those in the group who are vulnerable
- Listen to others' points of view
- Eliminate status or rank
- Value the learning in mistakes
- Set up win–wins
- Share the risk
- Assume it can be done
- Take on faith

The following are some of the actions that discourage:

- Pull rank
- Get angry
- React negatively
- Be cynical
- Correct others
- Point out only the flaws
- Blame
- Be impatient

The most popular technique in achieving group creativity is *brainstorming*. It works best on simple problems or those that are well defined. Other methods may include:

- *Checklisting* – this seeks to encourage creativity in a systematic way through Who?, What?, Why?, How?, Where?, When?
- *Delphi technique* – individuals are able to submit anonymous ideas to the group in writing
- *Gordon technique* – a topic for discussion is chosen that only loosely relates to the actual purpose of the study
- *Synectics* – this advocates the use of analogies in order to aid the creative process
- *PMI* – *p*lus, *m*inus and *i*nteresting points
- *CAF* – *c*onsider *a*ll *f*actors
- *TEC* – *t*arget and *t*ask thinking, *e*xpand and *e*xplore, *c*ontract and *c*onclude

19.10 OTHER ISSUES TO CONSIDER

19.10.1 Cost savings

Reports argue that where value management is carried out effectively then savings of between 5% and 25% are possible. With tightly designed projects the percentage is likely to be on the lower side. Traditional quantity surveying services are suggested to save clients between 5% and 10%, although this also depends upon the type of project and the nature of the client organisation. Quantity surveying services during the past 50 years have been helping to erode construction costs by evaluating design and construction solutions that have huge repeatable consequences throughout the industry. The introduction of new materials, methods of construction and methods of working is constantly changing the status quo in this respect.

19.10.2 The value management team

Members of the design team should be fully involved as a part of the value management team. This is then likely to avoid abortive design work that might otherwise be carried out in parallel to the value study. However, they will bring with them their own preconceptions of their design and may need persuasion to change their proposals.

19.10.3 Possible disruption to the process

The fear of many employed in the construction industry is that introduction of value management along with other processes will create more problems than it claims to solve. It is especially time-consuming on small projects and may not be valuable unless such projects offer repeatable benefits across a range of projects. Most will accept that value management is likely to effect changes which are of a beneficial nature.

19.10.4 Design liability

Design liability is a particularly sensitive issue. Designers will argue that the introduction of value management may derive design solutions for which they will need to accept the long-term liability and responsibility. Since designers remain ultimately responsible for the design, changes or recommendations must be accepted by them. It is not sensible to attempt to separate design liability from value management responsibility particularly at a time when many clients are calling for single-point responsibility for their construction projects.

19.10.5 Procurement

There is no reason why value management techniques should not be included in projects whether they are designed using a traditional model or a contractor-orientated approach.

19.10.6 Effects on project duration

An important question is whether the introduction of value management will extend the overall project duration from inception to completion. If design activities are suspended during the value management process then clearly the design process will take longer. The concept that value management will save costs (or improve value) should suggest that the contractor will be on site for a shorter period of time. There is some evidence to suggest that devoting more time to getting the design correct before construction starts on site will help to reduce the contract period and hence provide for possible savings in the contract cost.

19.10.7 Benefits of an independent review

There are overriding benefits of an independent review of a design, before it is constructed or manufactured. Value management provides such an independent review and where this can be included in a non-critical way then real benefits can usually be achieved simply by considering someone else's perspective.

19.11 SECURING VALUE

The briefing process sets out quality expectations from the client. It is inadequate in today's world simply to consider the aesthetics, durability and function of the project. It is important to consider the long-term value that the client can expect to derive from the construction project. Most major clients derive value from a development that contributes to their overall business aspirations. Without clear overall direction, project teams can focus more on design and construction issues than on really meeting the client's individual needs for the project. It has already been shown that disproportionate amounts can be spent on features of the project that overall have very limited worth.

Value can be derived in many different ways but will probably at least include the following:

- Better use of space
- Increase in worker productivity as a result of improved working conditions
- Attention to flexibility to allow for anticipated changes in future work and employment practices
- Reduced costs-in-use due to improved designs that consider future costs and the rising need for environmental improvements

19.11.1 Losing value

Since value can be derived from so many different project features, the greatest threat to providing added value is through poor overall and initial project definition. This might be due to an unbalanced design, through poor management of the processes involved and through unnecessary expenditure. Value management has often demonstrated that between 10% and 25% of the initial costs of a construction project can be attributed to unnecessary expenditure. This is commonly caused by the following factors:

- Prescriptive briefs that are too precise and do not allow the design team scope and flexibility to achieve an improved solution
- Inappropriately high standards of quality expected by the client
- Delays in the decision-making process
- Preference for bespoke design solutions rather than adopting standardised components that are able to achieve the same objective
- Lack of information on which to make sound decisions
- The supposed need for temporary works that make no real contribution to the final product
- Poor levels of communication within the project team
- Conflicting agendas or personalities within the project team providing solutions that satisfy them rather than meeting the needs of the client
- Mindsets that are fixed and rely on aged solutions that result in the recycling of old ideas
- A lack of collaborative working amongst the client's advisors, the design team and, where appropriate, the contractor or contractors involved

SELF ASSESSMENT QUESTIONS

1. Distinguish the differences between value engineering, value analysis and value management.

2. Selecting a project of your choice, identify where cost savings could be considered while at the same time maintaining the project's overall value.

3. Why has the construction industry only recently begun to introduce value management methods and why is the use of value management still so limited in practice?

BIBLIOGRAPHY

Adams, J., *Conceptual Blockbusting: A Guide to Better Ideas*. Addison Wesley 1986.

Ashworth, A. and Hogg, K., *Added Value in Design and Construction*. Longman 1999.

Ashworth, A. and Hogg, K., *Practice and Procedure for the Quantity Surveyor*. Blackwell Science 2002.

Connaughton, J.N. and Green, S.D., *Value Management in Construction: A Client's Guide*. CIRIA 1996.

Crum, L.W., *Value Engineering: The Organised Search for Value*. Longman 1971.

De Bono, E., *Lateral Thinking for Management*. Penguin 1971.

Dell'Isola, A.J., *Value Engineering in the Construction Industry*. Van Nostrand Reinhold 1982.

Green, S.D., *A SMART Methodology for Value Management*. Occasional Paper No. 53, Chartered Institute of Building 1992.

Kelly, J.R. and Male, S.P., *A Study of Value Management and Quantity Surveying Practice*. Surveyors' Publications 1988.

Kelly, J. and Poynter-Brown, R., *Value Management in Quantity Surveying Techniques: New Directions*. Blackwell Scientific 1990.

Kelly, J.R., Male, S.P. and Graham, D., *Value Management of Construction Projects*. Butterworth Publishing 2003.

Macedo, M.C., Dobrow, P.V. and O'Rourke, J.J., *Value Management for Construction*. Wiley Interscience 1978.

Miles, L.D., *Techniques for Value Analysis and Engineering*. McGraw-Hill 1972.

Mudge, A.E., *Value Engineering in Design and Construction*. McGraw-Hill 1971.

Norton, B.R. and McElligott, W.C., *Value Management in Construction: A Practical Guide*. Macmillan 1995.

O'Brien, J.J., *Value Analysis in Design and Construction*. McGraw-Hill 1976.

Peters, T. and Waterman, R.H., *In Search of Excellence*. G.K. Hall 1997.

Walker, P. and Greenwood, D., *Construction Companion to Risk and Value Management*. RIBA Enterprises 2002.

Zimmerman, L.W. and Hart, G.D., *Value Engineering: A Practical Approach for Owners, Designers and Contractors*. Van Nostrand Reinhold 1982.

CHAPTER 20

RISK ANALYSIS AND MANAGEMENT

LEARNING OUTCOMES

After reading this chapter, you should have an understanding about the principles of risk analysis and management as used in the construction industry. You should be able to:

- Distinguish between risk and uncertainty
- Understand the risk management process
- Identify the kinds of risk envisaged on a construction project
- Understand how to deal with risk
- Identify a number of techniques that are available for dealing with risk and uncertainty

20.1 INTRODUCTION

All activities include an element of risk and uncertainty. They are inherent in all future events, since we can never entirely predict what might take place. All future activities are influenced by a range of macro and micro issues, some of which can be anticipated and others which cannot even be imagined. For example, estimates of building costs are forecasts of future costs; if they were precise values then they would no longer be estimates.

20.2 RISK AND UNCERTAINTY

Risk arises when the assessment of the probability of a certain event is statistically measurable. It differs from uncertainty since it can be mathematically predicted, whereas uncertainty cannot. Risk relies upon an availability of previous known events for this purpose. Uncertainty occurs where there are no data on previous performance on which to base a judgement. Uncertainty arises in one of two ways. In the first case it can arise because it can be imagined or anticipated. Activities that involve the use of new materials or procedures may produce uncertain consequences.

Some of these can be defined. However, because new procedures may be being employed the possible knock-on effect of these may not always be known beforehand.

For example, on the basis of past performance, the developer of a new out-of-town office complex could assess the risks involved in building for lease on the outskirts of any major city in the UK. While risk is involved in making the decision there are sufficient data available by which to assess the risks. The forecasting of costs for a project deep within the Amazon jungle on which there are no data is one governed by uncertainty. On some occasions, while data are lacking, research may enable some information to be gathered to remove some of the aspects of uncertainty. The building of the Kariba Dam in Africa in the 1970s was instrumental in causing a major British contractor of that time to go into liquidation because aspects of risk and uncertainty were not properly evaluated.

In preparing a plan for the future, the associated risk and uncertainty increase the further these predictions are projected. Past experiences can be used to assess the confidence of our predictions. In reality, we routinely apply risk to everyday events. We make judgements on the basis of past knowledge and understanding. It is also possible to enhance the confidence in our expectations by examining and analysing data and other information. The forecasting of events at some distance in the future and the preoccupation with current issues and events often militate against thorough appraisals and subsequent decision-making.

Because capital investments often span a long period of time they are never made under conditions of certainty. Risk attempts to quantify events about which we have some knowledge. Uncertainty is concerned with events that either cannot be measured or cannot perhaps even be contemplated. Both risk and uncertainty may result in outcomes that are better or worse than expected. There is therefore an argument that to apply any form of analysis to problems of risk and uncertainty is unwise. These arguments tend to perpetuate the view that decisions are made without the support of all the facts or a rational evaluation of the information that is available. It must be stressed, however, that the analysis must always be supported by experience. The analysis should in many ways help to confirm what is already supposed, although this might not always be the case. Analysis frequently identifies more closely issues or events that may not have been properly considered in the assessment of the project. Techniques for dealing with risk and uncertainty must therefore always seek to combine both judgement and analysis, each relying on the other for support. A purely mathematical approach is unlikely to have many followers.

20.3 THE RISK MANAGEMENT PROCESS

While every construction project has its own independent attributes, the approach to the management of risk can be similar for all. Figure 20.1 identifies such a risk management process.

The identification of potential risks can be assisted through the use of a project checklist as shown in Table 20.1. The risk management process starts with the

Risk
Risks occur at the decision to build or
outset of the project

Identify
The potential risks envisaged are
initially identified

Assessment
The risks are then assessed
This aspect of the process may be repeated several times

Re-evaluation
Where it has been possible to manage out certain risks,
this may have identified further risks

Control
Some of the risks that can be identified are
considered beyond the control of the
those involved in the project

Fig. 20.1   The risk management process

Table 20.1   Simplified risk checklist

Client risks	Decision to build
	Team selection
	Briefing
	Time, cost and quality
Site risks	Choice of site
	Ground conditions
	Existing buildings
	Existing owners and occupants
	Access and boundaries
	Availability of services
Planning risks	Planning restrictions
	Environmental concerns
	Legal requirements
Design risks	Interpretation of brief
	Aesthetics and space
	Early price estimates
	Appointment of constructors
Construction risks	Site management
	Delays
	Increased costs
	Liquidation and insolvency
	Latent defects
	Health and safety
External risks	Market conditions
	Political considerations
	Government legislation
	Weather conditions

identification of all possible risks associated with a proposed project. It is desirable to start this process early in the project's life, since many of the major decisions are often determined at the outset. It is during the feasibility and viability stages that decisions that have the greatest impact on the project are made. It is also during these early phases that changes can be made to the process and the product that are likely to cause the least disruption. In identifying what the potential risks are it will be necessary to ascertain whether they will arise, what their effects might be and what measures can be taken to prevent their possible occurrence.

The assessment of risk, like many aspects of construction activity, is a combination of judgement and some form of numerical analysis. The qualitative assessment seeks to describe and classify the different risks involved, their relationship to each other, the potential impact on the project and the likelihood of occurrence. This assessment will also identify the risk responsibility and the extent that it can be removed or avoided. The different risks will be classified into categories and importance. These different categories may include, for example, environmental, contractual, financial and economic aspects.

The detailed quantitative assessment of risk is the one that is the most easily identified as risk analysis. A relatively simple financial approach is used in the allocation of an amount to cover contingencies on a typical project. The amount allocated has frequently been determined largely on the basis of rule of thumb or past experience. This simple assessment can be further enhanced by allocating probabilities, which take into account the likelihood of a particular risk occurring.

The method of probabilistic analysis relies upon the application of meaningful probabilities to estimates. Normally each important risk can only be realistically quantified in terms of optimistic, most likely and pessimistic. The sum of these probabilities will equal 100% (or 1).

Re-evaluation is the action that is required to reduce or eliminate the potential areas of risk that have been calculated. It may be desirable to redesign those parts of the project in an attempt to remove part or all of the risk. Alternatively a contingency plan should be prepared that will only be implemented where the identified risk occurs. The redesign may result in other potential risks being identified which will then need to be evaluated.

20.4 TYPES OF RISK

Many different types of risk are associated with construction projects and these affect the various parts of the construction process and the different parties involved. They may include, for example, items in the simplified checklist shown in Table 20.1.

More specifically some of the items from the checklist may materialise into typical risks as follows:

- Delays in providing the design by the agreed date
- Adverse ground conditions causing delays and increasing costs
- Exceptionally inclement weather

- Unforeseen increases in the costs of labour and materials
- Final costs in excess of the agreed budget amount

20.5 DEALING WITH RISK

The allocation of risk may be dealt with in several different ways, as follows.

20.5.1 Risks may be avoided

The total avoidance of all risks can only be achieved by non-activity. Risk avoidance may include a review of the overall project objectives leading to a reappraisal of the entire project. For example, the risks associated with building on poor load-bearing ground can be avoided by choosing to build the project elsewhere. In practice the site investigation report may recommend such a course of action, since the costs involved in building under these circumstances might be better justified in the finished project. Where the risk involved cannot be counterbalanced by the utility gained from the risky action, then the risk should always be avoided. Avoidance is only possible if that choice exists. If it does not, then the risk must be handled in some other way. Other examples of risk avoidance include the use of exemption clauses in contracts that seek to avoid the consequences of a particular course of action.

20.5.2 Risks may be accepted

The development of property is such that, if one wishes to be involved, then some risks must be accepted in one way or another. Sometimes risks are accepted in ignorance since the different liabilities have not been carefully considered. Contractors, for example, may undertake work for development companies unaware of the implications of the conditions of contract that are being suggested. Sometimes risks are recognised but the opportunity to either transfer or minimise them is neglected. A contract for a new development project may involve aspects of work of a highly technical or specialised nature. The contractor may recognise this but, instead of subcontracting the work to a specialist firm, decides to execute the work and therefore accept the risks that are involved. In other circumstances, risks may be accepted unintentionally, in the belief that the contract documents for a project mean one thing when they actually mean something rather different. In order for construction work to be carried out some risks will need to be accepted.

20.5.3 Risks may be reduced

It may be possible, usually by further investigation, to reduce the potential exposure to risk. This is more often achieved either by attempting to redesign parts of the work to lower the impact of risk or through improving the management process that is involved. In each of these situations additional investment is required to provide improved analyses.

Amounts included for contingencies in contracts represent items that will hopefully allow for any risks in order to keep the project within budget. In practice estimates of all known items can be calculated. The contingency sum represents the allowance for the unknown or uncertain elements of the project. Where the project still includes a large proportion of undecided work then the contingency sum will be higher than in those circumstances where all design decisions have been made. Even under these circumstances an amount will still be included for contingency items.

Contingencies have traditionally been calculated on the basis of past experience alone. Typically this was 5% for new build projects and 10% for refurbishment projects. By adopting a risk management approach, a more accurate prediction of contingencies can be made to enable the project to provide more realistic information to assist clients in their own planning.

20.5.4 Risks may be averaged

Averaging the risks means sharing the risks involved between the parties involved. The principle should be that risk should be allocated to the party that is best able to control it. Evidence suggests that where a building client places all of the risk with the contractor a worse deal is likely to be struck. This is especially true in times of tendering prosperity, since some of the risk will never materialise in practice. The traditional form of building procurement attempts to allocate the risks evenly between the contractor and the client. More aggressive clients have attempted through different contractual arrangements to place more of the risk with the contractor (see Ashworth, *Contractual Procedures in the Construction Industry*, Pearson Prentice Hall 2006). In some cases contractors have been content to accept this, particularly where it gave them greater control, as in the case of design and build projects. When averaging risks it is important to differentiate between the transference of risk itself and the allocation of risk responsibility.

20.5.5 Risks may be ignored

It is not recommended to ignore risks! The belief that the project will work out all right in the end is the precursor to many bankruptcies in the construction and property industries. Another view which suggests that all risks will be controlled and contained is equally unrealistic. Past evidence suggests the folly of this statement. Others will argue that their own intuition has been sufficient in the past in assessing the risks involved and they see no point in attempting to quantify risk. Risk is inherent in the construction of buildings. The opportunity to control it better should be welcomed by all of the parties concerned.

20.6 SOME TECHNIQUES FOR DEALING WITH RISK AND UNCERTAINTY

20.6.1 Brainstorming

Brainstorming is a most effective technique for attempting to identify the potential risks that are associated with a building project. Brainstorming involves an open discussion in order to generate a large number of different ideas. The process involves individuals with a variety of different backgrounds in order to capitalise on different points of view. In a typical brainstorming session the emphasis is upon generating a large number of ideas, in the anticipation of obtaining an excellent idea. It calls at times for thinking the unthinkable and for creativity and imagination. Its aim is to arrive at a workable solution, sometimes through unusual approaches to a problem.

Throughout the process the judgement of ideas is withheld so that potentially extreme solutions to a problem are not stifled. Ideas that are generated are criticised and evaluated at a later stage in the process. Large problems may need to be made more manageable by breaking them down into smaller parts.

Brainstorming used for risk analysis will seek to identify all the potential risks involved and the likelihood of their occurrence. It will also want to examine the potential impact that such risks may have on the parties involved and the response from the different participants who have been involved in the session.

It may be sensible to subdivide the project amongst its different phases and to provide potential checklists of the problems that have been envisaged on a typical project.

20.6.2 Synectics

Synectics is similar to brainstorming but the group selected includes only those individuals who are best equipped intellectually and psychologically to deal with complex problems associated with risk. The synectics process involves two steps. The first is *making the strange familiar*. It requires that a problem and its implications be clearly understood. The second is *making the familiar strange* which involves distorting, inventing and transposing the problem in an attempt to view it from a very different perspective.

20.6.3 Probability

Probability is an important concept when dealing with the analysis of risk. It is often helpful when dealing with uncertainties in cost and price forecasting to consider a range of results rather than a single outcome. It might be suggested that the development value of a proposed project is likely to be between £10m and £12m. It is also possible to attach to this range an estimate of its probability. Probabilities may be estimated from past results or statistical records of previous events, or they may be obtained by conducting experiments using a sampling procedure.

20.6.4 Decision tree analysis

Many investment decisions are not isolated events but are often a process regarding an overall strategy of development. A decision tree comprises a number of branches that originate from a first question, which might be, 'Should we carry out the development?' The analysis is characterised by a series of either–or decisions.

It is a pictorial method of showing a sequence of interrelated decisions highlighting possible courses of action and future possible outcomes. Where probabilities or the values of potential outcomes are known, they can be used as a method of quantification.

20.6.5 Sensitivity analysis

Sensitivity analysis has already been described in Chapter 13 (p. 312) and in an example on whole-life costing in Chapter 18 (p. 458). It is a practical method of showing the effects of risk or uncertainty on the project by applying different values to the variables and measuring the outcome. Essentially it is a method that is used to test the impact of changes in the values of the variables in a model. Such variables are capable of having a range of values attributed to them. For example, in a developer's budget the selling prices of houses are estimated. The selling price is affected by a number of different circumstances. Sensitivity analysis is used to test whether under extreme circumstances, if the development was to proceed, it would do so at a loss. Any of the variables can be changed to provide a worst and best scenario and the likelihood of either of these events actually taking place.

20.6.6 Simulation

Simulation assumes that the values of the different variables may be combined with each other on a random basis. See paragraph 16.8.

20.6.7 Portfolio theory

Portfolio theory or analysis is a technique that seeks to identify the efficient set of investment characteristics. Once these have been identified then the developer will choose those that best satisfy the overall objectives for the project. The analysis may require at the outset a long list of characteristics that may be of some importance. The list is often too extensive to explore fully and some of the characteristics may have a very minor impact upon the final solution. The important characteristics will then be evaluated, often numerically but not so in every case. Some of the characteristics will be considered as essential, others as required and a third group as desirable. Correlation coefficients can be measured between these characteristics.

20.6.8 Breakeven analysis

Breakeven analysis is discussed in Chapter 6, page 121. The breakeven point, or more correctly breakeven circle, can be used as a basis to assess the risk of adopting a particular decision pathway.

20.6.9 Scenario analysis

Scenario analysis essentially examines different scenarios as possible options. The aim is to consider the likely outcomes of a solution in a more carefully considered manner by examining the different variables involved in the decision-making process. Scenarios are chosen that represent the most likely, optimistic and pessimistic cases.

20.7 RISK ANALYSIS USING PROBABILITIES

The methods used for the preparation of early price estimates have been described in Chapter 14. The assessment of design and price risks in pre-tender estimates have already been considered in that chapter. It was suggested that the risks associated with both of these factors were greater at inception than when the project is more fully developed and costs have become more explicit. Table 14.5 indicates that it is no longer sufficient to provide clients with a single estimate of cost. It is now more appropriate to offer a range of values of a forecast tender sum. The technique of multiple estimating using risk analysis attempts to provide such a range of estimates. This procedure was originally devised by the former Property Services Agency within the Department of the Environment.

Traditionally, early price or approximate estimates have been prepared to provide clients with a budget of their expected costs. This was to avoid expensive design fees on schemes that they would not be able to afford. Traditionally clients were given a single lump sum figure. No measure of accuracy or confidence limits was given, although the estimate would have been prepared in accordance with appropriate skills and expertise. Estimate deviation was never much discussed and where estimates did not represent reasonably accurate forecasts then explanations for any discrepancy could usually be provided.

Since those days surveyors have become much more familiar with statistical analysis and clients have become much better informed and more demanding of their services. Table 20.2 provides an estimate for a proposed project using as a basis for the estimate the likely cost. This has been prepared using the simplified cost plan but any of the methods described earlier (see Chapter 14) could be used. From these costs, optimistic and pessimistic (worst cost) values would be calculated. The surveyor would then allocate probabilities to each of the three values for, in this case, the six sections of the simplified analysis. The amount included for preliminaries could be dealt with in a similar way. In this example 10% has been added for preliminaries and 5% for contingencies.

The estimate prepared using conventional means is described as the likely cost. After applying the above probabilistic process, in this example the expected cost is very similar to the likely cost. This is due to a large extent to the arrangement of the probability distribution. The estimate would be presented within the range of values of £495,650 and £672,750.

It is sometimes necessary to separate fixed risk items, i.e. items that may or may not be required in their entirety, e.g. air conditioning, from variable risk items which will be required but their extent is as yet undecided.

The method of calculation may appear to be subjective. Data and information from past projects will be used wherever possible, but it needs to be remembered

Table 20.2 Simplified cost plan based on probabilistic analysis

	Optimistic cost	Probability %	Likely cost	Probability %	Worst cost	Probability %	Expected cost
1 Substructure	55 000	20	65 000	60	75 000	20	65 000
2 Superstructure	140 000	20	178 000	70	190 000	10	171 600
3 Internal finishes	25 000	25	35 000	55	40 000	20	33 500
4 Fittings and furnishings	16 000	15	23 000	60	25 000	25	22 450
5 Services	120 000	15	145 000	65	165 000	20	145 250
6 External works	75 000	15	82 000	50	90 000	35	83 750
7 Preliminaries (10%)	43 100		52 800		58 500		52 155
8 Contingencies (5%)	21 550		26 400		29 250		26 078
Totals	495 650		607 200		672 750		599 783

Range of values   £495 650–£672 750
Likely cost   £599 783

that estimating and forecasting are not just applied sciences but also rely upon the application of sound judgement. As the project design evolves, the cost plan would be updated. As the design becomes more certain then the range of values would reduce.

CONCLUSIONS

The assessment of risk is now recognised as an everyday activity. It may be necessary on the basis of a risk assessment to avoid a construction project at almost any price – for example, where the risks involved outweigh any possible advantages, financial or otherwise. On other occasions, it may be essential that a construction project is completed, even though the potential risks may be considerable. In either case it is desirable that the risks involved have been fully investigated to safeguard both the interests of the client and the potential financial rewards that may be expected.

SELF ASSESSMENT QUESTIONS

1. Differentiate between risk and uncertainty, giving examples of each from the construction industry.

2. Identify a list of common risks encountered in the construction industry and rank these in order of importance for a typical project.

3. Using some of the risks that have been identified in question 1, explain how these could be dealt with in practice.

BIBLIOGRAPHY

Ashworth, A., *Contractual Procedures in the Construction Industry*. Pearson Prentice Hall 2006.
Byrne, P., *Risk, Uncertainty and Decision Making in Property*. Estates Gazette 1996.
Chapman, C. and Ward, S., *Project Risk Management*. John Wiley 1997.
Curtis, B., Ward, S.C. and Chapman, C.B., *Roles, Responsibilities and Risks in Management Contracting*. Construction Industry Research and Information Association 1991.
DAS, *Managing Risk in Procurement*. Commonwealth Procurement Guideline, Department of Administrative Services 1992.
Engineering Council, *Guidelines on Risk Issues*. 1993.
Flanagan, R. and Norman, G., *Risk Management in Construction*. Blackwell Science 1993.
Flanagan, R. and Stevens, S., *Risk Analysis in Quantity Surveying Techniques: New Directions* (ed. P.S. Brandon). Blackwell Science 1992.
Grey, S., *Practical Risk Assessment for Project Management*. John Wiley 1995.
Institution of Civil Engineers and Institute of Actuaries, *Risk Appraisal and Management of Projects*. Thomas Telford 1998.
Kingsmill-Velacott, A. and Bomforth, P., *Risk and Finance*. Blackwell Publishing 2004.

Newland, K., *Guide Project Risk Analysis and Management*. APM Publishing 2004.

Raftery, J., *Risk Analysis in Project Management*. Taylor and Francis 2003.

Sawczuk, P., *Risk Avoidance for the Building Team*. 1996.

Smith, N.J., Merna, T. and Jobling, P., *Managing Risk in Construction Projects*. Blackwell Publishing 1998.

Tweeds, *Guide to Risk Analysis and Management*. Laxton's 2006.

Uff, J. and Odams, R.V., *Risk, Management and Procurement in Construction*. Centre for Construction Law and Management 1995.

Walker, P. and Greenwood, D., *Construction Conpanion to Risk and Value Management*. RIBA Enterprises 2002.

CHAPTER 21

POST-CONTRACT COST CONTROL

LEARNING OUTCOMES

After reading this chapter, you should have an understanding about the principles and practice of post-contract cost control as used in the construction industry. You should be able to:

- Identify the different processes and techniques that are used
- Understand the principles involved with clients' financial reports
- Recognise the similarities and differences between clients' and contractors' objectives
- Develop income and expenditure S-curves
- Calculate project cash flows

21.1 INTRODUCTION

The financial control of any construction project commences at inception and continues until the issue of the final certificate. The time from the signing of the contract until the final certificate is termed the post-contract phase. The method used for controlling costs will depend on the following circumstances:

- The method used for contractor selection
- The method used for price determination for both the tender sum and the final account
- Whether the control is being exercised for the contractor or the client
- The role of the quantity surveyor in respect of budgeting and accounting

The methods used for contractor selection and price determination are constantly under review and change. While the client's and contractor's needs are similar, their objectives are different. Both wish to restrain their costs, but the aims of contractors are to increase their profits, while those of clients are to secure the highest possible value for money. Post-contract cost control includes the following:

- Interim valuations and certificates for payments
- Cash flow control and forecasts through budgetary control

- Financial statements showing the current and expected final costs for the project
- Final account, the agreement of the final certificate and the settlement of claims

The responsibilities in respect of post-contract services will vary depending on the terms of appointment of the surveyor and the provisions within the contract conditions. The following are some of the activities involved:

- Attendance at site meetings
- Preparation of documentation for subcontractors and suppliers, examination of quotations and invoices and making recommendations
- Advising on contractual implications
- Negotiation
- Confirmation of payments to nominated subcontractors
- Advising on the implications of extensions of the contract period
- Preparation of special reports on cost implications
- Completion of documentation which may be required for some clients, particularly government departments
- Working with auditors

21.2 MEASUREMENT CONTRACTS

The first duty of the quantity surveyor on the receipt of tenders is to examine them for any arithmetical or technical errors. This is then followed by the preparation of a written report on the tenders received and the quality of pricing applied in the contract documents. Advice will be offered on which contractor should be selected, and how any errors in the documents will be dealt with. These activities are strictly pre-contract, since the contractor and client will not yet have signed the appropriate documents. Post-contract activities tend to fall within two categories.

1. Interim certificates and statements

(i)   Preparing a budget and cash flow forecast.
(ii) Updating the financial statement and report.
(iii) Interim valuations which include
    - valuing the work completed on site
    - valuing any goods or materials manufactured on or off site which can be included under the terms of the contract
    - agreeing these amounts between the client's and the contractor's surveyors
    - notifying nominated subcontractors that payments can be expected

The amounts included in interim certificates should be fair and reasonable to both parties, representing an accurate picture of the financial position of the project. It needs to be remembered that the values calculated may not coincide with costs, since items of work undertaken at different times will incur different costs. For example, while a bill rate for building a brick wall in a multi-storey building will be the same for all floors, the contractor's costs will be less on the lower storeys.

The client is most vulnerable during the early stages of the works should the contractor become bankrupt or go into liquidation. However, the contractor must not be penalised through underpayment, since valuations are the lifeblood of the contractor's business.

2. Agreement of the final account

(i) Remeasurement of work
- as a result of variations authorised within the terms of the contract
- for provisional items or quantities which do not require a designer's instruction
- for the correction of previously undetected errors in the documents arising from mistakes in their preparation, but not contractor's errors
- for the execution of provisional sums undertaken by the contractor

(ii) Adjustment of prime cost sums
- on the basis of invoices from nominated subcontractors and suppliers or statutory undertakings which will include the original quotation as a basis together with variations which may have been authorised, increased costs, dayworks etc.
- for dealing with profit and general and other attendance

(iii) Provisional sums which form the basis of nominated work would be dealt with in a similar way to (ii).

(iv) Daywork accounts including their authorisation for applicability, time-material quantities and pricing.

(v) Increased costs for inflation, where this is a condition of the contract. The contract will identify the method to be used for their computation, either using actual increases or based on a formula adjustment.

(vi) Preparation or evaluation of contractual claims. The contractor may consider that full reimbursement under the terms of the contract has not been made. For example, the nature of the contract may have been changed, resulting in additional costs on preliminary items.

In addition to the adjustment of the above items, it will be necessary to consider the additional professional fees involved and the implications regarding value-added tax which are generally the matter of a separate agreement. A further item which will require adjustment in the final account and consideration throughout the contract is the expenditure or otherwise of contingencies. These are often described in the contract documents as 'a sum of money to be partially or wholly expended by the designer in the event of unforeseen circumstances'. Since the circumstances are unforeseen, it is never clear what this item is expected to cover, but it may include:

- Mistakes made by the design team
- Work which could not reasonably have been anticipated at the design stage, such as unknown ground conditions
- Work which could be considered only when construction work was in process
- Costs associated with new legislation
- Additional work required by the client

Contractors sometimes mistakenly believe that contingencies have been budgeted for by the client and should therefore be expended on the project. In discussing whether an item of work should be undertaken, the contractor's surveyor will sometimes enquire how much of the contingencies remains unspent. This is not the purpose of this item. In the final account, contingencies will always be omitted in their entirety, being replaced by the above where they have occurred.

In an attempt to allow for the possibility of mistakes on the part of the design team, it is practice to inflate some of the provisional quantities which will be automatically remeasured at some stage during the contract. This approach, however, can be misleading to the contractor and confusing to the client. It is a safeguard but an unprofessional procedure. The quantity surveyor should use skill and judgement in an attempt to present a fair and true financial picture at all times to all interested parties. The surveyor needs to strike a balance between being too cautious and overly optimistic. In order to be able to provide an efficient cost control service, full information regarding any intended financial expenditure on the project must always be made available. The quantity surveyor therefore needs to be an integral part of the strategic decision-making process, and to be prepared to instigate procedures appropriate to the project concerned by introducing a system that not only records the cost implications after the event but also forecasts and controls all possible changes in expenditure beforehand.

21.3 COST-REIMBURSABLE CONTRACTS

The procedures used with cost-reimbursable types of contractual arrangement are considerably different from those employed with measurement contracts. The degree of financial control is inferior, and a much greater reliance is placed on the integrity of the contractor. The first impression is one of a much fairer system, but the lack of control of the essential resources and any incentive to reduce costs prevent it being a popular method amongst clients. Also, the inherent element of human greed must not be underestimated. Contractors are selected through negotiation or competition, with the definition of 'cost' being of prime importance in the contract documents.

While it is not possible to predict with any certainty an estimate or final cost, the quantity surveyor may choose to provide some broad indications of cost for the various sections of work. These will have no contractual basis between the client and the contractor. During the progress of the works the quantity surveyor will check the workmen's time sheets and the invoices for materials on site. The clerk of works will have a daily record of all workmen, of either the main contractor or subcontractors on site, and this can be used for the verification of time sheets. A difficulty occurs of ensuring that the contractor uses the available resources efficiently and effectively. The client has little control over these aspects. The conditions of contract may state:

■ 'Only labour performing productive work can be included in the contractor's cost.' The definition of cost will already have identified how supervisory staff are to be charged to the contract. The question here relates to 'properly performing'. This is a value judgement, open to much interpretation, and difficult to enforce in practice.

■ 'Materials must be purchased competitively.' The quantity surveyor will need to see quotations and invoices, and may insist in some instances that competitive prices are obtained. This may not provide the desired solution, since on a measurement contract the contractor may be prepared to bargain with the suppliers to improve the discounts which are offered. On the more usual cost-reimbursement contract, a reduction in materials costs will directly reduce the contractor's profit, which provides the opposite incentive to reducing costs. A reconciliation between bought quantities and those in the finished structure must be carried out, but differences can remain inconclusive.

Valuations for interim payments are based on the cost sheets plus an assessment to value the work to date for sheets still to be submitted. No real forecast of expenditure can be made, and this has a serious disadvantage for the client. If the valuations which include the percentage for profit are reasonably accurate, this should aid the contractor's cash flow. It would appear on the face of it that there are no disadvantages to the contractor of using this system. Certainly in periods of recession, when too many contractors are chasing too few jobs, it serves the contractor well. However, there is no opportunity with the common method of cost plus for the contractor to make profits in relation to improved performance. In some ways the contractor may even be penalised for his own efficiency. The procedures for calculating the cost reimbursement are at best cumbersome and at worst tedious. Cost control as such is largely ineffective in real terms.

21.4 BUDGETARY CONTROL

Budgets are used for planning and controlling the income and expenditure in many different organisations. It is through the budget that a company's plans and objectives can be converted into quantitative and monetary terms. Without these a company has little control. The budget may represent a total sum divided among a number of subheadings or work packages. It is important that the various subheadings include a timescale, since the expenditure by both the contractor and the client needs to be matched against income or the availability of funds. While the contractor will have a work programme for the project and this can be costed, the procedure may be disrupted by delays on the part of all those involved and through changes (variations) to the original scheme.

This information will give a rate of expenditure and a rate of income throughout the project, and by deducting income from expenditure the amount of capital required at the different times can then be calculated. This is dealt with later in this chapter. The contractor will need to aggregate this information from all projects in order to determine the company position. For budgeting purposes these data are prepared in advance of work being carried out on site. The information will also be collected after the execution of the works in order to establish the 'as done' position, and to facilitate a comparison with the budget. This is known as budgetary control. In common with other control techniques, budgetary control is a continuous process undertaken throughout the contract duration. When variances from the budget occur, the contractor will need to assess the reasons for them. The client's budgetary control procedures are somewhat similar, although the control mechanism is different. The client's main concern is with the total project expenditure which has been forecast. The ability to control this depends on the sufficiency of pre-contract design, the number of subsequent variations, and the steps taken to avoid unforeseen circumstances and matters which fall outside this control, such as strikes. The client will also be concerned about the timing of expenditure, for funding purposes, but in this case, although control will be influenced by the above factors, the contractor's method of construction will also be influential. In the case of a large building client such as a county council, the interaction of the many different projects under construction at any one time will need to be aggregated to establish the total periodic payments required. Research has indicated that when such projects are grouped together, this often produces similar cash outflows for each month of the year. This obviously influences scheme approvals, starts and completions.

21.5 CLIENT'S FINANCIAL REPORTS

Throughout the construction phase the client will need to be advised on any changes to the probable final cost of the project. This is normally done through a financial statement at regular times throughout the contract period. The size and complexity of the project will determine how often these statements are prepared: typically every quarter. A typical example of a financial statement is shown in Figure 21.1. The report is in two parts: the first considers the current position and the second the likely final cost of the project.

The current position states how much money the client has already spent in terms of interim payments and the retention which has been withheld. This sum will be compared with an expenditure forecast to aid the client's cash flow. This is explained later. The second part of the report can be prepared in different ways and levels of detail. The importance of its contents depends on who will use it. Generally the client will wish to know only in general terms the anticipated total financial commitment. The report therefore seeks to forewarn the client of any future possible increases or decreases in the costs of the project. The technical details, even on a simplified report, will mean very little to the typical client. However, the designer may require more information in order to plan and rectify the remaining expenditure for the project. The information which is provided should be self-explanatory and easily referred to during the preparation of later statements. The financial statement updates the contract sum, which is the amount agreed between the contractor and client. The accuracy of the amounts will vary depending on the quality of the judgement provided. Adjustments to the contract sum can be made under the heading indicated in Figure 21.1. The financial statement will include:

CONTRACT	Architect's Ref:	Quantity Surveyor's Ref:

Original contract sum
Fluctuations/fixed price
Approved additional expenditure £   £

£   £

CURRENT APPROVED FIGURE £

Contract sum £   REMARKS

Deduct contingencies £

£

	A Savings £	B Extras £
1. P.C. sums ordered to date		
2. Provisional sums		
3. Provisional quantities		
4. Variation orders Nos. ....to.......		
5. Daywork sheets No.		
6. Claims		
7. Other extras (give description)		
Savings/extras	£	£

Add/deduct difference
between A and B   £

Add/deduct fluctuations on labour/materials
estimated to date   £

ESTIMATED FINAL COST £

Net amount in hand/overspent on approved figure £

Fig. 21.1   Form of financial statement
Source: Seaden Partnership

Table 21.1 Architect's instructions

	Add	Omit
1. Possession of site	–	–
2. Slingpile nomination	3 300	–
3. Gas main connection	–	4 000
4. Selection of facing bricks	12 600	–
5. Autobay size	–	16 000
6. Steelwork supply	6 750	–
7. Sewage connection	–	–
8. Opening sizes for main doors	800	–
9. Block finish to inspection room	–	250
Carried forward	23 450	20 250

- Quotations set against prime cost and provisional sums already accepted although the work may be incomplete or not even started.
- Variation orders which have been issued to the contractor. This information can be recorded in a form similar to Table 21.1. The designer should obtain the appropriate cost advice prior to the issue of any instruction. Once it has been issued, the quantity surveyor can estimate its likely effect on the contract, and this can be amended when the work has been carried out and agreed with the contractor.
- Provisional quantities remeasurement, even though these may represent only an approximate calculation or still have to be agreed with the contractor.
- Daywork sheets are often not submitted as requested, but some allowance needs to be made for them by way of an estimate.
- Claims for delays or disruptions may already have been intimated, and it would be foolish to discount these from the statement. An amount will need to be included together with an assessment of the likelihood of their acceptance.
- The quantity surveyor may also be aware that the designer is considering issuing instructions on other matters pertinent to the cost of the project. For example, the quantity surveyor will be aware of where the bill is inadequate or generous in the light of current knowledge. These aspects will need to be considered carefully. They can be recorded in a similar manner to architect's instructions. Once they have been dealt with by the architect they are removed from this list and added to the architect's instructions schedule.

The adjustment of all of the above items will be either added to or deducted from the contract sum (less contingencies). At all stages during the progress of the works the adjustments required will be based on a combination of fact and judgement. One final adjustment is required for projects which include the fluctuations provision. This is probably the most difficult assessment of all, since it will be necessary not only to include the increased costs which have already taken place, but also to estimate those up to the end of the contract. The client's forecast of cost will need to include all possible items. The only exceptions are professional fees, VAT and furnishings, which are unlikely to be part of the construction contract.

21.6 CLIENT'S CASH FLOW

In addition to the client's prime concern with the total project costs, the timing of cash flows is important. Equal monthly instalments cannot be assumed; indeed, as the project proceeds, a peak in activity is achieved about two-thirds of the way through the contract period. The client's advisers will therefore need to prepare an expenditure cash flow which is linked to the contractor's programme of activities. On large and complex projects and in periods of high inflation, the timing of payments might result in higher tender sums being a better economic choice for the project as a whole. Table 21.2 represents the cash flows for two contractors based on their own programmes for executing the works. Contractor A intends to set up on site a highly mechanised system that will produce cost benefits and savings later in the contract. Contractor B intends a more traditional approach, typifying much of the scenario in the UK. The client's opportunity cost of capital is 10%. On the basis of submitting the lowest tender sum, contractor A is the logical choice, since both contractors meet the requirements in terms of quality and time constraints. However, when cash flows and the costs of finance are taken into account, contractor B is the better alternative. If the interest rates were higher this would make contractor B an even better choice. In practice such cash flows would be calculated on a monthly basis.

The Department of Health has developed a method of expenditure forecasting for clients using a formula. This has been developed on the basis of plotting the

Table 21.2   Client's cash flow

Contract period (years)	Cash flow	Discount factor @ 10%	Time value of payment
Contractor A (high mechanisation)			
1	850 000 ×	0.90909	772 727
2	610 000 ×	0.82644	504 128
3	800 000 ×	0.75131	601 048
4	970 000 ×	0.68301	662 520
5	510 000 ×	0.62092	316 669
6	450 000 ×	0.56447	254 011
Tender sum	4 190 000	Net present value	3 111 103
Contractor B (traditional)			
1	300 000 ×	0.90909	272 727
2	500 000 ×	0.82644	413 220
3	750 000 ×	0.75131	563 482
4	1 200 000 ×	0.68301	819 612
5	1 200 000 ×	0.62092	745 104
6	350 000 ×	0.56447	197 565
Tender sum	4 300 000	Net present value	3 011 710

totals of interim valuations for a number of contracts. The curves of best fit are drawn, giving an 'S-curve' of approximately the same shape for each contract. Therefore, within a range of cost and over a limited contract period, a standard S-curve can be used to help predict the expenditure flow for future contracts. This can then be used as a comparison against actual expenditure from valuations. The former PSA also developed a similar system. References to both systems are given in the bibliography at the end of this chapter.

21.7 CONTRACTOR'S COST CONTROL

The contractor, having priced the project successfully enough to win the contract through tendering, must now ensure that the work can be completed at most for the estimated costs. One of the duties of the contractor's quantity surveyor is to monitor the expenditure, and advise site management of action that should be taken. This process also includes the costs of subcontractors, since these form a part of the main contractor's total expenditure. The contractor's surveyor will also comment on the profitability of different site operations. Where loss-making situations are encountered, decisions need to be taken to reverse this position if at all possible. Wherever an instruction suggests a different construction process to that originally envisaged, details of the costs of the site operations are recorded. The contractor's surveyor will also advise on the cost implications of the alternative construction methods which could be selected by site management. For example, the following items were included as a part of the drainage bill. The items were not identified separately from the remainder of the work.

a. Trench excavation, average 750 mm deep
   for drainpipe   80 LM @ £30.00 = 2 400.00
b. Granular filling around 150 mm pipe   80 LM @ £15.00 = 1 200.00
c. 150 mm drain pipe   80 LM @ £20.00 = 1 600.00
Total (£)   = 5 200.00

This drain to be laid was 2 m away from, and ran parallel to, an existing building within an industrial complex of structures. The contractor had intended to use a machine (JCB or similar) for excavating. It was quickly realised that this would not be possible because of the considerable number of drains and services which crossed the proposed line of the drain. A cost record was kept of the operation, which included mainly hand excavation with the machine working at a reduced efficiency due to the nature of the work. These actual costs are given in Table 21.3.

In order to make a direct comparison with the price charged, the contractor would need to include a sum for profit. On this project, according to the contractor's pricing notes, this represented 4%.

Contractor's cost   8 581.76 (see Table 21.3)
Profit 4%   343.27
Total   £8 925.03

Table 21.3   Contractors' record sheet: excavation at Goose Mill

JCB excavator 8 hours	@	£90.00	= 720.00
Granular filling 18 m³	@	£30.00	= 540.00
150 mm drainpipe 8 m	@	£10.00	= 80.00
Labourer 362 hours	@	£18.00	= 6 516.00
Dumper 6 hours	@	£40.00	= 240.00
			= 8 096.00
Head office overheads 6%			= 485.76
Cost of operation (£)			= 8 581.76

This represents an under-recovery against the work included in the documents of £3,175.03 or 72%.

Discounting the fact that estimators can sometimes be wide of the mark when estimating, even with common items, the contractor would seek reasons for such a wide variation between costs and prices. This will be done for two reasons: first, in an attempt to recoup some of the loss; second, to avert such errors in future work. The above situation may have arisen for one of the following reasons:

■ The character of the work is different from that envisaged at the time of tender.
■ The conditions for executing the work have changed.
■ Adverse weather conditions severely disrupted the work.
■ There was inefficient use of resources.
■ There was excessive wastage of materials.
■ Plant had to stand idle for long periods of time.
■ Plant had been incorrectly selected.
■ Delays had occurred because of a lack of accurate design information.

This list is not, of course, exhaustive, and often when the project is disturbed by the client or designer this can have a knock-on effect on the efficiency and outputs of the contractor's resources. Contractors may suggest that they always work to a high level of efficiency. This is not always the case, and the loss is sometimes due to their own inefficiency. However, in the above case the contractor appears to have a good case for the recovery of the loss. The client's quantity surveyor, in accepting the argument, would then try to see how reasonable the quantified data were, and whether they had been independently checked or recorded by the clerk of works. A prudent contractor would always draw such work to the attention of the clerk of works to ensure that at least the recording was fair and reasonable.

Circumstances will arise where an inefficient use of resources is entirely the fault of the contractor. Costing which shows that a project has lost money is of limited use where the contractor cannot remedy it. The contractor needs to be able to ascertain which part of the job is in deficit and to know as soon as it starts to lose money. The objectives, therefore, of a cost control system are:

- To carry out the works so that the planned profits are achieved
- To provide feedback for use in future estimating
- To cost each stage or building operation, with information being available in sufficient time so that possible corrective action can be taken
- To achieve the benefits suggested within a reasonable level of administration charges

Ideally, therefore, a cost control system should provide for reports on a daily basis. Since this could become an extremely expensive procedure, the cost of work done is checked weekly in an attempt to allow some corrective action to be taken. If the information is appropriately recorded, it can then be used as a basis for bonusing and valuations for interim certificates. The measurement of the work done should be undertaken by someone who correctly understands the demarcation between the various operations, since misrecording of costs can easily occur. The costs can then be properly compared with the value of work concerned.

21.8 CONTRACTOR'S CASH FLOW

Contractors are not, as is sometimes supposed, concerned only with profit or turnover. Other factors need to be considered in assessing the worth of a company or the viability of a new project. For example, the shareholders will be primarily concerned with the measurement of their return on the capital invested. Contractors have become more acutely aware of the need to maintain a flow of cash through the company. Cash is important for day-to-day existence, and some contractors have suffered a downturn not because their work was not profitable but due to an insufficiency of cash in the short term. In periods of high inflation, poor cash flows have resulted in reduced profits which in turn have produced an adverse effect for the shareholders' return. It is necessary therefore to strike the correct balance between cash flow and return on capital.

21.8.1 Contractor's income and expenditure curves

Table 21.4 represents the costs and payments for a project with a contract period of ten months and a defects liability period of six months. These are shown diagrammatically in Figure 21.2.

Expenditure

This is a combination of all the contractor's costs, and will include wages, materials, plant, subcontractors and overheads (head office charges). Expenditure is an ongoing item which will occur at irregular intervals other than for payment of wages. The expenditure S-curve is therefore shown as a continuous curve.

The expenditure on this contract peaks at month 7 and then begins to decline up to the completion of the contract. Any expenditure beyond month 10 is likely to represent minor items of work which had still to be carried out after the certificate

Table 21.4   Contractor's income and expenditure (£)

Month	Expenditure	Cumulative expenditure	Cumulative valuation	Net income
1	15 100	15 100	21 100	20 045
2	21 200	36 300	47 200	44 840
3	28 700	65 000	80 000	76 000
4	47 700	102 700	125 000	118 750
5	40 500	143 200	177 400	168 530
6	52 800	196 000	240 000	228 000
7	61 500	257 500	309 100	296 212
8	57 500	315 500	366 500	353 612
9	47 500	363 500	408 000	395 112
10	31 500	395 000	428 000	421 556
11–15	11 000	406 000	428 000	421 556
16	2 500	408 500	436 200	436 200

Retention is 5%, with a limit of 3% (similar to ICE Conditions of Contract). Contract sum was agreed at £429 600 on a fixed price basis; therefore the retention fund maximum is £12 888.

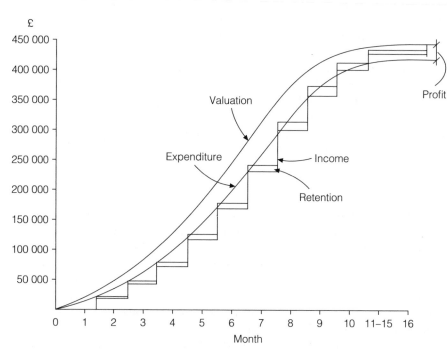

Fig. 21.2   S-curve

of practical completion and the making good of defects. The contractor's income and expenditure data provide a good indication of the progress of the works.

Valuation

This is also shown as a continuous curve since, although it may be impracticable to value the work in this way, this is how the value of the work changes. For convenience and clarity it is shown as a smooth curve even though in practice this will not strictly be the case. The comparison of expenditure and valuation will show profitability for that point in time. Some care of course needs to be exercised here, since this may not be a true representation due to unevenness in the way that the works have been priced.

Income

The third line on the graph represents the amount and timing of receipts paid to the contractor. Payment under the majority of forms is now almost standard at within fourteen days of the date of certification to the client. For this reason income is shown as the midpoint of the month. The two lines represent the income received and the amount of retention. Retention is normally released to the contractor in two equal parts, at completion and with the issue of the final certificate.

21.9 COST COMPARISON

In practice it is always difficult to make comparisons between costs and valuations, since either the full items of expenditure are unavailable or the valuation has only been approximately prepared. However, the contractor does need to determine which contracts are profitable and which are not, and also to determine which operations gain or lose money. The information which is then generated may be used to form the basis of contractual claims or to assist in future tendering and the contractor's selection of projects for which to tender.

Both the actual profit and percentage on cost can be calculated from Table 21.5. This offers the contractor an indication of the financial trend, although in order to measure this realistically these figures need to be compared with their respective budgets. It is unclear from these data alone whether the contract was successful. By inference it can be suggested that the project reached its most successful stage in terms of profit alone at month 7. The profit–expenditure ratio, however, had been decreasing since the commencement of the contract, indicating a possible front loading of the contract. This is also a typical feature of fixed price contracts if the anticipated profit has been distributed evenly throughout the project. Towards the end of the contract the project was probably losing money. For example, compare the expenditure with the valuation for month 9. This may suggest that the work has been deliberately overvalued during its early stages.

Table 21.5   Contractor's profitability

Month	Actual profit (valuation − expenditure)	Profit % on expenditure
1	6 000	39.73
2	10 900	30.02
3	15 000	23.07
4	22 300	21.07
5	34 200	23.88
6	44 000	22.44
7	51 600	20.03
8	51 500	16.34
9	44 500	12.24
10	33 000	8.35
11–15	22 000	5.42
16	27 700	6.78

21.10 PROJECT CASH FLOW

Using the data given in Table 21.5 and shown in Figure 21.3, the contractor can also establish the cash flow and borrowing requirement. Although in this case the information is retrospective and therefore of historical use only, the forecast amounts could be used to build a model of these requirements. The contractor expends £15,100 in month 1, and this is indicated as a negative cash flow. A valuation is then prepared and agreed, but 14 days may elapse before the contractor receives any payment. One half of month 2's expenditure is therefore spent before this payment

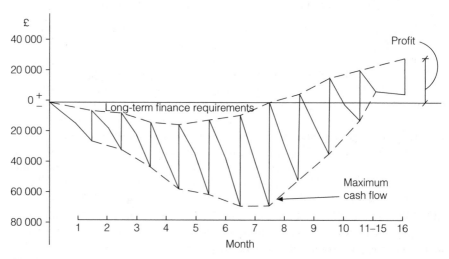

Fig. 21.3   Project cash flows

is received. Payment is indicated as a positive cash flow. This same procedure is then repeated throughout the contract. The expenditure is shown progressively from month to month since expenditure on wages, materials etc. will be made throughout. Income is a single payment made once per month and is represented by the vertical line. The diagram is often referred to as a sawtooth diagram.

Inflation and interest charges will therefore influence the actual profits received from a contract. Although the actual profit measured as a ratio between income and expenditure is shown to be 6.78%, this percentage will be reduced by the size of each cash outflow and the interest rates applied by the banks for borrowing short-term finance. The adverse effects of a negative cash flow can be mitigated to some extent by:

- Profit margins. The amount of these and the fact that they have been evenly applied will ensure that the project is self-financing more quickly. It will also bring into benefit the secondary factor of reducing bank charges.
- Interim payments. The period is traditionally one month. If this can be reduced to provisional amounts paid weekly, as occurs on some of the large projects, then this will improve the cash flow.
- Retentions. The more of this that is retained and the greater the delay in its release, the worse will be the cash flow.
- Delay in receipt of payments. Where the client does not honour certificates promptly, this will cause the cash flow to deteriorate.
- Delays in making payments. The greater the time between receiving goods or services and paying for them, the better will be the cash flow, even if it results in loss of discounts for normal trade terms. This is especially so in times of high inflation.
- Overvaluations. This has the effect of improving cash flows, but the private quantity surveyor's caution is more likely to result in the contractor's valuations being undervalued, creating the opposite effect.

Many contractors have in the past faced bankruptcy or liquidation because they overstretched their commitments, which resulted in cash flows that became unsustainable. It is always tempting to take on every project available and expand too quickly in an unplanned fashion, but the apparently lucrative contract, unless based on sound financial analysis and planning, often results in disaster.

21.11 COST MANAGEMENT

The management of the overall cost of the project is ultimately the responsibility of the project manager. This function is usually carried out by a cost manager or quantity surveyor. The main tasks that are required to be carried out include:

- Management of the base cost estimate
- Allocation and control of the risk allowance
- Operation of change procedures

- Production of cost reports, estimates and forecasts
- Reporting on the cost consequences of any decisions and responsible for initiating any corrective actions that might become necessary
- Maintaining up-to-date estimated final costs
- Advising on cash flow scenarios
- Management of all items of expenditure
- Advising on ways of avoiding a cost overspend
- Issuing cost update statements at regular intervals
- Ensuring that full and proper accounts are monitored of all transactions, payments and changes

The principal areas of cost management are in controlling expenditure, advising on cash flow and payments. In order for this to be done effectively the scope of the works must be clearly defined allowing for no ambiguities in interpretation. Estimates of cash flow will be prepared that are consistent with the agreed contract programme from inception through to completion of the works. Wherever practicable, cost inputs to the design should be made to find smarter ways of achieving the same design objectives and thus create added value solutions. A part of this process is ensuring that all orders are properly authorised. Contracts should be prepared that provide full control, and materials and works should be properly specified in order that the work can be procured effectively. An appropriate allowance for risk should be included and this should be allocated to the party that is best able to control it. The planning of cash flows based on the budgetary allocation to the programme should be prepared in order that all transactions can be forecasted and properly recorded, authorised and if necessary justified.

21.11.1 Financial reviews

Each financial review should ensure that:

- The latest estimate of final cost and previously agreed budget should be compared, and any differences should be clearly explained.
- Risk allowances should represent identified risks and not be absorbed by some assumed contingency provision.
- The project is still affordable.
- Funds are available to meet any interim payments by the specified date.

Risks and risk allowances should be reviewed on a regular basis, especially when formal estimates are prepared but also through the design and construction stages. As more firm commitments are agreed then the amounts for risk will be reduced accordingly.

21.11.2 Change management

The design at the outset of the project should be as complete as possible. Projects that include only vague concepts cannot be properly planned and adherence to budgets will be difficult, if not impossible. Changes should be minimised as far as is practicable to allow contractors to be able to source the work fairly and within timescales. Discussions with planning authorities should take place early in order that their requirements can be anticipated. Site investigations and, where appropriate, condition surveys of existing buildings should be completed as early as possible to remove as much uncertainty as possible. During the design stage elemental cost planning will be used to evaluate possible changes and design solutions. Any variance from the current cost plan can then be noted and remedial action taken if this is required.

Many of the changes during the construction period will arise from instructions to the principal contractor from the design team. These will follow formal practices and be in writing. These instructions should be issued with delegated authority and costed before they are issued. Each instruction should be justified in terms of added value and its overall impact on the project.

SELF ASSESSMENT QUESTIONS

1. Describe the procedures involved in keeping clients informed about changes in costs that might occur during the construction phase of a project.

2. Describe the different processes that are used for controlling the costs of measurement and cost-reimbursable contracts.

3. 'Cash flow is the lifeblood of the contractor.' Discuss this statement, suggesting ways in which contractors can create positive cash flows.

BIBLIOGRAPHY

Ashworth, A., 'How to cut your losses', *Building*, February 1978.

Ashworth, A., *Contractual Procedures in the Construction Industry*. Longman 1994.

Ashworth, A. and Hogg, K., *Willis's Practice and Procedure for the Quantity Surveyor*. Blackwell 2007.

Barnes, N.M.L., 'Profit, cash flow or return', *New Civil Engineer*, 1974.

Barrett, F.R., *Financial Reporting, Profit and Provisions*. Chartered Institute of Building Technical Information Service No. 12, 1982.

Gobourne, J., *Site Cost Control in the Construction Industry*. Butterworth 1982.

Hudson, K., 'DHSS expenditure forecasting method', *Chartered Quantity Surveyor*. B & QS Quarterly 1978.

Nisbet, J., 'Post contract cost control, a sadly neglected skill', *Chartered Quantity Surveyor*, January 1979.

Oxley, R. and Poskit, J., *Management Techniques in the Construction Industry*. Blackwell Science 1996.

Ramus, J., Birchall, S., and Griffiths, P., *Contract Practice for Surveyors*. Elsevier 2006.

RICS, *The Surveyor's Construction Handbook*. RICS (annual publication).

Seeley, I.H., *Quantity Surveying Practice*. Macmillan 1997.

Somerville, D.R., *Cash Flow and Financial Management Control*. Chartered Institute of Building Surveying Information Service No. 4, 1997.

Trimble, E.G. and Kerr, D., 'How much profit from contracts goes to the bank?', *Construction News*, March 1974.

CHAPTER 22

FACILITIES MANAGEMENT

LEARNING OUTCOMES

After reading this chapter, you should have an understanding about the role of facilities management and its relevance to the study of building costs in the construction industry. You should be able to:

- Appreciate the development of facilities management
- Identify the various aspects that are associated with facilities management
- Recognise the importance of building costs within facilities management
- Understand the various aspects of facilities operations management

22.1 INTRODUCTION

Building cost appraisal over 50 years ago was restricted to a concept relating to initial building costs alone. The emphasis was directed towards approximate estimating, which was largely required for a client's budgeting and financial purposes. The practice also included the preparation of contractual information and documentation for tendering purposes and the preparation and agreement of final accounts. These procedures were applied to the actual building works or building contract alone. No consideration was given to the other costs for which a client would be involved. These were frequently described as the land costs, professional fees, furniture and fittings and maintenance. The importance of total costs and how initial construction costs might be better considered within the principles of overall total-cost appraisal were not envisaged. The consideration of future costs or costs-in-use was an idea that was not yet even being proposed by the industry. However, during the high levels of construction activity in the 1960s, it was recognised that the existing procedures were inadequate for meeting the needs of clients.

Clients' expectations of their buildings, in line with expectations for other goods and services, were also increasing. Approximate estimating was frequently too approximate and some schemes had to be redesigned after tenders had been received. Hence the concept of cost planning was introduced. Cost planning is explained in Chapter 15. Much later whole-life costing, which considered the costs of a project from inception through to demolition, helped to change and inform the philosophy of initial design solutions (see Chapters 17 and 18). But even this philosophy was restricted to those items that were traditionally thought of as building works.

As long ago as the beginning of the twentieth century it was recognised that spatial design was perhaps the most important single variable influencing the costs of buildings. This acknowledgement was the basis of the much used superficial floor area method for calculating the approximate initial costs of buildings. It was also recognised that buildings providing the same function and for the same number of occupants often resulted in different costs for both their construction and future maintenance. Procedures were therefore introduced to rationalise building design by setting parameters that identified just how much space should be allowed. The application of cost limits sought to reinforce these ideas.

Facilities management goes much further than the above important ideas alone. It considers not just the initial and future costs associated with the building, but the entire costs that are incidental in a client's business. Its aim is to manage the total facilities provided, in addition to the costs that are more traditionally related to buildings. This procedure includes examining the amenities of the business to see if they can be undertaken more effectively, efficiently and economically. This information is often set within a context of national patterns and trends, as well as in the particular organisation concerned. Benchmarking techniques are often applied in this respect.

22.2 OVERVIEW OF FACILITIES MANAGEMENT

If you invite a group of individuals to describe the meaning of facilities management, you will obtain as many definitions as there are members of the group. The many and varied functions that are involved in facilities management are not new but the way that these are managed have become more integrated and extended. The role of a facilities manager is recognised as being more proactive and different from that of a traditional estates manager. Facilities management also differs from property management in that the range of activities that it performs far exceeds just buildings and property.

Facilities management is concerned not just with the building structure, services and finishings but also with the activities that go on within the building. It is much more than a combination of the traditional disciplines of estate management and building maintenance. It includes a range of activities that can be broadly summarised under seven headings, as shown in Figure 22.1. This figure indicates the extent to which these services are carried out in-house or contracted out to external consultants. This division of activities will vary depending upon the type, nature and size of the organisation concerned. The headings can be further subdivided into the key facilities management services shown in Table 22.1. These can also include pest control, porterage, laundry, environmental testing, furniture management, etc.

The professional role of facilities managers has been increasingly recognised over the past few years. The emphasis has been on improving value for money in

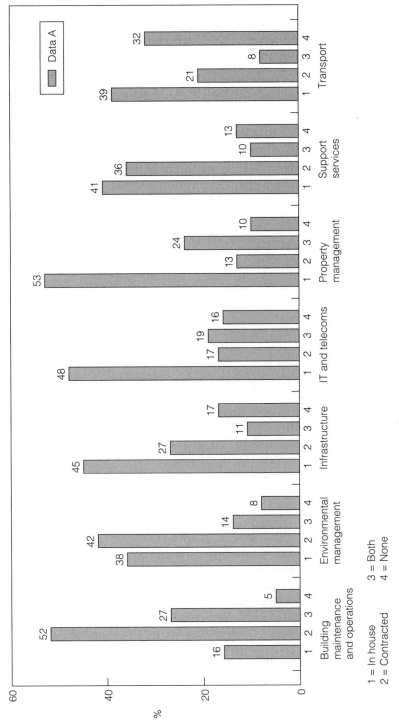

Fig. 22.1 How services are procured by sector (percentage allocation).
Source: Premises and Facilities Management Journal

1 = In house   3 = Both
2 = Contracted   4 = None

Table 22.1   Key facilities management services

Building maintenance and operations	Electrical
	Mechanical
	Fabric
Environmental management	Energy management
	Health and safety
	Waste management
Infrastructure	Utilities
IT and telecommunications	Telecommunications services
	Systems administration and management
	Customer response support
Property management	Asset management
	Space planning
Support services	Catering and vending
	Cleaning
	Office support services
	Security
Transport	Fleet management

Source: Premises and Facilities Management Journal

the provision of traditional facilities services. The use of external consultants (outsourcing) for this purpose has largely been the result of their promise to reduce costs while at the same time maintaining or improving quality. This has been achieved through the wider experience gained by consultants in managing different clients and by their having a clearer focus on their work, not being distracted by in-house politics. Consultants are better able to disseminate good practices because of their contacts with a range of different clients. External consultants also claim to remove some of the hassle from the client. The use of external consultants is partially threatened by the demands of accelerating organisational change. New work practices such as hot desking, hotelling and remote working call for changes in the way that the workplace is being used.

Facilities management has grown out of a range of different disciplines, most of which came from general management, as shown in Figure 22.2. About one-third of practising facilities managers are members of the British Institute of Facilities Management (BIFM). As many as 65% hold other professional qualifications but less than 10% have any academic qualifications in facilities management. Qualifications and skills development are therefore two important priorities for facilities managers. The BIFM's national professional qualification and training programme is being used by major public and private sector organisations.

22.2.1 A strategic discipline

The BIFM has been proactive in marketing this profession to the business community especially in the UK but also overseas. There has been a heightened

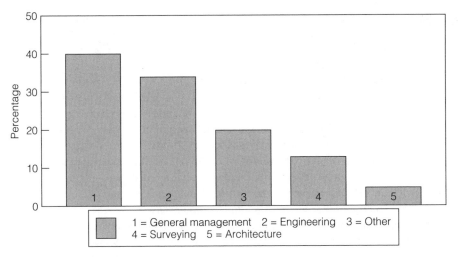

Fig. 22.2   Background of facilities managers

awareness of the discipline that has been driven by a number of different factors including:

- The general interest in outsourcing as an important management topic
- The relatively heavy media coverage of the Private Finance Initiative (PFI), now more commonly referred to as Public–Private Partnerships (PPP)
- The increasing importance of facilities management as viewed by the financial community

Whilst it is now commonly understood that a well-designed and well-managed workplace can directly impact on staff morale, job satisfaction and productivity, other perceived benefits of facilities management are perhaps less well known. Facilities management is a strategic discipline because it translates the high-level changes required by senior mangers and decision-makers into the day-to-day reality of the workforce, their working space and expected performance.

Facilities management is able, amongst other things, to:

- Deliver effective management of an organisation's assets
- Enable new working styles and processes which are essential in our current technology-driven age
- Enhance and project an organisation's identity and image
- Assist in the integration of processes associated with change, post-merger and acquisition
- Deliver business continuity and workforce protection in an era of heightened security threats

The facilities management sector today is large and complex, comprising a combination of in-house departments, specialist contractors, large multi-service companies, and consortia who are delivering a full range of design, build, finance and management projects.

In the UK alone the sector is estimated to be worth over £55bn, and the BIFM is one of the fastest growing professional bodies.

22.3 RELEVANCE OF BUILDING COSTS

The annual costs of buildings in any firm or organisation make a significant contribution towards the firm's annual overhead charges. They are often the second highest cost centre after wages and salaries. For example, in colleges of education these costs typically represent about 10% of the overall annual expenditure. The annual expenditure on plant and equipment accounts for a further 5% of the budget. Making better use of the premises, reducing the costs or attempting to do more for less are therefore commonsense approaches in a world that places quality and value for money at the top of its lists of objectives.

Table 22.2 lists some of the activities with which a facilities manager will need to be conversant in respect of the buildings or premises requirement. Considering these aspects and the fact that many of these services are currently contracted out (see Figure 22.1) a facilities manager needs a breadth of understanding but is unlikely to have an in-depth knowledge in all the areas of activity. A good facilities management team is therefore likely to comprise a breadth of expertise rather than simply a group of similar disciplines. When deciding which activities to contract out an examination of the available in-house expertise will be essential. Consideration will also need to be given to the respective advantages, costs, convenience and control. In order of priority, a survey of facilities managers listed quality, value for money, price, capability, service provided, past performance and financial stability as important factors to consider when contracting out aspects of their work to others.

Table 22.2   Building aspects of facilities management

Procurement management	Purchase of land and buildings
	Management of leases
	Preparation of design briefs
Property maintenance	Repairs, running and replacement costs and their activities
Modernisation	Planned improvements and developments
Space utilisation	Reviews depending upon changes to functions and practices
	Planning and management of relocation
	Interior layouts
Equipment and plant	Rolling programmes for repair and replacement
	Utilisation expectation
	Specifications
	Inventories
Communications	Analysis and modernisation of communications systems
Environmental management	Health and safety at work legislation
Security	Fire and theft, storm damage, equipment failure

22.4 OPERATIONS MANAGEMENT

The operations management aspect of facilities management considers the activities within the building that are required to allow it to be used effectively. They include a range of services that are incidental to the core work of the business but are necessary to enable it to function properly. The emphasis placed upon these and other activities will vary depending upon the type of building and the purpose of its use.

22.4.1 Space management

Spatial planning and control are an important activity for facilities managers. Providing the correct amount of space for a particular function and activity is to some extent limited by the spatial arrangements provided in a building. It should be possible to prescribe this space in a better way in new buildings than where an existing building is to be adapted. Open-plan designs in theory have the opportunity to offer the best solutions, but this does not necessarily mean that the space is used in the best possible way (see Chapter 5). The arrangement of space must also take into consideration the ease of communication between different employees or groups of individuals.

Spatial planning requires set standards to be developed for particular activities. For example, in hospital design, a defined amount of space is allocated per 'bed' not just in terms of its adjacent space but also in connection with the space provided in ancillary areas. Standard units of space are not just some theoretical area but are derived from a study of the organisation's past experience, the individuals involved and through comparison with benchmark standards in similar environments. The provision of the correct amount and arrangement of space enables efficiency and productivity to be improved in a company's core business. Spatial management is not a one-off solution since work patterns and activities are evolving. The adaptation of appropriate space, when the opportunity arises, for changes in working patterns must also follow.

An example of spatial management issues is considered towards the end of this chapter.

22.4.2 Maintenance

Maintenance work is required to preserve buildings and other items of equipment in a state of fitness for purpose. Such practices are encouraged to increase the life expectancy of an asset, reduce the energy consumption that is required and avoid emergency work should the asset fail to perform the functions that are required. The main concern of maintenance is directed towards the building structure or fabric, the internal and external appearances which are usually cosmetic in nature and periodic servicing of the engineering services.

Maintenance work is often described as planned and preventative. The former must relate to a company's strategic plan and an overall, often long-term, approach since most buildings are maintained on this basis. The alternative approach is to completely refurbish a building in its entirety. This procedure is only carried out at relatively long time intervals. Some of the issues relating to maintenance costs are described in Chapter 17.

22.4.3 Engineering services

Engineering services are those aspects that help to bring a building structure to life. They include an increasing list of items such as plumbing and water supply, heating, ventilation and air-conditioning, lifts, escalators and conveyors, electrical power and lighting, gas, security and protective installations and communication systems.

The approach adopted for their management will vary with the type of client, the building, its age and use. For many of these systems, particularly the complex ones, their routine maintenance and repair will be carried out by specialist contractors. On-call contracts will be required where for instance breakdown may have a serious effect on the continuation of the business until the repair can be made. An acceptable performance in use will be the main goal for all of the engineering services.

The application of engineering services provides scope for the facilities manager to develop cost-effective solutions. These solutions will be based upon an analysis of past performance and projected improvements. Long-term solutions may require rapidly obsolete provision being replaced by more reliable plant and equipment that are also less expensive in terms of their energy usage and other running costs. It is also tempting to put off until tomorrow what could be done today, resulting in repair costs out of all proportion to the savings achieved today. The modern engineering services use technologies that are more reliable and over time may be shown to be more cost-effective. An unexpected breakdown not only has a cost for its repair but will result in the knock-on effect of lost production.

22.4.4 Asset registers

A part of the facility manager's work is maintaining and updating the register of the company's inventory of assets, e.g. plant and equipment. The inventory should list all of the items that have been purchased by the company, their date of acquisition, initial value, current value (if appropriate), age and expected renewal date. The location of the asset and, if appropriate, the manufacturer's serial number should be included. The register will remain an active document and be constantly updated. There are more obvious direct financial benefits to be achieved where a computer-assisted facilities management program is adopted. This will provide tighter management controls, and annual audits of plant and equipment should be simplified. The company will also have improved data on which to plan for the future. The inventory provides an accurate checklist for insurance purposes and to substantiate claims should these be necessary. Items of equipment that are missing, in an incorrect location or damaged can be easily traced through the inventory.

The asset register can also be employed for taxation purposes in respect of depreciation and replacement funding. It acts as a financial register for the company's assets.

22.4.5 Equipment servicing

The plant and equipment provided will vary from the heavy equipment needs of manufacturing industry to small items of office equipment such as information technology, communications technologies, shredders, photocopiers etc. Wherever such equipment exists some form of servicing arrangement will need to be provided. Short-term service agreements usually come with the purchase of equipment and always with leasing and hire purchase arrangements. Long-term provision for maintenance and repairs will be essential so that breakdowns, should they occur, can be rectified very quickly. New EU regulations may mean that initial guarantees will need to have a two-year warranty rather than, as at present, only twelve months.

Where equipment uses consumables such as paper in photocopiers, cheap supplies of paper may cause breakdown failures. This inconvenience may be increased if such supplies also invalidate warranty agreements. This may cause an additional delay in arranging for repairs by the photocopier supplier.

The regular servicing of equipment should result in fewer unexpected breakdowns. This preventative maintenance is also often a part of any warranty agreement, whether in the short term for new equipment or where long-term maintenance and repair contracts have been formulated. The warranties normally insist that the equipment is maintained by a reputable or named contractor. The work must conform with the manufacturer's instructions using genuine parts of equipment.

22.4.6 Systems and software

Information technology and computers are now an integral part of commerce and industry. Few items of equipment are now available that do not in some way or other include computer technologies. In selecting computer systems it is advisable to decide on a single system, particularly in respect of compatibility. This is especially important where manufacturing and office tasks are to be linked together, where, for example, computer-aided design (CAD) is to be linked with computer-aided manufacture (CAM).

The factors that help to dictate the choice of particular systems include: tasks to be performed, familiarity, recommendation, need for flexibility and the requirements for future expansion. The level of information technology literacy is growing all the time but there still remains a constant need for training and updating. The life expectancy of a building's structure can be a hundred years, engineering services in buildings fifteen years and finishes and fittings ten years. By comparison hardware and software systems are becoming outdated after a period of only three years.

The following should be considered in choosing an information technology system:

- Compatibility: ensure that equipment is compatible with other industry standard hardware and software
- Supplier: choose a reputable supplier and where necessary take up references with existing customers of the products
- Clarity: be clear on what is actually required within the company

- Costs: consider not only the purchase and setting-up costs but the charges for annual updating, training etc.
- Maintenance: most companies require an on-site maintenance agreement and thus a local supplier is preferable to one at the other end of the country

The issues of maintaining programs, integration of processes, cost containment, recruitment and meeting project deadlines remain the top five problems of using information technology and computers in practice.

22.4.7 Energy management

The three largest users of energy in modern society are transportation, industry and buildings. The demands from heating, cooling and lighting systems in buildings are responsible for roughly one-third of the energy that is consumed worldwide. Buildings in developed countries account for 50%–60% of energy use.

An examination of energy use in buildings together with an analysis of the costs involved is always likely to result in potential savings in expenditure. The most common approaches that are adopted in improving energy efficiency in buildings have typically taken the form of providing additional thermal insulation through a variety of measures. These applications have little to do with energy efficiency but only the efficient use of energy. Energy efficiency requires that the primary source of energy is also considered since this is measured at the consumer's meter.

22.4.8 Security

The provision of security will range from a basic commonsense attitude towards crime prevention to high security arrangements requiring personnel, digital entry points, alarm systems and closed circuit television. Regulating staff access to certain areas and controlling public access are other features to consider. Receptionists are frequently the first point of contact on entering a building. The need for high security for the protection of personnel or goods and the need for confidentiality in the case of government departments and some commercial and industrial companies require particular considerations. Organisations that might be the target of terrorist attacks require even more stringent systems to be employed. The size of the organisation and the frequency of visitors will determine the features that need to be considered, such as allowing controlled access into different parts of the building and communications with staff within the building.

The following should be considered:

- Assess the level of security that is required and should be provided.
- Consider the overall security rating of the premises.
- Examine the building or site's curtilage, noting all points of entry, both actual and possible.
- Decide whether different levels of security are required in the building.
- Anticipate the possible types of attack or intrusion.
- Devise methods that will frustrate intruders.
- Liaise with other members of staff to determine their concerns.

22.4.9 Communication systems

Communications is now a byword for good and effective management. Communication can be oral or written; it may require the use of internal memoranda, a telephone, fax machine, the use of electronic mail or video links. It is an area within most firms and organisations that has been the subject of irregular but frequent change and evolution. Modern electronic communication systems provide for a range of options that are easy to use. Modern systems are linked to computers to provide easy access dialling, call-back, call-waiting, interrupt messages etc.

22.4.10 Cleaning

The provision of a healthy and clean working environment is essential in most types of buildings. General cleaning and caretaking services are routine items carried out on a daily or weekly basis with more detailed work being done at less frequent intervals. The routine requirements will be dictated by the business processes being employed. In some circumstances specialist cleaning and decontamination will be required. In some manufacturing processes a lack of cleanliness at the correct level can result in the industrial process being disrupted and the goods that have already been manufactured being destroyed because dirt has made them worthless. The safe disposal of waste materials has increased company costs in countries that have become conscious of the environment and overall public safety. Legislation also restricts the way in which the disposal of certain materials may be carried out. It should be remembered that some waste products have a scrap or resale value and it is therefore essential to separate recyclable materials from ordinary waste products. Environmentally conscious firms will do this even for no monetary gain.

The following should be considered:

- The frequency of cleaning that is required
- The standard of cleanliness demanded by the business or industrial processes
- The need to provide safe methods of disposal for dangerous materials
- Provision in the case of emergency cleaning requirements
- The relative costs of different kinds of cleaning procedures and their impact on the design or refurbishment of new buildings and structures
- Schemes to train operatives in more efficient and economic ways of cleaning
- Raising the profile within a company of the importance of carrying out all work in an environmentally friendly manner

22.4.11 Staff welfare

Under the heading of staff welfare is included catering, creche provisions, first-aid, salaries and pensions, counselling etc. The amount of staff welfare provisions will depend upon the size, type and location of the organisation. In the larger ones these are sometimes referred to as central company services. They are provided incidental to the main core business but are now considered to be essential within a good organisation. The origins of staff welfare can be traced back to the industrial corporations of the nineteenth century where it was realised that in spite of the industrial revolution and the greater dependence on machinery, labour still remained a key element in the process. Some employers also possessed social consciences and in spite of the need to make profits recognised that this should be done with an attitude of dignity towards those whom they employed.

Staff welfare now often extends beyond the provision of catering and similar facilities to sports and other leisure pursuits. It may encompass industrial chaplains and staff development and training beyond that which is normally expected from an employee. Some employers recognise that if employees are content then they are likely to serve the firm better and to remain as good company people. Cost-effectiveness and value for money remain key criteria but research has shown that, by having an ethical approach and some concern for people, companies can remain profitable.

22.4.12 Quality assurance

The general purpose standard, to which quality management systems can be required to conform, was initially designed for the mass production manufacturing and engineering industry (originally BS 5750 but now replaced by the international standard ISO 9000). Each step of the process is controlled by means of procedures manuals. The aims of these are to help to achieve product consistency. This is a common difficulty with service industries where product quality also depends on interactions between customer and supplier.

Properly implemented, the system has proved beneficial in many companies, establishing a discipline in the operation of quality-producing procedures. It can also be a potentially powerful marketing tool. The main features of ISO 9000 are:

- A well-documented but somewhat bureaucratic system
- Well-defined procedures and processes
- Clearly defined standards for the quality system
- A product consistency, which in practice may be difficult to achieve
- A system that is audited by a third party

Total quality management (TQM) is a system that seeks to realign the mission, culture and working practices of an organisation by means of pursuing continued quality improvement (see Chapter 6, p. 116). The main features of TQM are:

- A company-wide commitment to quality
- A focus on satisfying customers' needs
- A commitment to continuous quality improvement
- All staff are responsible for achieving quality outcomes

The key to a successful quality assurance system is in the interpretation of correct quality. This should not be interpreted as necessarily meaning the best possible quality, since this might be a wasteful and uneconomic policy to follow.

22.4.13 Health and safety

Employers have been responsible for health and safety for a very long time and have statutory obligations to fulfil which include:

- Securing the health, safety and welfare of persons at work
- Protecting others against risks to health and safety arising out of work activities
- Controlling the keeping and use of dangerous substances
- Controlling the emission into the atmosphere of noxious or offensive substances

The two particular categories of risk for an employer to consider are:

1. The health and safety of employees whilst they are at work
2. The risk to persons, not in the employer's employment, but arising from the employer's activities

Facilities managers often have a direct responsibility for health and safety and particularly so where they have been designated as such within a company. Their interest in health and safety covers the following:

- Ensuring that all personnel adopt safe working practices
- Keeping plant and equipment in a safe state of repair
- Identification of dangerous areas of working
- Identification of potentially dangerous industrial practices
- Adequacy of storage, testing and maintenance of equipment
- Provision of appropriate safety equipment
- Execution of safety audits
- Risk analysis of health and safety matters

22.4.14 Post-occupancy analysis

Post-occupancy analysis (POA) is a process of evaluating buildings after they have been in use for some time. Rather than focus solely on the maintenance of the building structure, POA considers the users of the building. Practices and procedures change over time, so the need for space changes. In some cases the only resolution to the problem is to offer some building adaptation in order to improve the building's performance for its users. The knowledge and information that can be gained from a POA provide a better understanding of how buildings can be designed for better utilisation in the future. Far too often, however, designers fail to seek out this information in any sort of methodical way, relying instead only on an owner's perspective, hearsay and a rule of thumb approach.

The briefing process to clients, under the RIBA Work Stages, is complex but highly fragmented in its approach. Sometimes clients may appear to get in the way of a designer, who feels that he or she has a better understanding of the client's needs. Clearly a designer is selected to bring in new ideas and fresh perspectives to the design. Also sometimes functionality is sacrificed to aesthetics. POA evolved in response to the growing concern that completed buildings are not performing to the expectations of the occupants. The Latham report (1994) compared buildings to automobiles and found that the latter scored high marks for fitness for purpose. Buildings, on the other hand, received a much less complimentary evaluation.

POA was initially developed to relate the design to the building's use. A POA is particularly helpful to designers of future buildings and is essential in complex building design and in relation to functional arrangement. In practice, of course, the different users of a building bring in many different perspectives and these are not always in agreement. POA recognises this and seeks to develop a broad agreement. POA encourages users to think carefully about their workplace, for example, and this in itself helps to generate new ideas to inform future practices.

It also needs to be recognised that buildings in themselves usually have a much longer life span than the business itself. They represent a durable asset and when a business changes and no longer requires its premises for work, other uses will need to be found. Buildings designed solely for a specific commercial or manufacturing process will be much less adaptable and their onwards value reduced since considerable adaptations may be required. A building's value is represented by its function. When that function ceases, in some cases it may be unsuitable other than for a similar purpose. Buildings involve a number of owners and firms throughout their operational life. Each has its own perspective and some form of adaptation is usually required in the case of existing buildings that may have been in use by previous owners for some time.

Ultimately, the effective use and management of buildings is about their fitness of purpose to the users. Where there is not a close fit, then users may require considerable adaptation or even new premises. This fitness for purpose is made up of a number of elements, not only the activities of the users, which may be appropriate, but also the size of the premises, which may no longer match the functions that are being carried out, or there may even have been a growth in the number of occupants.

Table 22.3   Virtuous features of office buildings

Convenient location	Proximity to transport and shops
Shallower plan forms	Access to natural light
Lower occupant densities	People can spend long hours at their desks
Cellularisation	Especially for meeting rooms
Reliable heating	Comfortable and predictable conditions
Ventilation	Controlled without unwanted air infiltration
Openable windows	
Views out	
Quiet areas	Especially at break times
Efficient security	Particularly where people work late into the evening
Vending facilities	Convenient and subsidised
Cleaning	Regular cleaning and maintenance
Usable controls	For heating, blinds, ventilation
Occupancy patterns	That are predictable
Management	Well-informed, responsible and diligent

Source: Adapted from Institute of Directors (2001)

Table 22.4 Vicious features of office buildings

Location	Inconvenient and difficult to access
Deeper plans	Poor natural lighting and ventilation
Open-plan areas	More interruptions
Circulation routes	Illogical layouts
Workgroups	Too large
Working hours	Too long
Immobility	People tied to one place
VDU	Long amounts of time spent at VDUs
Technology	Complex arrangements
Noise	Irrelevant and intrusive
Facilities management	Ineffective, indifferent or arrogant
Quiet areas	Non-existent
Working environment	Insecure
Staff facilities	Poor
Work space	Untidy and poorly maintained

Source: Adapted from Institute of Directors (2001)

Tables 22.3 and 22.4 indicate virtuous and vicious characteristics of work spaces in offices. Some of these features can be achieved with minimal expense but others will be more costly. The facilities manager will be interested in these criteria since they are likely to have an impact on the productivity of the workforce. The occupants are most concerned with what might be termed the boring basics and not with image, aesthetics or the design quality. The vicious is the reverse of the virtuous. What often makes the difference between a good and bad building is the attitude of the management and the organisational culture and resources.

Cost studies of buildings, to be fully effective, must now consider the bigger, holistic picture and not just initial building costs in isolation from the other factors. Building costs are nevertheless very important, since an incorrect solution initially may become difficult or too expensive to remedy at a later time.

22.5 SPACE MANAGEMENT EXAMPLE

22.5.1 College of further and higher education

The initial impression of any building project is gained by the layout of the site and its architectural attributes. It is also concerned with the arrangement of space that is provided within the accommodation that is available. It is necessary to carefully distinguish between the physical characteristics and the way in which the premises are arranged and managed. The primary aspects of the former are:

- Location
- Type
- Amount
- Condition

Within a college of further or higher education it is necessary to determine whether:

■ The size of rooms being used matches the number of students
■ The rooms being used are appropriate for the intended activity
■ The rooms are located adjacent to other related activities

The management of the accommodation will also need to be appraised in respect of how well the space is used, e.g.

■ Deployment through sensible timetabling
■ Weekly room utilisation

22.5.2 Capacity

The Department for Education and Employment has set down guidelines for the appropriate amount of space required. These recommend areas varying between approximately 7 m² and 13 m² per full-time equivalent student (FTES). The areas required are dependent upon the activities being undertaken and the type and level of programmes offered. However, the recommendations are apt to become quickly outdated as new methods of teaching and study evolve. In most cases the overall spatial requirements are now much less since the formal class teaching hours have been reduced. This is a result of a move towards more student-centred activities. Guidelines are available for the amount of teaching space that is required for each student. For example, in a typical classroom, each student requires about 2 m² of floor space. This compares with as little as 1 m² in a purpose-built lecture theatre and as much as 5 m² in a laboratory. In heavy craft workshops, such as engineering or construction, the recommended area is approximately 8 m² per student. Table 22.5 suggests some guide figures for typical room area ratios.

   After staff costs, premises-related expenditure is the next most significant item in a college's budget. These costs include rates, heating, lighting, cleaning and the general running costs associated with the maintenance of the buildings and site. Many of these elements are fixed costs bearing little relationship to their actual usage. Thus, if the volume of activity can be increased, and in most colleges this is occurring, the unit cost or cost per full-time equivalent student (the typical unit of measure) decreases.

Table 22.5   Typical room area ratios

Location	Typical area per student (m²)
Lecture theatres	1.0
Classrooms	1.8–2.1
Laboratories	3.0–4.6
Art and design studios	3.2–5.6
Construction and engineering workshops	7.5–8.4

Source: Department for Education (1992)

22.5.3 Utilisation

The utilisation of the accommodation can be monitored against national norms or the college's own target figures. In order to compare this, it is first necessary to prepare an inventory of the available accommodation and then measure the actual usage.

- What room space is available (seats)?
- When is this space available (frequency)?
- How is this space used (occupancy)?

Example

A lecture theatre capable of seating 100 students is available for 40 hours each week. During a typical week it is in use for 30 of these hours and on average is occupied by 70 students. The utilisation factor is:

$$\frac{30 \times 70}{40 \times 100} = 52.5\%$$

Some college managers suggest that their available space should be used for about 80% of the time at 80% capacity, i.e. 64% utilisation. Accommodation plans provided by colleges are often based upon this indicator, but achieving this over a long period of time is difficult. Students who prematurely leave courses or fail to attend as required or the cancellation of lectures result in utilisation rates that are perhaps closer to 49% (i.e. 70% × 70%).

22.5.4 Efficiency

The colleges that provide for a large proportion of part-time students, adult education or evening short courses operate their premises on a three-session, five-days-per-week basis. Days are then typically of twelve hours' duration. Some colleges will also open parts of their premises, such as libraries, at weekends, work to an extended college year and encourage summer schools to use their accommodation in order to achieve greater efficiency.

An examination of room utilisation data shows that there is a dramatic reduction in the need for teaching accommodation between the hours of 1200 and 1400 and that a significant decline in use occurs after 1500 hours, particularly later in the week. This trend is less pronounced in colleges where a significant amount of adult education occurs. Figure 22.3 represents a day's activity in a typical college showing the hourly needs for accommodation.

22.5.5 Condition

A large proportion of the accommodation used by colleges was built either at the turn of the twentieth century or during the large educational buildings programme of the 1960s. Only a limited amount of systematic maintenance has since been undertaken

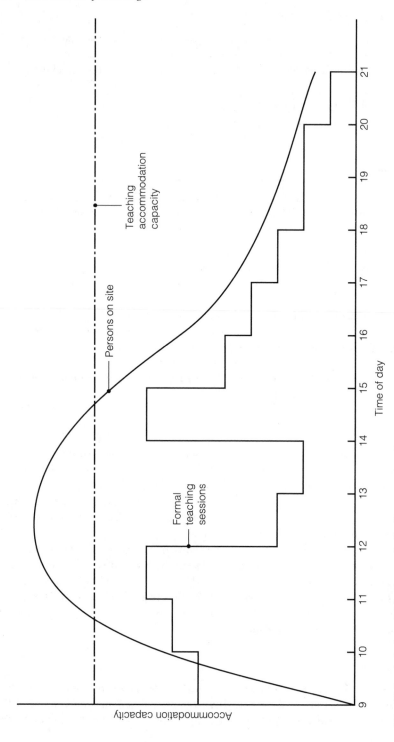

Fig. 22.3   A day's activities in a typical college or university

and consequently a good deal of building stock is in a less than satisfactory condition. Condition monitoring is a concept which concerns comfort for those who use the premises, and includes decorative order, heating and ventilation, lighting, sound and acoustics, type of furniture, cleanliness, ambience, ease of access and location, and displays and artefacts to help create a stimulating learning environment.

22.5.6 Future developments

In assessing the suitability of the accommodation, it is necessary to take into account the probable and possible developments for its use. These may include:

- Changes in teaching and learning styles
- Movement towards large group teaching
- The 'reading' for a degree and the importance of self-supported study
- The needs for study skills and open-access workshops
- An extension of the opening times of the premises, especially the library and other learning resources
- Resource-based learning developments

Colleges have accommodation policies and systems that are used for reviewing and evaluating the accommodation. The policies include:

- Information on changes in health and safety regulations
- Maintenance programmes
- Suitability for current use
- Links with other accommodation for flexibility
- Innovation
- Cost
- Overall effectiveness

CONCLUSIONS

The emerging discipline of the facilities manager has helped towards providing a new focus for the study of building costs. It is no longer sufficient to examine costs in isolation from other activities carried out by clients. The rationalisation of building designs in order to provide the space that is required rather than the space that was wanted has been a positive way forward. Changes in working practices will continue to have implications on the provision of this space. The consideration of whole-life costs further enhanced the advice that was provided to clients. Facilities management has moved yet one stage further by seeking to place building costs within a context of overall client costs.

SELF ASSESSMENT QUESTIONS

1. Describe the kinds of activities that are undertaken by facilities managers and explain why companies are now organising their businesses in this kind of way.

2. What kinds of knowledge and experiences should an individual ideally have to take on the role and responsibility of a facilities manager?

3. Describe the relationship between facilities management and the whole-life costs of buildings.

BIBLIOGRAPHY

Alexander, K., *Facilities Management: The Professional Guide to the State of the Art in International Facilities Management*. Blenheim Business Publications 1995.

Armstrong, J., *Facilities Management Manuals: A Best Practice Guide*. CIRIA 2002.

Ashworth, A. and Harvey, R.C., *Assessing Quality in Further and Higher Education*. Jessica Kingsley 1994.

Barrett, P., *Facilities Management: Towards Best Practice*. Blackwell Publishing 1998.

Barrett, P. and Baldry, D., *Facilities Management: Towards Better Practice*. Blackwell Publishing 2003.

Becker, F.D., *The Total Workplace: Facilities Management and the Elastic Organisation*. Van Nostrand Reinhold 1990.

Best, R., Langston, C. and De Valance, G., *Workplace Strategies and Facilities Management*. Butterworth Heinemann 2003.

BIFM, *Rethinking Facilities Management*. British Institute of Facilities Management 2003.

Department for Education, *Further Education and Sixth Form Colleges: Developing Strategies for Accommodation*. Department for Education, Architects and Building Branch 1992.

Evans, K., *Facilities Management: Practical Guide to Facility Management*. Spon 1996.

Hinks, J., Kelly, J. and McDougall, G., *Facilities Management and the Chartered Surveyor*. RICS 1999.

Institute of Directors, *Buildings that Work for Your Business*. Director Publications Ltd 2001.

Langston, C., *Sustainable Practices: ESD and the Construction Industry*. Envirobook Publishing 1997.

Langston, C. and Lauge-Kristensen, R., *Strategic Management of Built Assets*. Butterworth-Heinemann 2002.

Latham, M. *Constructing the Team*. HMSO 1994.

McGregor, W. and Then, D.S.S., *Facilities Management and the Business of Space*. Arnold 1999.

Park, A., *Facilities Management: An Explanation*. Macmillan 1998.

Powell, C., *Facilities Management: Nature, Causes and Consequences*. CIOB Technical Information Service No. 34 1991.

Robertson, D., *Auditing Facilities Management: An Overview*. McGraw-Hill 1994.

SJT Associates Ltd, *Facilities Costs and Trends Survey*. University of Strathclyde 1992.

Spedding, A., *The CIOB Handbook of Facilities Management*. CIOB 1994.

University of Strathclyde, *An Overview of the Facilities Management Industry*. Centre for Facilities Management, University of Strathclyde 1992.

Williams, B., *Facilities Economics including Premises Audits*. Building Economics Bureau 1996.

CHAPTER 23

THE FUTURE DIRECTION OF COST STUDIES OF BUILDINGS

LEARNING OUTCOMES

After reading this chapter, you should have an understanding of the future direction of cost studies and building. You should be able to:

■ Recognise how important developments from the past have influenced the future
■ Identify the shift that is taking place in focusing more on the process rather than on the product
■ Appreciate the principles of best-value solutions
■ Recognise the importance of adding value to the process and the product
■ Distinguish other techniques that are important in evaluating construction solutions

23.1 INTRODUCTION

A former colleague of mine used to argue that there was really no such thing as cost optimisation. He was, of course, quite correct. We can only optimise within the constraints of our current knowledge and experiences. In attempting to discover where the future direction of cost studies in buildings lies, this book follows a similar argument. What is certain is that the discipline will not stand still. But we have not yet acquired all of the knowledge and experiences for all time and in our lifetime we probably never will.

It must not be forgotten that quantity surveyors have for generations been assisting designers, architects and civil engineers to find ways in which the same result can be achieved for a lower cost. This has in practice meant identifying and then reducing waste and efficiencies in the design. During the process of taking off quantities for bills of quantities, for example, there are numerous examples that can be quoted where a surveyor has made recommendations to the architect that a design detail could be simplified without altering the aesthetics or function of the design in any way. Most architects have been grateful to receive such advice in enabling them to achieve a form of design optimisation. Such simplification has inevitably resulted in a reduction in initial construction costs and in some cases in the ongoing costs-in-use. This is a primitive form of value engineering whilst not generally following the principles as outlined in Chapter 19.

The study of building costs is still barely half a century old. The first substantive textbooks published on the subject did not appear until 1961 (see Chapter 2). The emphasis then was on the cost planning of buildings and the parallel development of the two methods used for this process that were outlined in Chapter 15. It was realised at that stage that very similar buildings, in terms of their function, were being constructed around the United Kingdom, since this is where the study of this subject was first developed. The best examples that can be cited relate to public buildings such as schools, houses and hospitals. It is surprising now that such comparisons were not made much earlier in time. That, of course, relates to current knowledge and experiences, and to hindsight, which is a tool that everyone would like to acquire.

These comparisons were discussed in Chapter 15 with reference to school buildings and the variable amount of space and subsequent costs that were being allocated during the era shortly after the Second World War. During this time there was also a period of austerity but with a massive building programme to replace war-damaged schools and schools that had fallen into serious disrepair during the war years, and the need to provide new schools for an aspiring population. Necessity is the mother of invention. The true importance of building assets had yet to be realised – there was, rather, a philosophy of make do and mend.

Three factors were particularly significant as the importance of cost control and cost planning during the early design of new building projects emerged at that time. There was the need for the large-scale public building programme referred to above. There was also a trend towards using non-traditional building materials and industrialised systems of building to meet the demands of early completion. These tended to increase the complexities of both design and construction, at least initially until designers and contractors acquired familiarity with such systems. Clients in both the public and the private sectors were becoming increasingly aware of the necessity for improved cost advice and reliable evaluations prior to committing the project and its associated finances to working drawings.

This pattern of studying the costs associated with buildings became standard good practice during the ensuing years. Cost planning grew as a subject and other titles such as cost modelling, costs-in-use, whole-life costing and value management were added to recognise and acknowledge the importance of costs during both development and construction.

Much later, it became recognised that to evaluate the different types of construction technologies or even to make comparisons in the floor space provided for buildings with similar functions and similar occupants was insufficient in itself. Greater attention needed to be given to the way in which the floor space was used. For example, in 1992 in the UK the management of further education colleges was removed from local authorities and given to centralised funding councils. Comparisons were made between colleges in England, Scotland and Wales that revealed considerable differences in provision. Even with the considerable expansion of student numbers in these colleges over the following ten years, over one million square feet (100 000 m²) of floor space was able to be disposed of. This not only reduced the capital costs required for the buildings' expected future replacement but also gave considerable savings in their ongoing whole-life costs.

23.2 A PARADIGM SHIFT IN PRACTICES

The construction industry worldwide is currently undergoing a phase of massive change and redirection. In the UK this is the result of a drive to re-engineer the construction process that has been stimulated by a government-driven agenda through reports such as *Constructing the Team* (Latham 1994) and *Rethinking Construction* (Egan 1998; Crane 2002). These provide a focus for change in the way in which capital construction projects are procured and the need to improve quality, time and cost characteristics. They offer a new agenda to enable an increasingly dynamic and responsive industry to address the way in which it operates and organises its activities, improves value for money and focuses more clearly on the needs of the client and on getting it right first time.

Much of the important and valuable work that has been undertaken to improve the economics of construction has focused on the building product, such as the finished school building or supermarket. It has previously been discussed that building costs are normally quantity-related, whereas civil engineering costs are more process-determined (see Chapter 4, p. 53). The gains that have undoubtedly been achieved through the correct application of cost planning techniques and cost studies of different components (see Chapter 9) cannot be easily refuted by anyone who understands them. However, such gains may be considered to be small when set in the context of a changing construction industry and the demands made by some clients for a reduction in capital building costs alone of some 30%. Whilst the emphasis has quite rightly been focused on the product, the time is now right to pay attention to other factors, such as the way that building projects are conceived through procurement routes and the way in which the construction industry performs its tasks. Also, much of the practices of cost planning and cost studies have been, for want of a better term, professional quantity surveyor-orientated. The industry has already done a considerable amount of refocusing by appointing the contractor as the lead consultant on a great many schemes.

At the beginning of the twentieth century, a family house cost approximately the same as a family car. At the beginning of the 21st century, the ratio between the two had increased to approximately 10 : 1. If the costs of manufacturing motor vehicles had stayed the same, these would now be beyond the reach of most individuals and a scarce commodity. The car manufacturing industry has had to examine its cost–value ratios to help it remain in business. It has also addressed the issues of manufacturing costs, defects and whole-life costs. All sorts of comparisons have been made between the manufacture of cars and the construction of buildings, and generally it can be shown that the progress that has been made with car manufacture has not yet been matched in the construction of buildings.

23.3 ADDED VALUE

Added value is not new. It has been around for over 200 years, although until recently little seems to have been written about it. A United States of America Treasury official, Mr Tenche Cox, supposedly devised the idea. The technique is useful in measuring company performance and in monitoring manpower productivity or work outputs. It is claimed by some authors to be the real key to prosperity. It is usually referred to as added value rather than value added, since the latter description gives the impression of an afterthought, but the two terms are often thought of as synonymous with each other.

Added value is a term that is used to describe the contribution that a process makes to the development of its products. At a primitive level man goes into the forest and cuts down a tree and converts this into furniture. In so doing he adds value to the raw materials he has used. In a more complex industrial society, a manufacturing business purchases materials, components, fuel and various other services. It converts these various resources into products that can be sold for more than their combined cost. In so doing it adds value. The amount of added value may be insignificant to make the process worthwhile. Different individuals and firms will assess the significance in different ways. The construction industry uses plant, materials, people and other resources. Its completed projects are usually greater than the sum of the various parts. This represents added value. Clients may be prepared to invest £10,000 in professional fees for £100,000 of demonstrable client benefit or added value. There is an increasing demand on the part of some clients to assess the payment of professional fees in this way. It is believed by many that where benefits, or added value, cannot be demonstrated then the associated professional services may disappear.

An alternative method of increasing added value without changing prices or products is to reduce the costs of the production involved. The techniques of value engineering, analysis and management have been shown to be effective in this way. The organised approach to cost reduction goes beyond conventional methods by questioning the function of the product, component or process.

In the context of business, activities are considered to be repetitive actions that are performed to fulfil a business function. These activities can be judged to be either value-added or non-value-added. An activity may increase the worth of a product or service to the client, and where a client is willing to purchase such an activity then it is considered to be added value. Some activities simply add time or cost to a service but do not increase its worth significantly enough to the client. In these situations they are described as non-added value. The additional time or costs here are considered to be unnecessary and should therefore be eliminated.

It should be noted that the measure of added value is not the effort that has gone into this activity. Added value is defined by the satisfaction of the customer or client and not by the producer. The manufacture of quill pens, for example, may require considerable effort on the part of the producer, but since nobody now wishes to purchase these products then no added value has been created.

An added value analysis is not merely a quantitative analysis. It is a management tool that assists in the identification of strengths and weaknesses by the interpretation of the issues raised by the added value analysis.

23.4 ADDING VALUE IN CONSTRUCTION

The Quantity Surveyors Division of the Royal Institution of Chartered Surveyors published in 1998 a report titled *The Challenge of Change*, which considered the changing business world, the needs of clients and the skills that will be required by practitioners in the future. It emphasised the importance of lean production methods and of removing from the construction process anything that did not add value. This would enable a reassessment of the concepts of value and the needs of clients. In this report value was described as 'a capability provided to a customer at the right time and at an appropriate price, as defined by the customer'. The report identified several ways in which added value might be achieved. These included:

- Facilitating earlier trading by helping clients to gain market advantage by selecting appropriate procurement routes
- Reducing maintenance costs through the use of whole-life costing
- Minimising disruptions to businesses during repair and maintenance works
- Providing a point of single responsibility for clients
- Reducing the focus on the cost of project components and increasing the focus on the benefit that the elements of the project can bring to a client's business
- Offering a capability of calculating the added value to the client's business of different alternatives
- Understanding how a client's profits might be increased or business costs reduced
- Being prepared to challenge designs by suggesting new ways of undertaking work
- Suggesting, developing and advising on different contractual and procurement arrangements
- Understanding how construction work may be differently sourced, procured, supplied and installed
- Gaining an understanding of how a client's competitors source work elements and carry out projects
- Establishing sector benchmark data to facilitate inter-project comparison on design periods, lead times, installation methods and specifications for comparable projects

Comparisons that have already been made, and continue to be made, with other industries suggest that the targets set by this report are not overly optimistic. The increased use of information technology at all levels in industry is likely to play a major role in their attainment. The report draws on achievements now being realised in the engineering industry.

23.4.1 Construction industry attitudes towards added value

Tables 23.1, 23.2 and 23.3 are the result of a survey of different companies representing a spectrum of firms in the construction industry. The response from these firms was encouraging, in that all respondents felt that some reductions in construction costs could possibly be achieved without affecting the value. However, none of those surveyed felt that a reduction anywhere near a 30% target (Construction Industry Board, 1996) was realistic, unless more radical and

Table 23.1   Possible percentage reductions in costs that could be achieved without affecting value

	Percentage
Consultants	0–5
Contractors – General	10–20
Subcontractors	5–10
Housing	5–15
Component manufacturers	5–10
Clients	0–40

Source: Ashworth (1996)

Table 23.2   Areas where savings are likely to be achieved

	Percentage
Design	30
Component manufacture	25
Procurement	20
Construction	40
Management	10

Source: Ashworth (1996)

Table 23.3   Some suggested areas where savings could be achieved

	Percentage
Design readiness	85
Single point responsibility	60
Standardisation of components	70
Standardisation of designs	45
Off-site manufacture	55
Continuity of workloads	90
Increased mechanisation	45
Teamwork	85

Source: Ashworth (1996)

fundamental changes to the industry and its products were made. They also felt that such fundamental changes were likely to be unacceptable to clients. Whilst some contractors believed that savings of up to 20% were possible, this was set against a background where they were in greater control of the building process as the lead consultant. Also under these circumstances, they expected single point responsibility and little interference or variations during construction operations on site from the client. House builders indicated that they had already achieved considerable savings and their main problem was the consistent demand for their products and its subsequent effect upon the volume of units that were able to be constructed. Some suggested that during the recent recession in the industry, construction costs were somewhat artificial anyway. Both suppliers and subcontractors had been prepared to offer quite low prices for their products and services simply to maintain their business activities.

Consultants, especially designers (mainly architects), on the other hand, were rather more pessimistic about possible long-term cost reductions, although they did accept that savings could always be made. They tended to suggest that as far as possible these issues were always examined carefully during the design process. There was naturally some resistance amongst designers towards further standardisation, in the use of both designs and building components.

Component manufacturers were eager to find better and more economical solutions. They believed that the introduction of new technologies into the manufacturing process would not only improve product specifications but also enable their products to be manufactured more easily and economically. They were very much aligned to the progress that had been achieved in manufacturing industry in general. Long-term investments in expensive plant and equipment that have since been made have had an effect on volume manufacture, and this was an important consideration in respect of reductions in cost.

Whilst clients thought that costs could be reduced without detrimentally affecting quality, they expected real costs to increase, especially after a recession in the industry. They did accept that there was always likely to be a better way of achieving the same objectives, but that consequent cost savings would be balanced against prevailing tender conditions. Many still insisted that competitive tendering provided the best opportunity to reduce costs against defined specifications in standards and quality. Any possible upturn in building activity would have the effect of artificially increasing prices. Some of the larger clients of the industry have already suggested that the '30% reduction in costs' would be insufficient to meet their own needs and expectations.

Inevitably, the different organisations surveyed expected cost savings to be achieved by one of the other parties involved in the process. Consultants wanted a more precise brief from clients and better organisation from contractors. Contractors wanted more completeness in the designs with less interference from consultants, and through consultants from clients during the construction process on site. Overall, the suggestion that there is scope for design cost reductions indicated that value analysis applications have still some way to go and that the lessons suggested from buildability have not been fully realised.

All of the different parties involved wanted more ideal circumstances and situations to exist in order to effect acceptable cost reductions. A more buoyant industry with longer-term future prospects would encourage investment in the industry from all quarters. Consultants and contractors would become more profitable and this general feeling of well-being would effect greater efficiencies in practice. The stop–go nature of the industry was incompatible with achieving the best buildings. Others argued that such a requirement was incompatible with the nature of the industry and an aim that would not be easily achieved. Any attempt at smoothing out the workload pattern would be welcomed, even if this meant an industry of a reduced capacity.

Many of the specific areas where cost reductions could be achieved have been well rehearsed in recent times and are considered more fully in Chapter 2. All of the parties involved want to reduce unnecessary waste and its consequent expenditure. A co-ordinated industry-wide approach would help to solve the problem in the quickest and most effective way. Trust between parties is a key ingredient. Some contractors argued that in the past some of their good ideas and suggestions for reducing the costs of building and shortening the contract period had frequently been dismissed by designers working under traditional contractual arrangements.

23.5 LEAN CONSTRUCTION

The lean construction process is a derivative of the lean manufacturing process. This concept has been popular since the early 1980s in the manufacturing sector. The original thinking was developed in Japan, although it is now being considered and introduced worldwide. It is concerned with the elimination of waste activities and processes that create no added value. It is about doing more for less.

Lean manufacturing is based on the elimination of waste, including time lost waiting for missed or delayed material supplies, unnecessary storage and the value that is tied up in large stocks or components waiting for assembly. The Just-in-Time philosophy forms an integral part of this concept, and lean manufacturers have needed to develop a network of suppliers and to form a robust supply chain to meet their requirements at the appropriate time. Significant efforts are made to encourage all such firms to adopt the same principles. Lean manufacturers have thus moved away from traditional relationships with suppliers to partnering arrangements with a smaller number of clearly focused suppliers.

Lean production is the generic version of the Toyota Production System. This system is recognised as the most efficient production system in the world today. Incidentally, Toyota's activities in the construction industry are larger than those of its more well-known automobile business! It also needs at the outset to be acknowledged that construction production is different from that of making motor cars, although it is possible to learn from and adapt successful methodologies from other industries. In the automobile manufacturing industry, spectacular advances in productivity, quality and cost reduction have been achieved in the past ten to fifteen years. Construction, by comparison, has not yet made these advances. It also remains the most fragmented of all industries, but this can be seen as both a strength and a weakness.

The application of lean production techniques in motor vehicle manufacturing has been a huge success and is associated with three important factors:

- The simplification of manufacturing dies
- The development of long-term relationships with a small number of suppliers, to allow just-in-time management
- Changes in work practices, i.e. in the culture and ethos of practices, most notably the introduction of team-working and quality circles

Lean thinking is aimed at delivering what clients want, on time and with zero defects. Lean construction has identified poor design information which results in a large amount of redesign work. Several organisations around the world have established themselves as centres for lean construction development. The Lean Construction Institute in the USA, for example, is dedicated to eliminating waste and increasing value.

Because it is a philosophy, lean construction can be pursued through a number of different approaches (Figure 23.1). The lean principles have been identified as follows:

- Eliminate all kinds of waste. This includes not just the waste of materials on site, but all aspects, functions or activities that do not add value to the project.
- Precisely specify value from the perspective of the ultimate customer.
- Clearly identify the process that delivers what the customer values. This is sometimes referred to as the value stream.
- Eliminate all non-added-value steps or stages in the process.

Fig. 23.1 Lean construction in practice
Source: Ashworth (2006), p. 176

- Make the remaining added-value steps flow without interruption, through managing the interfaces between the different steps.
- Let the customer pull, do not make anything until it is needed, then make it quickly. Adopt the philosophy of *just-in-time management* to reduce stockpiles and storage costs.
- Pursue perfection through continuous improvement.

23.6 BENCHMARKING

Benchmarking owes much of its relatively recent development to the success achieved by the Xerox Corporation. In the late 1970s, Rank Xerox were severely challenged by Japanese competitors who were able to manufacture photocopiers of a better quality and at a lower cost. In response, Xerox successfully implemented a programme of benchmarking which incorporated the comprehensive examination of their entire organisation. Better practice, in any relevant process used by another company, was identified and translated into new Xerox practice. This facilitated a significant improvement in company performance and an improvement in their competitive position.

Benchmarking has been described as a method for organisational improvement that involves continuous, systematic evaluation of products, services and processes of organisations that are recognised as representing best practices. It is a system that uses objective comparisons of both processes and products. It may make internal comparisons within a single firm, by, for example, comparing the performances of different building sites. It can be an external system, where comparisons of performance are made against similar and dissimilar firms. The enterprises that are recognised, internationally, as being the world leaders are described as having the best practices and are the most efficient. The Construction Industry Board in the UK has published a list of *key performance indicators* that are updated regularly and can be used by any firm to measure its performance.

There is considerable interest in and practice of benchmarking techniques across a wide range of industries. It has been suggested that probably all of the top firms and companies use it in one way or another. It is also worth remembering that benchmarking is not just practised by those firms that are lagging behind and need to improve their performance, but by the international world leaders, as a tool for maintaining their competitive edge. The focus of benchmarking is on the need for continuous improvement.

The aim of benchmarking is to improve the performance of an organisation by:

- Identifying the best known practices relevant to the fulfilment of a company's mission. As will be discussed later, this may be found internally or externally to the company.
- Utilising the information obtained from an analysis of best practice to design and expedite a programme of changes to improve company performance.
  It is important to stress that benchmarking is not about blindly copying the

processes of another company, since in many situations this would fail. It requires an analysis of why performance is better elsewhere, and the translation of the resultant information into an action suited to the company under consideration.

In theory, therefore, benchmarking is a very simple process involving a clear pattern of thought:

- Who does it better?
- How do they do it better?
- Adapt and/or adopt best practices
- Improve performance

23.6.1 The advantages of benchmarking

The advantages of benchmarking include:

- Opportunities to avoid complacency, which may be nurtured by monopolistic situations and result in the loss of market leadership
- Improvements in the efficiency of a company leading to an increase in profits
- Improvements in client satisfaction levels resulting in an improved reputation, leading to more work, better work and an increase in profits
- Improvements in employee satisfaction in working in a more efficient, profitable and rewarding environment

Benchmarking may be performed by access to several reference points. In the case of a large national contracting organisation, a multi-tiered approach may be possible, using, for example:

Internal comparisons
- Comparisons within a section, region or division
- Comparisons within the entire company including all regions and divisions

External comparisons
- Comparisons with similar competitors in the UK construction industry
- Comparisons with non-competitors, which might include those outside of the construction sector

23.6.2 Factors critical to the success of benchmarking

The following are important factors that are critical to the success of benchmarking.

- *The support of senior management.* Token support will not be enough. It is probable that in most organisations, benchmarking will be carried out without relief from other responsibilities and in this sense is an added burden. A great commitment to benchmarking through ownership, active involvement and leadership will therefore be required.
- *An understanding of the processes of an organisation.*

- *Adequate resources.* Although a study may be carried out internally, without the use of an external consultant, staff time, often highly paid staff time, is likely to be an essential ingredient.
- *Staff development.* Staff participating in a benchmarking study should be familiar with the methods used to collect and analyse data that are necessary to identify and examine best practice.
- *Team participation.*
- *Adequate planning of the study.* A timetable of the study and a subsequent action plan should be established and monitored.

23.6.3 Benchmarking and the construction industry

The low standing of the construction industry within the UK suggests that there is significant opportunity for benchmarking to be used with great effect. As discussed in other chapters of this book, clients too frequently criticise the industry because buildings are delivered late, they exceed budget costs and are of poor quality. This provides three broad areas for improvement. This view has recently been enforced by the Rethinking Construction movement. In construction, the need to improve is clear. Clients need better value from their projects, and construction companies need reasonable profits to assure their long-term future. Both points of view not only increasingly recognise that there is plenty of scope to improve, but they also have a powerful mutual interest in doing so.

23.7 RETHINKING THE CONSTRUCTION INDUSTRY

Rethinking the construction industry is a movement in the UK about implementing the radical change identified in the Egan Report, which was published in 1998. This report identified inefficiencies and waste in construction which result, for example, in high construction costs, low productivity, poor safety and excessive material wastage.

The principles established by the task force charged with implementing change were to place the client in the position of leadership, and provide integrated teams for the delivery of projects through partnering and supply chain management to improve the respect given to people. These objectives are aimed at achieving radical improvements in design, quality, customer satisfaction and sustainability of UK construction. This would enable the industry to recruit, train and retain a skilled workforce at all levels by improving its employment practices and health and safety performance.

The task force proposed seven targets for improvement which underpin the Rethinking Construction movement:

- Reduced capital costs
- Reduced construction time
- Better predictability
- Fewer defects

- Fewer accidents
- Increased productivity
- Increased turnover and profit

An important characteristic of Rethinking Construction is for it to involve all sectors of the industry. Its success to date has been variable, but it is especially good amongst the larger contracting organisations.

The Rethinking Construction movement has an important effect on the cost studies of buildings, especially in the context of helping to drive down costs. A further feature of this movement is its commitment towards continuous improvements in quality, efficiency, sustainability and added value. It has an important focus on best value principles and whole-life costing.

In an attempt to illustrate the advantages that Rethinking Construction can have it has developed an initiative known as demonstration programmes. These provide the opportunity for leading edge organisations to promote projects that demonstrate innovation and change that can be measured and evaluated. There are in excess of 500 such projects in the programme. These have indicated that:

- Clients are happier (11% increase in client satisfaction)
- Quality is increasing (30% more projects are reporting fewer defects)
- It's a safer place to work (figures suggest that demonstration projects are 25% safer than the construction industry at large)
- Keeping promises (15% more demonstration projects are ahead of programmes and on budget)
- More productive workforce
- Quicker projects (demonstration projects are completing schemes 10% quicker than a year ago)

Some of the larger contracting organisations are claiming £30 million annualised cost savings through streamlining their programmes.

23.8 SUSTAINABLE CONSTRUCTION

The design, construction and maintenance of buildings has an important impact on the environment and our natural resources. As was noted in Chapter 1 (p. 22), buildings consume large amounts of the energy produced and designers therefore have some responsibility to reduce the global demand. The challenge is to build smart. This will enable such buildings to use a minimum amount of non-renewable energy, produce minimum pollution and reduce the associated costs involved to a minimum. At the same time such buildings should increase the comfort, health and safety of the people who live and work in them. It was also noted in Chapter 1 that buildings are a major source of the pollution that causes urban air problems and the pollutants that contribute to climate change. Traditional building practices often overlook the interrelationships between building components, their surroundings and their occupants.

Sustainable development can be defined as the development needs of today being compatible with and not compromising the needs of future generations. The philosophy of sustainable development borrows freely from the science of environmental economics. There is the basic recognition of a relationship between environmental issues and economics in general – the one affects the other. The government has outlined its views on sustainable development in *Building a Better Quality of Life* (2000). This strategy seeks to achieve four broad aims:

- Social progress which recognises the needs of everyone
- Maintenance of high and stable levels of economic growth and employment
- Protection and, if possible, enhancement of the environment
- Prudent use of natural resources

In recognising the importance of sustainable building development, the construction industry is challenged to provide built assets that:

- Regenerate and revitalise town centres and other urban areas where decay has taken place
- Plan communities to place less reliance on the motor car
- Use energy more efficiently
- Minimise mineral extraction
- Cause minimum damage to the natural and social environments
- Provide training through schemes such as Welfare to Work and the New Deal
- Enhance the quality of life
- Will be acceptable to future generations

Some ideas on how to implement the above are shown in Table 23.4. Many of these ideas make good business sense to all the parties that are involved in the project. They provide a less contractual claims culture with a more predictable contract outcome. They reduce the risk of environmental liabilities and provide more efficient buildings with a more positive image. These in turn attract better quality clients for designers and contractors and indicate that they are proactive firms, fostering good relationships with repeat projects. In a more environmentally conscious world these approaches reassure potential employees that such projects and organisations are not an environmental liability.

Green buildings or sustainable construction projects can often involve more expense than conventional construction, especially in the soft costs for additional design, analysis, engineering, energy modelling, commissioning and certification to relevant standards. Studies carried out in both the UK and the USA show that higher levels of sustainable construction may involve additional capital costs. Such additional costs have traditionally been assumed to be compensated for in the payback or whole-life cost return. However, the increased costs associated with, for example, indoor environmental quality and the incorporation of more environmentally friendly materials may find investments more difficult to substantiate on the value of such savings alone. It may also be argued that employee productivity improves with green designs, although the evidence for this has yet to be evaluated in longitudinal studies over a much longer timescale. Others will argue

Table 23.4 Sustainable construction action list

Clients	Designers	Contractors	Suppliers
Demand forms of contract which foster a partnership, non-adversarial approach	Ensure that employees are up to date with the most recent guidance on how to design with consideration to environmental efficiency	Operate contracts in a non-adversarial way with clients, subcontractors and suppliers	Find sources of materials with low environmental impact. Use materials that would otherwise be waste
Specify BREEM (Building Research Establishment Environmental Assessment Method) concepts		Plan construction activities to minimise negative environmental impact	Source as many of the raw materials as possible from local suppliers and manufacturers
Encourage innovation	Attempt to educate existing clients on the benefits of considering a sustainable approach to construction	Engage and train all those involved with the project	Assess, review and audit activities and put into place appropriate environmental management systems
Include on-site environmental performance requirements in tender documents		Protect the health and safety of those involved on the project and third parties	
Speak with the local community about the proposed project			

Source: Construction Best Practice Programme (a programme funded by the DTI and steered jointly by the Government and Industry)

that green buildings can only be really cost-effective where governments are prepared to offset some of the costs through grants and subsidies to encourage owners, designers and constructors to opt for sustainable designs.

Sustainable construction is one of the important areas today where cost studies can play a significant part for the long term. The part that whole-life costing has to play in this context is considerable. Sustainable construction practices offer the opportunity to create environmentally sound and resource-efficient buildings by using an integrated approach to design. Such buildings promote resource conservation, consider the environmental impacts and waste minimisation, create a healthy and comfortable environment, reduce operation costs and help to address issues such as historical preservation, transportation and infrastructure systems. The entire life cycle of the building and its components are considered together with the economic and environmental impact and performance. Sustainable development is therefore viewed as a complete package with the aspects of the environment and economics being of paramount importance. The principles of sustainable construction apply equally to new development, refurbishment and to the solution of ongoing maintenance problems.

23.9 PROCUREMENT

Perhaps one of the greatest changes the construction industry has seen in the past twenty-five years is the way in which buildings are procured (see Chapter 10). The reasons for change are many but in part due to dissatisfaction with the adversarial aspects of earlier methods and the expectations from new procurement practices that have originated from research, overseas comparisons, government intervention and industry evolution. As noted in Chapter 10, there is no panacea for all projects and no final solution to procurement arrangements. The methods used will continue to evolve to incorporate new technologies, proven practices and improved expectations to suit modern methods of construction. The methods used for procurement, of course, have a great impact on cost, especially since they determine how risk will be apportioned and allocated.

Some have argued that research in construction economics has been more prevalent at the construction stage than at the design stage. Certainly, if research papers in journals are examined they might support this argument. But much of the research on cost studies of buildings has been carried out during the design stage, since it is understood that this can offer the best opportunities for the industry's clients to achieve efficiencies that then help to reduce costs and increase added value. As the process of design evolves, develops and is finalised, opportunities for change become less possible as the design becomes etched in tablets of stone.

Regardless of where a project originates, either from designers or from constructors, cost studies of buildings will remain important. This is to ensure not only that the best practices are adopted but that this is done in the context of the added value that is produced. This may mean looking at the wider picture of whole-life costs, comparing design alternatives or finding more economical solutions to construction site problems through modern methods of construction.

Much has already changed over the past 50 years, but this is likely to represent only a small fraction of what will be achieved in the future. Some will continue to focus their knowledge, skills and understanding on the design, others will find a greater application of possible improvements at the construction stage, whereas some will continue to focus on the project in-use phase. There are continuous improvements to be made at every stage, through the application of new technologies and elegant ideas that will in the future make today look old-fashioned. Care must be taken to consider the wider picture and to ensure that this is not forgotten. Attempting to safeguard any uneconomic practices and procedures is doomed and they will eventually be lost anyway. Sustainable methods of procurement (see Chapter 10) are assumed to be the best way forward at the present time.

23.10 MODERN METHODS OF CONSTRUCTION

'Modern methods of construction', or MMC, is the term used to embrace a range of technologies and processes involving various forms of supply chain specifications, prefabrication and off-site assembly. The rising costs of traditional methods of construction are unlikely to meet the demands for projects in the future. Industry decision-makers are recognising the significant benefits of MMC. MMC is about improved products and better processes. It aims to improve business efficiency, quality, customer satisfaction, environmental performance, sustainability and the predictability of delivery timescales. Some of the advantages claimed for MMC include:

- Reduction in on-site time
- Reduction of on-site labour requirements
- Off-site work can be taken off the overall critical path, provided it is planned properly
- Shorter total development time
- Design and manufacture of off-site elements can start as soon as outline planning permission is obtained
- Makes a more effective use of materials, goods and components
- Supply chain management can interface with planning authorities and with manufacturers
- An early design freeze is critical because design changes after manufacture has started will then become expensive
- MMC compares well with the relative cost-effectiveness of different construction methods
- Faster completion results in a shorter borrowing period
- Quality control in off-site manufacture results in less snagging and less on-site inspection
- The risk profiles of MMC projects migrate from later stages to earlier stages
- Health and safety on site should be improved

MMC responds well to durability (will the building last as long?), whole-life costs (will it cost the same to maintain?) and performance (will it operate as well?). MMC is about finding efficient ways to deliver high-quality, well-designed buildings, quickly, efficiently and economically.

Whilst the above advantages are claimed for MMC, cost studies of buildings should evaluate them in terms of construction economy.

23.11 RESEARCH

The link between research and practice is fundamental. Sometimes the latter has constrained research, especially that by academic researchers, by wanting solutions to tomorrow's problems today. Research is a long process and potential solutions today may be found to be false tomorrow, as the industry continues to evolve to meet the needs of its customers. Some research is also about thinking out of the box: looking for the less obvious solutions. The different directions taken by the construction industry and some researchers seem, at times, not to be in tandem. This can be seen as beneficial in driving forward new ideas and solutions. Unfortunately, industry is sometimes suspicious of those involved in research because they cannot provide either immediate or relevant solutions. Researchers, on the other hand, sometimes feel that practitioners are too wedded to current practice to allow new ideas to flourish. Each side has a contribution to make and a common language needs to be used to ensure inclusivity. Sometimes research is presented in a way that fails to connect with practice. In other cases practice ignores potentially good solutions because it fails to see the connection. We are all aware of research in other industries that has provided solutions in unexpected areas, where the problem itself was not actually being targeted. We are all also equally aware of research that has provided spin-offs to improve practices and products almost as a by-product of the main thrust of research.

CONCLUSIONS

Over the past half-century there has been considerable interest and an advance in understanding of how building costs are generated. There can be no doubt that such studies have contributed considerably to helping the construction industry develop more economic products and thus add value to a client's investment. There are many factors that have influenced and continue to influence the achievement of yet further added value in the construction industry. The appetite for such change remains considerable.

Evidence indicates that an emphasis on the product alone will not satisfy the client's long-term objectives. Attention is rightly being focused on the process of development from design through to construction and into use. An examination of the practices used in the industry will further add to continuous improvements that must be our long-term goal for the future.

Having set out the key principles and elements that should inform the direction of *Cost Studies of Buildings*, it is hoped that all those who are involved will see that this is a journey rather than a destination. The key questions of time, cost and quality will always remain even if they are in their different guises. The challenge is, of course, to create a coherent framework that can inform, challenge and guide the future direction of allowing us to build more, and more that is of the desired quality, in a more economical manner. We have achieved much through examining manufacturing processes, such as off-site construction, through comparing the costs of similar function buildings and through focusing not just on the needs of today but by looking to the future and the legacy that will remain for future generations.

An improved built environment provides a feel-good factor for everyone. This must be achieved in the future within a sustainable framework in as many parts of the world as possible. This might mean an increase in costs, but by exploring different ways of achieving the aims and aspirations of clients, this should be able to be done as economically as possible. We need to be able to see what has been achieved, what has been possible and what foundation principles have been laid. The important issues are likely to continue to be important themes, and they include providing a sustainable built environment.

It is worth concluding with a quotation from John Ruskin:

There is hardly anything in the world that man cannot make a little worse and sell a little cheaper, and the people who consider price only are this man's lawful prey.

This has never been the aim of *Cost Studies of Buildings*. Most importantly, it is also not about acquiring the cheaper things in life, but about having the more expensive things that cost less. This aim can be achieved if we examine carefully what we build and how we build it.

SELF ASSESSMENT QUESTIONS

1. Explain why there should be a shift in building economics focus from the product to the process.

2. Explain the main principle behind added value and where this can be applied to construction industry products.

3. In what ways is lean construction a form of added value?

BIBLIOGRAPHY

Ashworth, A., *The Shift from Cost to Value*. Proceedings of the CIB W55 conference, Zagreb 1996.
Ashworth, A., *Contractual Procedures in the Construction Industry*, 5th edition. Pearson Education 2006.
Ashworth, A. and Hogg, K., *Added Value in Design and Construction*. Pearson Education 2000.
Best, R. and De Valance, G., *Building in Value*. Arnold 1999.
Construction Industry Board, *Towards a 30% Productivity Improvement in Construction*. Thomas Telford Publishing 1996.
Crane, A., *Rethinking Construction: 2002*. Rethinking Construction Ltd 2002.
Department of the Environment, Transport and the Regions, *Building a Better Quality of Life: A Strategy for More Sustainable Construction*. 2000.
Egan, J., *Rethinking Construction*. Department of the Environment, Transport and the Regions 1998.
Gray, C., *Value for Money*. Reading Construction Forum 1996.
Kelly, J., Morledge, R. and Wilkinson, S., *Best Value in Construction*. Blackwell Publishing 2002.
Langston, C., *Sustainable Practices: ESD and the Construction Industry*. Envirobook Publishing 1997.
Latham, M., *Constructing the Team*. HMSO 1994.
National Audit Office, *Using Modern Methods of Construction to Build More Homes Quickly and Efficiently*. National Audit Office 2005.
Powell, C., *The Challenge of Change*. Royal Institution of Chartered Surveyors 1998.
Spencer, N.C. and Winch, G.M., *How Buildings Add Value for Clients*. Thomas Telford Publishing 2002.
Strategic Forum for Construction, *Accelerating Change*. Department of Trade and Industry 2002.
Womack, J. and Jones, D., *Lean Thinking*. Simon & Schuster, New York 1996.
Wood, E.G., *Added Value: The Key to Prosperity*. Business Books 1978.

INDEX

accounts method 292
accuracy 56, 64, 334–7, 355, 382–3, 398
added value 35, 540, 541
aesthetics 18, 91, 96, 119, 201, 417
agents 306
air conditioning 206
amount of £1 per annum table 296, 297
analysis,
    and evaluation 476
    principles of 167
annual equivalent 442
annual sinking fund 297
annuity £1 will purchase 299
appearance 18, 90, 118
appraisal costs 119
approximate quantities 325, 358
architect's instructions 505
ARCOM 272, 276
asset registers 524
AUBEA 272
Audit Commission 9, 10, 225
average rate of return 310
awareness 30

BCIS 6, 27, 54, 80, 81, 83, 84, 347, 382,
    397
    cost analysis 163, 167, 202, 211, 347, 382,
    397
    indices 134, 136–7
    NRM 147, 149, 153, 155
benchmarking 10, 29, 143, 156, 545–8
Best Practice Programme 110
best value 9, 10, 224
bills of quantities 54–5, 66–8, 166, 346, 382,
    388

brainstorming 491
BRE 113, 270, 382
breakeven analysis 121–2, 493
British Standard 3811: 1974 407
BS 5750 115
BSI 115, 420
BSIRA 420
budget 340, 343, 379–80
budgetary control 502
buildability 100, 543
buildings,
    bulletin 26, 29, 164
    cost studies 228
    costs 134, 211, 522
    economics 3, 21, 25–6, 28–9, 31, 33, 35
    green 22, 550
    grouping of 99–100
    healthy 107
    index 132, 134, 136
    industrial 236
    life of 14, 409, 450
    prices 82
    research organisations 269–70
business orientation 281

calculations 398, 447–51
capacity 532
capital 234
    allowances 231, 234–5, 457
    expenditure 357
    investment 487
carbon tax 255–6
case studies 268, 476
cash flow 248, 385, 457–8, 506, 509, 512–13
CEBE 276

CEEC 273
change management 514–15
chronology 29, 383
CIB 28, 271, 276
CIOB 274–5
CIRIA 271, 272, 420
civil engineering 55, 101–2, 169, 539
cleaning 446, 527
client 5, 14, 20, 89, 104–5, 217, 345, 503,
    506, 520
CNBR 276
COBRA 276
collaboration 259–60
college of further and higher education 531
communication installations 209
communication systems 522, 527
comparative
   cost planning 348, 352
   method 291
comparison 86
   cost relationships 129
   other industry 227
component,
   failure 418, 421, 422
   life 418, 421, 450
   manufacturers 542
compound interest 293
computer-aided design (CAD) 280, 334, 402
computers 32, 65, 333, 399, 525
condition 289, 533–5
conference method 320
consistency 56
constructing an index 128
Constructing Excellence 225
Constructing the Team 539
construction 15, 212
   adding value 541
   budget 340, 343, 379
   costs 86, 302
   development process 43
   economic regulator 43
   economics 3, 13
   firms 44
   industry 37, 38, 43, 269, 277, 441, 542
     Europe 39
   life cycle costing 82
   method 170
   modern methods 552–3
   project 21

sectors 40
services 234
statistics 137
constructional details 101
consultancy 266
contingencies 306, 505
contract,
   conditions 329
   period 168, 170
contractor 5, 551
contractor's,
   cost control 507
   cost models 384
   method 291
contractual,
   arrangements 11
   claims 505
   procedures 12, 308
corporation tax 234
cost 6, 86, 89–91, 168
   advice 3, 10, 401
   analysis 68, 69–75, 81, 156, 163, 166
   benefit analysis 29, 313–16
   breakdown 213
   changes 467
   checking 349, 352–3
   codes 58, 389
   comparison 398, 511
   control 3–5, 398, 538
   data 51, 53–5, 60, 67, 85
   engineering 327
   estimates 152–3, 155, 159
   indices 132
   information 66, 127, 168
   limit 354, 356, 472
   management 434, 513–14
   modelling 32, 381, 384, 386, 538
   models 328, 383–6, 402
   optimisation 121, 385, 537
   plan 152, 349–50, 363–6
   planning 128, 161, 345–8, 538
   planning process 348–9
   practice 287
   ratio 171
   reimbursable contracts 501
   research 256–8
   savings 482
   sensitive 353
   studies 3, 537

costs-in-use 11, 29, 76–8, 195, 423–5, 443, 538
creative work 266
creativity 103, 480–82
CRISP 276–7
cube rules 323
cyclical movement 139

data 17, 333
    classification 421
    sources of lifespan 420
Davis Langdon 137, 211
daywork 505
decision tree analysis 492
decision-making changes 467
definitions 165
demonstration projects 280
depreciation 238–40
design 15, 440
    economics 88, 329
    liability 483
    method 3, 17
    optimisation models 385
    risk 331
    sustainable economic 106–7
    team 153, 157
designer 7, 14, 19, 88, 551
designers' cost models 384
deterioration 410–11
    physical 412–13
developers',
    budgets 300
    profit 303
development 33, 43, 157, 212, 243, 262, 476
    appraisal 287
    density 213
    financial assistance for 243
    technical 160
    value 288
dilemmas 106
dimensions 118
discount rate 427–8
discounted cash flow 122, 122–3
discounting 311
disruption 105, 482
distortion 67, 423
diversification 280
drainage 189, 210, 377
drivers for change 279
duty of care 13

ECAM 275
ECCE 273
economic,
    analysis 16
    assessment 120
    life 416
    regulator 43
economic life 415–16
economics 21, 25–6, 112, 416, 539
economy 8, 288–9
EDCON 104
EDICOM 280
EDICON 280
education 31–2, 313
effectiveness 8, 125
efficiency 8, 93, 533
Electrical installations 208, 376
element 107, 162
elemental,
    cost planning 160–61, 350, 515
    cost studies 168
    cost synthesis 357
    estimating 326
'Elsie' 399, 399–400
embellishment 102
empirical methods 387
employer's requirements 218
employment 40, 280
energy 446
    conservation 442
    efficiency 107
    management 526
engineering services 98, 203, 330, 524
enterprise zones 236, 245
environmental,
    impact 21
    management 522
EPSRC 271
equipment servicing 525
equities 251
errors 55, 64, 67, 360
estate management 446
estimate,
    classification 319
    preparing 339
estimating 55, 58, 152, 153, 496
    accuracy 334
    methods 320
European Union 39, 45, 248

evaluation 476
evolution 107, 217
exclusions 330
expert systems 334, 399, 402
external,
    doors 193, 370
    services 379
    walls 190, 362, 369–70
    works 210, 330

facilities management 15, 67, 517–18
failure costs 120, 122
fashion change 466
feedback 55, 57, 393, 398, 509
fees 157, 303, 505
final account 164, 500
finance 304
financial,
    assistance 246–8
    reports 503
    reviews 514
finishes 119–201
    ceiling 202, 374
    floor 201, 373
    internal 199
    wall 202, 372
fire fighting 209
    protection 95, 164, 181
fittings 203, 374
fluctuations 62
forecasting 129, 403, 408, 466, 487, 496
frames 101, 164, 176, 194
function 17, 19, 91
functional life 415
furniture 203, 374
future 34, 46, 279, 423, 537
fuzzy set theory 432

general rates 446
GIFA 94, 166, 172–3, 177, 202, 208, 350–1, 363, 377
government 41, 289
grants 230, 245–6, 248
gross development value 301
ground conditions 92

health and safety 42, 529
heating 205, 375, 452
hedonic price modelling 293

height 97
heuristics 397
hidden costs 423
history 25, 30–4, 216, 347, 468–9
HMRC 231
HMSO 275
hotels 236
housing 137, 254

inception 15, 383, 440
independent review 483
index constructing 128
index numbers 128
indices 127, 130, 132, 135
inflation 159, 425–7, 513
information technology 30, 280, 333
infrastructure 289
initial costs 406
Inland Revenue 235
innovation 23, 47, 257
inspection 10, 363
insurance 446
integrated approach 43
interim certificates 499, 509
internal,
    doors 198–9, 371
    finishes 199–200
    rate of return 311
    walls 196, 371
international 42, 86, 281
interview surveys 267
investment 41, 230, 249
    appraisal 287, 289–90, 308
    method 292
    yield 301
ISO 9000 114–15, 281
ISO 15686–5 82

just-in-time 544

Key Performance Indicators 143–5
knowledge 30, 281–2
    management 23

labour,
    costs 52, 63, 131
    output 59, 64
Laspeyre 130
Latham report 529

lean construction 544–6
legal,
    aspects 338
    life 415, 417
letting 306
life cycle costing 82
life expectancy 83
lifespan data 420
lift installation 208, 376
linear regression analysis 141
location 62, 67, 92, 168, 170, 289
loose fit 95

maintenance 105, 421, 423, 445, 523
manufacture 104, 194
manufacturing industry 543, 545
market conditions 63, 129, 168, 329
market intelligence 84–5
market surveys 81
material costs 63, 64
materials 14, 118, 176–82, 194
    low-impact 107
mathematical models 385, 403
measurement contracts 499
method selecting 331
methodology 475–6
minor new works 65, 445
Monte Carlo 392, 432
mortgage repayment tables 299
moving average 139–40

Nationwide Building Society index 137
necessity 311, 538
net present value 311, 442
NRM 147, 149, 160, 161
    cost analysis 211
    purpose 150
    structure 150

obsolescence 409–10, 414, 415
OGC Gateway Process 147, 151, 152–3, 160
operational,
    bills 167
    research 268
optimal investment criterion 312
optimisation 121, 537
optimum envelope area 21
orientation 457, 475
overheads 63, 157

overseas property 254
overseas work 37

Paasche 130
Parry's valuation tables 294
partitions 196–8, 371
partnering 217, 220, 545
pay-back method 308, 311
performance 89, 119, 194
Performance Indicators 226
plan shape 95, 160, 192
    index 21
planning efficiency 93–4
plant 52
    and machinery 237
plumbing installation 204
policies 421
policy-making changes 467
population 289
portfolio theory 493
post-contract cost control 349, 354, 498
post-occupancy analysis 529
postponability 311
practitioners 33, 258, 407
pre-assembly 103
pre-tender price estimating 11, 318, 331, 339
precautions 62
predictability 144–5
    future maintenance costs 423
prefabrication 102, 103, 545
preliminaries 67, 156, 169, 379
preliminary,
    cost plan 349
    estimate 349
present value of £1 per annum table
    (dual rate) 298
present value of £1 per annum table
    (single rate) 297
present value of £1 table 296
prevention costs 119
price 6
    analysis 68
    books 60, 61, 64, 81, 358
    indices 132
    risk 331, 350
pricing conditions 358
private finance initiative 41
probabilities risk analysis and 494–5
probability 431–2, 492

procurement 11, 34, 63, 82, 157, 216, 228, 440, 483, 552
  evolution of systems 217
  selection and options 220
  systems 219
production cost models 384
professions 30, 45–6, 257, 402
profit 5, 52, 63, 157
profits method 292
programme 340
project cash flow 512
property investment 249–54
proportion 358
proposed cost plan 366
protection 181
provisional quantities 504
public sector price guides 65
Public Private Partnerships 221–4, 521

QS2000 280–82
quality 102–3, 112–14, 281
  assurance 114, 528
  considerations 124, 329, 359
  control 114, 116
  costs 119, 122
  design 212
  standards 117
  systems 114
quantifying 332
quantities priced bills 66
quantity,
  considerations 358
  surveying 260–3, 282
  surveyors 10, 13–14, 280–1, 352, 499, 505, 507–8, 537

rates of interest 429
  changes in 451
recession 6, 261
redecorations 445
refurbishment 104
regional,
  initiatives 245–6
  variation 137–8
regression analysis 141, 388
regulations 95
reinstatement method 292
reliability 56, 64, 398
renewal 107, 448

repair costs 103, 124, 243
replacement costs 103, 124, 407, 422
research 33, 47, 246, 257, 553
  applied 266
  basic 266
  dissemination 275
  historical 268
  methods 266–9
  organisations 269–71
  process 263–4
  strategic 266
  university 274
  wheel 265
research and development 237–8
research,
  experimental 267
residual method 291
resource,
  analysis 327
  models 386
Retail developments 211, 214
retail prices index 135–6
Rethinking Construction 9, 548–9
revenue 234
RIBA Plan of Work 147, 151, 152–3, 160, 347, 529
RICS 148–51, 164, 220, 272–5, 281, 347
risk 157–8
  analysis 486, 494
  dealing with 490–1
  management 486, 487–9
roofs 108–9, 184–90, 368–9
running costs 105, 358

s-curves 498, 509, 510
sanitary appliances 374
scenario analysis 493
scholarship 266
seasonal adjustment 141
security 170, 195, 522, 526
selection 63, 476
sensitivity analysis 312, 430–1, 458, 493
simple interest 293
simulation 391, 493
site,
  area 153
  considerations 67, 91, 170
  works 376–7
size 62, 92, 172, 194

social life 415–16

space management 523, 531

stability 118

staff welfare 527

staircases 190, 369

standardisation 103

statistics 131

   housing and construction 137

storey,

   enclosure method 324

   height 99

storeys 173

strength 118

structural morphology 102

subcontractors 52, 67

substructure 162, 171, 173, 366

superficial,

   area method 322

   perimeter method 323

superstructure 176

suppliers 67, 104, 279, 551

supply chains 552

surveys 267

sustainable construction 549–51

synectics 492

systems and software 525

tax-efficient design 241

taxation 230–2, 242, 428–9

technological,

   change 466

   life 414–15

technology 18

tender 11, 168

   price index 133, 135

   reconciliation 349, 352

   sum 346, 349, 385

terminology 473–5

textbooks 26, 55, 538

theories 21

time 89, 90, 145, 422

time series analysis 143

total project information 116, 167

Total Quality Management 114, 116, 528, 545

town-centre retail premises 212–13

trench-fill foundations 101, 169, 174–5

trends 28, 127, 138–41, 387

Trust companies 270

uncertainty 402

   accounting for 430

   risk and 486

   techniques 491

understanding 281

unit,

   method 321

   rate 52

unnecessary costs 479

upper floors 182–4, 367

urban renewal 244

usage 332, 398

uses of index numbers 128

utilisation 533

valuation 511

   methods of 290–3

   tables 293–300

valuations 498–9

value 6, 9, 288–9, 433

   Added Tax 153, 159, 232, 505

   engineering 473–4, 538

   losing 484

   management 472, 473–7, 482–3, 538

   for money 5, 7, 33, 34, 90, 109, 355, 401

variations 67, 398, 418

ventilation systems 205, 375

wall-to-floor ratio 95–6, 173

water installations 375

'what' scenarios 477–9

whole-life costs 98, 105, 211, 385, 406–8, 434, 439, 442, 538

Wilderness Study Group 32

windows and external doors 193–6, 370

workmanship 67

world of facts 268

World War II 30, 281–2, 354, 538